MATERIALS MODELING FOR MACRO TO MICRO/NANO SCALE SYSTEMS

AAP Research Notes on Nanoscience and Nanotechnology

MATERIALS MODELING FOR MACRO TO MICRO/NANO SCALE SYSTEMS

Edited by
Satya Bir Singh, PhD
Prabhat Ranjan, PhD
A. K. Haghi, PhD

AAP APPLE ACADEMIC PRESS

First edition published 2022

Apple Academic Press Inc.
1265 Goldenrod Circle, NE,
Palm Bay, FL 32905 USA

4164 Lakeshore Road, Burlington,
ON, L7L 1A4 Canada

CRC Press
6000 Broken Sound Parkway NW,
Suite 300, Boca Raton, FL 33487-2742 USA

4 Park Square, Milton Park,
Abingdon, Oxon, OX14 4RN UK

© 2022 Apple Academic Press, Inc.

Apple Academic Press exclusively co-publishes with CRC Press, an imprint of Taylor & Francis Group, LLC

Library and Archives Canada Cataloguing in Publication

Title: Materials modeling for macro to micro/nano scale systems / edited by Satya Bir Singh, PhD, Prabhat Ranjan, PhD, A.K. Haghi, PhD.

Names: Singh, Satya Bir, editor. | Ranjan, Prabhat, editor. | Haghi, A. K., editor.

Series: AAP research notes on nanoscience & nanotechnology.

Description: First edition. | Series statement: AAP research notes on nanoscience and nanotechnology | Includes bibliographical references and index.

Identifiers: Canadiana (print) 20210376848 | Canadiana (ebook) 20210376899 | ISBN 9781774630198 (hardcover) | ISBN 9781774639528 (softcover) | ISBN 9781003180524 (ebook)

Subjects: LCSH: Nanostructured materials.

Classification: LCC TA418.9.N35 M38 2022 | DDC 620.1/15—dc23

Library of Congress Cataloging-in-Publication Data

....

CIP data on file with US Library of Congress

....

ISBN: 978-1-77463-019-8 (hbk)
ISBN: 978-1-77463-952-8 (pbk)
ISBN: 978-1-00318-052-4 (ebk)

ABOUT THE BOOK SERIES: AAP RESEARCH NOTES ON NANOSCIENCE AND NANOTECHNOLOGY

AAP Research Notes on Nanoscience & Nanotechnology reports on research development in the field of nanoscience and nanotechnology for academic institutes and industrial sectors interested in advanced research.

OTHER BOOKS IN THE AAP RESEARCH NOTES ON NANOSCIENCE AND NANOTECHNOLOGY BOOK SERIES

- **Nanomechanics and Micromechanics: Generalized Models and Nonclassical Engineering Approaches**
  Editors: Satya Bir Singh, PhD, Alexander V. Vakhrushev, DSc, and A. K. Haghi, PhD

- **Materials Modeling for Macro to Micro/Nano Scale Systems**
  Editors: Satya Bir Singh, PhD, Prabhat Ranjan, PhD, and A. K. Haghi, PhD

- **Carbon Nanotubes: Functionalization and Potential Applications**
  Editors: Ann Rose Abraham, PhD, Soney C. George, PhD, and A. K. Haghi, PhD

- **Carbon Nanotubes for a Green Environment: Balancing the Risks and Rewards**
  Editors: Shrikaant Kulkarni, PhD, Iuliana Stoica, PhD, and A. K. Haghi, PhD

ABOUT THE EDITORS

Satya Bir Singh, PhD
Professor, Department of Mathematics, Punjabi University, Patiala, India

Satya Bir Singh, PhD, is a Professor of Mathematics at Punjabi University Patiala, Patiala, Punjab, India. Prior to this, he worked as an Assistant Professor in Mathematics at the Thapar Institute of Engineering and Technology, Patiala, India. He has published about 125 research papers in journals of national and international repute and has given invited talks at various conferences and workshops. He has also organized several national and international conferences. He has been a coordinator and principal investigator of several schemes funded by the Department of Science and Technology, Government of India, New Delhi; the University Grants Commission, Government of India, New Delhi; and the All India Council for Technical Education, Government of India, New Delhi. He has 21 years of teaching and research experience. His areas of interest include mechanics of composite materials, optimization techniques, and numerical analysis. He is a life member of various learned bodies.

Prabhat Ranjan, PhD
Assistant Professor in the Department of Mechatronics Engineering at Manipal University Jaipur, Jaipur, Rajasthan, India

Prabhat Ranjan, PhD, is an Assistant Professor in the Department of Mechatronics Engineering at Manipal University Jaipur, India. He is the author of the book *Basic Electronics* and editor of *Computational Chemistry Methodology in Structural Biology and Materials Sciences*. Dr. Ranjan has published more than 10 research papers in peer-reviewed journals of high repute and dozens of book chapters in high-end research edited books. He has received the prestigious President Award of Manipal University, Jaipur, India, given for the development of the university; a Material Design Scholarship from Imperial College of London, UK; a DAAD Fellowship; and a CFCAM-France. Dr Ranjan has received several grants and also participated in national and international conferences and summer schools.

He holds a Bachelor of Engineering degree in Electronics and Communication and a Master of Technology in Instrumentation Control System Engineering from the Manipal Academy of Higher Education, Manipal, India; and a PhD in engineering from Manipal University Jaipur, India.

A. K. Haghi, PhD
Professor Emeritus of Engineering Sciences, Former Editor-in-Chief, International Journal of Chemoinformatics and Chemical Engineering and Polymers Research Journal; Member, Canadian Research and Development Center of Sciences and Cultures

A. K. Haghi, PhD, is the author and editor of over 200 books, as well as over 1000 published papers in various journals and conference proceedings. Dr. Haghi has received several grants, consulted for a number of major corporations, and is a frequent speaker to national and international audiences. Since 1983, he served as professor at several universities. He is the former Editor-in-Chief of the *International Journal of Chemoinformatics and Chemical Engineering and Polymers Research Journal* and is on the editorial boards of many international journals. He is also a member of the Canadian Research and Development Center of Sciences and Cultures.

Dr. Haghi holds a BSc in urban and environmental engineering from the University of North Carolina (USA), an MSc in mechanical engineering from North Carolina A&T State University (USA), a DEA in applied mechanics, acoustics and materials from the Université de Technologie de Compiègne (France), and a PhD in engineering sciences from Université de Franche-Comté (France).

CONTENTS

CONTRIBUTORS

Saranya Babu
Laboratory for Molecular Electronics and Photonics, Department of Physics, National Institute of Technology, Calicut, Kattangal, Kerala 673601, India

Savita Bansal
Department of Mathematics, Punjabi University, Patiala, India

Dipshikha Bharali
Department of Chemical Sciences, Tezpur University, Napaam, Assam, 784028, India

Ramesh Chandra Deka
Department of Chemical Sciences, Tezpur University, Napaam, Assam, 784028, India

Swarup S. Deshmukh
Department of Mechanical Engineering, National Institute of Technology Durgapur, Durgapur, West Bengal, India

Rasna Devi
Department of Chemical Sciences, Tezpur University, Napaam, Assam, 784028, India

Arjyajyoti Goswami
Department of Mechanical Engineering, National Institute of Technology Durgapur, Durgapur, West Bengal, India

A.K. Haghi
Canadian Research and Development Center of Sciences and Cultures, Montreal, Canada
Emeritus of Engineering Sciences, University of Guilan, Rasht, Iran

T. A. Shahul Hameed
Department of Electronics and Communication, TKM College of Engineering, Kollam, Kerala, India

Vijay S. Jadhav
Mechanical Engineering Department, Government College of Engineering Karad, Karad, Maharashtra, India

Parveen Lata
Department of Basic and Applied Sciences, Punjabi University, Patiala, India

Tinu Lowerence
Department of Chemistry, University of Calicut, Malappuram, India

Prateek Mekkat
Laboratory for Molecular Electronics and Photonics, Department of Physics, National Institute of Technology, Calicut, Kattangal, Kerala 673601, India

K. Muraleedharan
Department of Chemistry, University of Calicut, Calicut, Kerala 673635, India

K. Nusrath
Department of Chemistry, University of Calicut, Calicut, Kerala 673635, India

M. Pratheek
Laboratory for Molecular Electronics and Photonics (LAMP), Department of Physics, National Institute of Technology, Calicut, Kerala, India

P. Predeep
Laboratory for Molecular Electronics and Photonics (LAMP), Department of Physics, National Institute of Technology, Calicut, Kerala, India

Prabhat Ranjan
Department of Mechatronics Engineering, Manipal University Jaipur, India

Pradip Deb Roy
Department of Mechanical Engineering, National Institute of Technology Silchar, Assam, India

K. Sabira
Department of Chemistry, University of Calicut, Malappuram, India

Shivdev Shahi
Department of Mathematics, Punjabi University, India
Emeritus of Engineering Sciences, University of Guilan, Rasht, Iran

Ramakant Shrivastava
Mechanical Engineering Department, Government College of Engineering Karad, Karad, Maharashtra, India

N. V. Sindhu
Department of Chemistry, University of Calicut, Calicut, India

Satya Bir Singh
Department of Mathematics, Punjabi University, Patiala, India

Sukhveer Singh
Punjabi University APS Neighbourhood Campus, Dehla Seehan, Sangrur, Punjab, India

ABBREVIATIONS

AVAC	antisolvent vapor-assisted crystallization
BBC	bottom boundary condition
BET	Brunauer–Emmett–Teller
BTNPs	barium titanate nanoparticles
BTO	barium titanyl oxalate
CNT	carbon nano tubes
CTAC	cavitation triggered asymmetrical crystallization
CV	crystal violet
CVS	chemical vapor synthesis
DCM	dichloromethane
DFPT	density functional perturbation theory
DFT	density functional theory
DMF	dimethylformamide
DMSO	dimethyl sulfoxide
DOS	density of states
DSC	differential scanning calorimetry
DSSCs	dye sensitized solar cells
EML	emissive layer
ETL	electron transport layer
FESEM	field-emission scanning electron microscope
FT-IR	Fourier transform infrared
FWO	Flynn–Wall–Ozawa
GA	genetic algorithm
GBL	gamma-butyrolactone
GGA	generalized gradient approximation
GS	graphene sheets
HI	hot injection
HTL	hole transport layer
ITC	inverse temperature crystal
KAS	Kissinger–Akahira–Sunose
LARP	ligand-assisted reprecipitation
LB	Luria Bertani
LDH	layered double hydroxide
LHP	lead halide perovskite

LMCT	ligand-to-metal charge transfer
LWT	linear wave theory
MAI	methylamine iodide
MB	methylene blue
MHP	metal halide perovskite
MMCT	metal-to-metal charge transfer
MMO	mixed metal oxide
MPTF	microcrystalline perovskite thin film
NC	nanocrystal
NWT	numerical wave tank
PAW	projector augmented wave
PC	powder concentration
PCE	power conversion efficiency
PL	photoluminescence
PLQY	photoluminescence quantum yield
POFFT	pulse off time
PONT	pulse on time
PSC	perovskite solar cell
PV	photovoltaic
QD	quantum dot
RhB	rhodamine B
SAED	selected area electron diffraction
SDS	sodium dodecyl sulfate
SEM	scanning electron microscope
SLITC	space limited inverse temperature crystal growth
SV	servo voltage
SWl	still water level
TEM	transmission electron microscope
TG	thermogravimetric
UV	ultraviolet
VASP	Vienna Ab-initio Simulation Package
VCA	virtual crystal approximation
VOF	volume of fluid
w-EDM	wire electric discharge machine
WF	wire feed
XRD	X-ray diffraction
YE^2A^2	Yu–Emmerich extended averaging approach

PREFACE

This book was written for a broad audience with a prior and basic knowledge of modeling of advanced materials.

The present volume offers a state-of-the-art report on the various recent scientific developments in the theory of engineering materials and the basic theoretical concepts in advanced mechanics of materials as well as the wide range of experimental and numerical applications.

Following this, the volume does not only address the sophisticated reader but also, for the interested beginner in the area of materials and composites, a collection of research-oriented chapters.

The book is addressed to a wide readership, and it will be useful for undergraduate and graduate students and as a reference source for professionals including engineers, applied mathematicians, and others working on different application of nanomaterials in engineering.

This new book also offers an introduction to numerical methods by employing a readily accessible and compact format, and it demonstrates an overview of new methods, for advanced students in mechanical engineering and mechatronics.

It also provides step-by-step descriptions of how to formulate numerical problems and develops techniques for solving them. A number of engineering case studies are also intended for academics, including graduate students and experienced researchers interested in state-of-the-art computational methods for solving challenging problems in engineering.

CHAPTER 1

NANOCRYSTALLINE PEROVSKITES: FROM MATERIALS TO MODELING

SARANYA BABU, PRATEEK MEKKAT, and P. PREDEEP*

Laboratory for Molecular Electronics and Photonics, Department of Physics, National Institute of Technology, Calicut, Kattangal, Kerala 673601, India

Corresponding author. E-mail: predeep@nitc.ac.in

ABSTRACT

Besides the exciting application potential of perovskites in photovoltaics, organic and inorganic lead halide perovskites are intensively investigated for various applications such as light-emitting diodes and photodetectors due to their unparalleled optoelectronic properties. However, the photoluminescence quantum yield of bulk perovskite is limited due to mobile ionic defects, small exciton binding, and so on. Here, the role of nanocrystalline perovskites comes into play. Nanocrystalline perovskites show excellent detection performance due to high exciton binding energy and quantum confinement. They are used in photodetectors due to high crystalline quality and large surface area. Perovskite nanocrystals (NCs) show high photoluminescence quantum yield as well. The morphology of the perovskite NCs is important since they affect its optoelectronic properties. The composition structure and size of the NCs can be tuned during synthesis and post-synthesis transformations. Various methods, such as hot injection method, ligand-assisted reprecipitation strategy, one-pot reaction, ultrasonic method, microwave-assisted ball milling, and so on, are used for its synthesis. Each method has its own advantages and disadvantages. The challenges in the practical use of nanocrystalline perovskites are nontoxicity, stability, and mass production. In the synthesis methods mentioned above, ball milling achieves the best yield, whereas hot injection leads to the best morphology. Morphology is very much important in the fabrication of semiconducting devices. This chapter discusses the structure, synthesis, and advantages of nanocrystalline perovskites in detail.

1.1 INTRODUCTION

Solar photovoltaic (PV) has already been accepted as the major sustainable energy source in the globe's survival efforts against the stark reality of fast depletion as well as the warming effects of fossil fuels. The currently largest deployed solar cell installations use silicon solar cells, which are having a power conversion efficiency (PCE) of 24.7% [1]. However, due to the limitations imposed by the Auger recombination, the extent to which this efficiency can be further increased is almost exhausted, and one has to look for other options such as multijunction solar cells and also new materials and device architectures. The latest entry into this quest for materials and techniques is perovskite absorbers. In a short period of its first introduction as a light harvesting material, its PCE showed a phenomenological increase year by year: to 3.8% in 2009 [2], 6.5% in 2011 [3], 20.1% in 2017 [4], and 22.1% in 2017 [5].

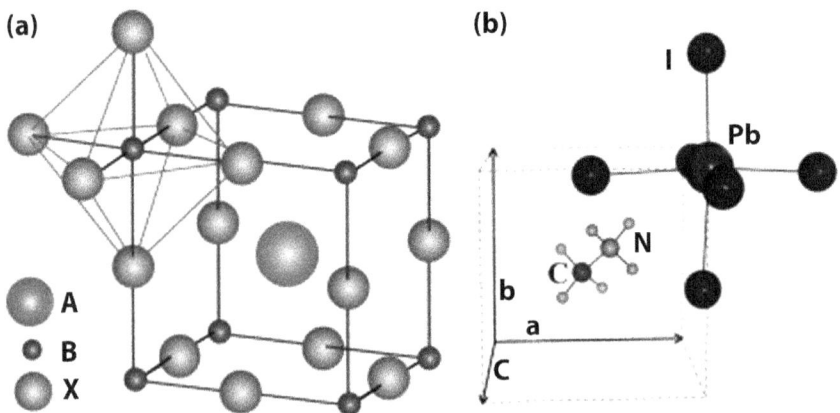

FIGURE 1.1   (a) ABX_3 perovskite structure showing BX_6 octahedral and larger A cation occupied in cubo-octahedral site. (b) Unit cell of cubic $CH_3NH_3PbI_3$ perovskite. Original figure in (a) reprinted with permission from [9] (Copyright 2016 Elsevier). Original figure in (b) was reprinted from [10] (Copyright 2013 American Chemical Society).

A perovskite [6] is an organic–inorganic hybrid with a formula ABX_3, as shown in Figure 1.1. Actually, the word perovskite is specific to a mineral, that is, calcium titanate ($CaTiO_3$). It was first discovered in the Ural Mountains of Russia back in 1839. In the case of organic–inorganic hybrid perovskites, A and B sites in the crystal structure are smaller metal cations, which are charge balanced with X site anions. Depending on the ionic or elemental radii of the A, B, and X elements, the crystal structure could change from a high-symmetry cubic phase to a tetragonal phase or to a low-symmetry orthorhombic phase. Cesium lead halide perovskites (LHPs) were first discovered in 1892 [7], and hybrid organic–inorganic methylammonium LHPs ($CH_3NH_3PbX_3$) were first synthesized by Weber in 1978 [8].

The rapid increase in the PCE of perovskite solar cells (PSCs) is due to its unique optoelectronic properties. $CH_3NH_3PbI_3$ has high absorption coefficient in the visible region and low-energy direct bandgap (~1.5 eV) [11]. It has small effective masses [12] for electrons and holes, long photogenerated carrier diffusion length (100–1000 nm) and lifetime, high dielectric constant, high carrier mobility, and high defect tolerance [13]. The band of metal halide perovskites (MHPs) can be tuned by altering [14] the ratio of halides or metal. This is useful in extending the absorption to longer wavelengths without changing the absorption coefficient. Bandgap can also be tuned by changing the organic cation in the A site of ABX_3. By changing the ratio of X site anion from lead to tin, the bandgap decreases from 1.55 to 1.17 eV [15]. The properties of the perovskite structure affect the performance and stability of perovskite devices.

Single-crystal solar cells have great advantages in PVs due to their charge carrier lifetime, light absorption region, and carrier diffusion length [16]. This has been proved in the case of silicon. However, Si has the advantage that its diffusion length is high enough that single-crystal wafers can be of thickness even up to millimeters. This makes wafer processing easy out of large single-crystal ingots of Si. Wafers could be cut out using sawing or wire cutting. However, perovskites have comparatively shorter diffusion lengths, and the thickness of the wafers should be within a few tens of micrometers. Cutting out such thin wafers using conventional methods is extremely difficult and not at all plausible. This makes the use of single-crystal perovskite in PV application in the form of wafers just like in the case of Si solar cells. Obviously, there are not many reports on highly efficient perovskite single-crystal wafer solar cells. Of course, there are many reports about using perovskite single crystals for forming thin films of perovskite for highly efficient solar cells. However, these are not about using single-crystal wafers directly as the absorbing layers. These reports are about evaporation single-crystal perovskite material to form thin films.

Also, there are reports about dissolving single-crystal perovskites to form thin films by solution processing through spin coating. This naturally raises the question as how to take advantage of single-crystal state of perovskites in solar cell applications. One possible solution is to grow single-crystal wafers of perovskites. However, growing large-area single-crystal wafers of thickness in the range of few tens of micrometers is a big challenge. Another possible solution is going for nanoscale. First, grow perovskite single-crystal nanocrystals (NCs) and then explore their use in device applications such as solar cells through films formed through polymer composites. Perovskite NCs are competitive with their single bulk single crystal since they can reduce the intragrain and intergrain defects that affect device performance [17].

Perovskite NCs have improved quantum efficiency of up to 90% [14] for PV applications, greatly reducing the loss mechanisms prevalent in bulk films. Even though the properties of NCs are similar to bulk counterparts, they show quantum confinement effects [18]. This helps in improving PV performance. This has generated much interest in nanocrystalline perovskites especially that of $MAPbI_3$ and has become an emerging area of research.

One of the advantages that make the perovskite absorbers attractive for PV applications is the tunability of the bandgap. This aspect is more in hand with perovskite NCs. Perovskite NCs exhibit bright and narrow-band photoluminescence (PL) [19]. We can easily tune the bandgap from ultraviolet to near-infrared region by changing the halide composition or the size of the NCs [20, 21]. Also, PL lifetime of perovskite NCs increases with decreasing bandgap energy [22]. The PL of the perovskite NCs covers [23] the entire visible range with narrow linewidths. The increase in exciton binding energy of the NCs contributes to the enhancement in photoluminescence quantum yield (PLQY) [21]. An important aspect is that the size of the NCs has a little effect on the emission spectrum. This is due to the small Bohr radius [24] of the perovskites in comparison with their bulk counterparts. Defect tolerance is an enabling factor for the efficient PV properties and bright PL [21].

Bulk perovskites exhibit [21] cubic, tetragonal, and orthorhombic structure in order of decreasing symmetry. Cubic phase is the highest temperature phase, and the phase transitions occur at well-defined tempera-tures. Like the bulk counterparts, perovskite NCs crystallize in tetragonal, pseudocubic, and orthorhombic structures. Because of its ionic character, we can synthesize perovskite NCs even at room temperature. The shape size and morphology of the perovskite NCs can be controlled by changing the synthesis routes. The most common morphology is cubic. They are highly crystalline in nature. The electronic and optical properties of the material can be controlled by doping it with external impurities. Magnetic dopants can enhance quantum mechanical spins, spin-orientation-dependent transport, and so on [19]. These versatile properties of perovskites in nanoscale can be useful in various electronic and photonic applications. Before discussing the reported applications of perovskite NCs, a discussion on their possible routes of synthesis would be helpful.

1.2   SYNTHESIS OF PEROVSKITE NANOCRYSTALS

MHPs are ionic in nature. Thus, the material undergoes faster nucleation, and highly crystalline NCs can be easily synthesized even at room temperature. The development of size and morphology controllable perovskite is a challenge in the field of nanotechnology. Various methods were adapted for the synthesis of perovskite NCs. Among them, colloidal synthesis is the most important one. A number of colloidal synthetic procedures are available [25] for the synthesis of MHPs having different chemical nature and morphologies.

Rivesta and Jain [26] pioneered the wet chemistry synthesis of hybrid perovskite NCs in 2014. They obtained $CH_3NH_3PbBr_3$ (MAPbBr$_3$) perovskite nanoparticles by using octyl ammonium bromide as the capping ligand. This is a versatile technique for the synthesis of chalcogenide NCs with high quality and is carried out by the addition of methyl ammonium bromide, oleic acid, and a noncoordinated solvent. This synthesis was based on the hot injection (HI) method.

Nedelcu et al. [27] developed a room-temperature-based ligand-assisted reprecipitation (LARP) technique to synthesize highly luminescent colloidal MAPbBr$_3$ nanoparticles. In this method, MAPbBr$_3$ precursors were mixed in a good solvent (polar solvent) dissolving organic salts. This solution is added to a poor solvent (nonpolar) under vigorous stirring in the presence of long-chain organic ligands to promote the reprecipitation process. By adjusting the concentration of the precursor solution and regulating the precipitation temperature [14], we can tune the size and bandgap of the prepared nanoparticles. The first demonstration of a size-tuned bandgap in MAPbBr$_3$ was first reported by Nedelcu et al. [27]. The reaction temperature of the LARP process was controlled for obtaining NCs with emission peak in the range of 475–520 nm. Akkerman et al. [28] applied the concentration tuning in the LARP approach to synthesize MAPbBr$_3$ with adjustable size. The octylamine function as a catalyst in the LARP process. Yantara et al. [29] prepared NCs in the size ranging from 6.6 to 13.3 nm by regulating the amount of octylamine in the solution.

NCs with fine-tuned composition and optical properties can be synthesized by post-ion exchange transformation of NCs. This has been demonstrated as a highly effective strategy for the synthesis of colloidal MHP NCs [26]. Nedelcu et al. [27] and Akkerman et al. [28] applied the anion exchange procedure to obtain colloidal cesium halide NCs. By tuning the anion ratio in the colloids highly luminescent CsPbX$_3$ (X = Cl, Br, I), NCs were prepared. The emission of the CsPX$_3$ NCs can be regulated over the visible range from 410 to 700 nm. MAPbBr$_3$ NCs with tunable bandgap cover over the wide range 1.5–3.1 eV can be synthesized by the anion exchange method.

Solvothermal synthesis is another strategy used in the preparation of NCs with well-controlled size, shape, and high crystallinity [29]. In solvothermal reaction, the precursors and a nonaqueous solvent are taken in a steel container known as autoclave. The precursors were premixed, and the solution is maintained at a certain temperature for a few hours. The product obtained from solvothermal reaction is highly pure and is of good quality. By dynamically controlled seed-mediated growth, the shape of the structures can be varied from nanocubes to nanowires. This variation is due to predissolution of precursors.

Jang et al. [30], Aharon et al. [31], and Bai et al. [32] developed the ultrasonication-assisted synthetic procedure for the preparation of high-quality MHP NCs. The advantage of this method is that it can be used for large-scale production, and one can avoid the use of polar solvents. MHP NCs with a wide range of compositions were synthesized [33] by the direct ultrasonication of the precursors in the presence of coordinating ligands.

High-quality MHP NCs with tunable compositions and high throughput are synthesized by microwave-induced preparation. Liang et al. [34] synthesized multiple colloidal CsPbX$_3$ NCs with tunable properties and morphologies by one-step microwave irradiation in a heterogeneous solid–liquid reaction system with no precursor preparations. The anion exchange process can be used to tune the PL emission color over the entire visible spectrum. Figure 1.2 shows the NCs of various morphologies prepared [35] using this method. The NCs prepared by this method were stable up to two months with only a minor reduction in their PLQY.

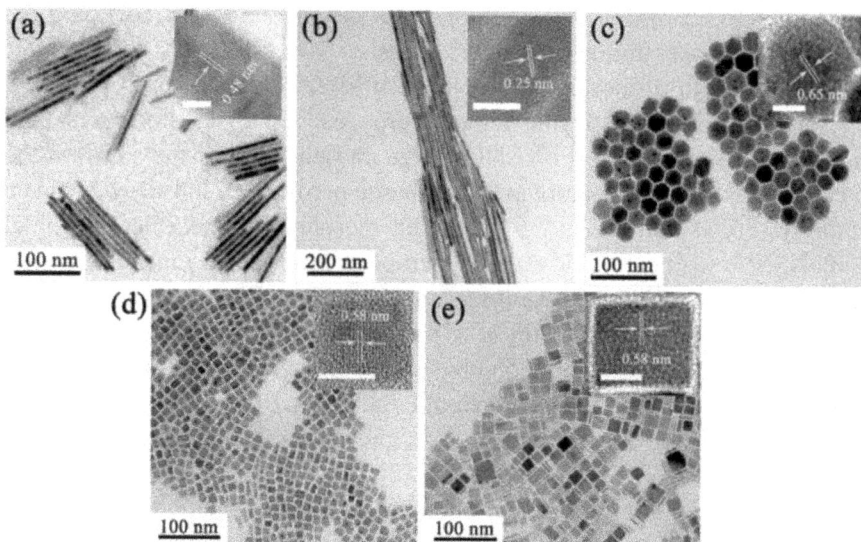

FIGURE 1.2 TEM images of the colloidal CsPbX$_3$ NCs with different morphologies: (a) CsPbI$_3$ nanorods, (b) nanowires, (c) hexagonal nanoplates, (d) CsPbBr$_3$ nanocubes, and (e) nanoplates. (Reprinted with permission from [36]. Copyright 2017 the Royal Society of Chemistry.)

The morphology of the perovskite NCs is of prime importance since it affects the optoelectronic properties and quality of NCs. In colloidal synthesis precursors, ligand-employed reaction temperature and reaction time influence the morphology. All the synthetic routes discussed in this section are capable of controlling the morphology of MHP NCs.

1.2.1 HOT INJECTION METHOD

This method was developed two and half decades back. In this method, one of the precursors is rapidly injected into the solution of remaining precursors, ligands, and a high-boiling-point solvent [37]. A rapid nucleation burst occurs after the injection, leading to the simultaneous formation of small nuclei. The nuclei start to grow when there is a decrease in the number of monomers. NCs evolve over time characterized by narrow size distribution. We can separate the growth stages in time. The key parameters that enable to control the size, size distribution, and shape of colloidal NCs synthesized by the HI method are (1) the ratio of the surfactants to the precursors, (2) the injection temperature of the cation or anion precursor, (3) the reaction time, and (iv) the concentration of the precursors.

A wide range of morphology-controlled MHP NCs have been produced by various groups of varying temperature, reaction time, precursors, and ligands. All inorganic $CsPbX_3$ NCs were chosen for the study due to the fact that hybrid organic–inorganic-lead-based perovskites show low intrinsic stability. Protesescu et al. [36] prepared morphology-controlled colloidal $CsPbX_3$ NCs with oleic acid and oleylamine as the ligands. Ligands are used to stabilize the NCs by preventing agglomeration. By adjusting the reaction temperature, the size of the NCs can be tuned in the range of 4–15 nm. The crystals obtained exhibited cubic morphology. The $CsPbX_3$ NCs other than $CsPbI_3$ remained stable in the cubic phase when it was stored at room temperature. When the feeding ratio of precursors and ligands is slightly changed in the works by Deka et al. [37] and Park et al. [38], improved-quality NCs with strong PL were obtained. In another method, NH_4X (X = Cl, Br, I) and PbO was used. PbO was used as a substitute to PbX_2 (X = Cl, Br, I). This provided a halide-rich atmosphere [39] that helped in enhancing the quality and durability of the products. The chemical stability of $CsPbI_3$ nanocubes improved remarkably when tri-octylphosphine and PbI_2 were used as the precursors.

Colloidal $CsPbBr_3$ nanocubes, nanoplatelets, and nanosheets were prepared by Liang et al. [34] and Almeida et al. [35] by controlling the amount of oleic acid and oleylamine. NCs with nanowire morphology and uniform growth direction were obtained by varying the reaction temperature, as shown in Figure 1.3 [40]. $CsPbBr_3$ nanoplatelets could be prepared by increasing the ratio of aliphatic ammonium ions. The competition among aliphatic ammonium ions and Cs^+ for lattice sites helps in promoting the formation hybrid layered structures [35]. In some cases, alkyl ammonium bromide (oleylamine-HBr) functioned as dimension-modulating reagent in the preparation of $CsPbBr_3$ nanocubes and nanoplatelets. The size tunability of the $CsPbBr_3$ NCs depends on oleylamine-HBr concentration.

Protesescu et al. [42] used acids and amines with short alkyl chain instead of oleic acid and oleylamine. Monodisperse colloidal $CsPbBr_3$ nanowires with diameters in the range of 10.1±1.6 nm were synthesized by Zhang et al. [41] by using octylamine and oleylamine as the ligands. Preparation using octonic acid and octylamine led to the formation of nanosheets [44]. By adjusting the ratio of octanoic acid and octylamine to the long-chain ligands of oleic acid and oleylamine, the lateral size of the resulting nanosheets can be tuned from 300 nm to 5 µm. A higher ratio of short- to long-chain ligands yields larger

NCs. The shape, size, and surface properties of the $CsPbBr_3$ NCs depend on the chain length variation [39] of alkyl chain ligands including acids and amines. In addition to the temperature effect, the chain length variation of carboxylic acids and amines showed independent correlation to the shape and size of the NCs. The effect of chain length on the shape and size of the NCs is summarized in Figure 1.4 [39].

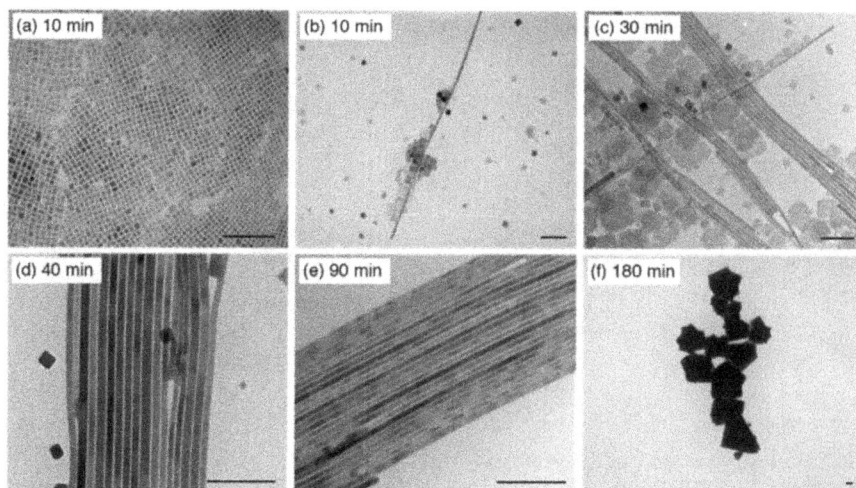

FIGURE 1.3 Shape evolution of the as-prepared $CsPbBr_3$ nanostructures synthesized with oleic acid and oleylamine as the ligands with different reaction times. (Reprinted with permission from [41]. Copyright 2015 American Chemical Society.)

Colloial hybrid organic–inorganic LHP NCs were synthesized by several groups. Tyagi et al. [45] reported the initial work on the synthesis of colloidal methyl ammonium lead bromide ($MAPbBr_3$) nanoplatelets. $PbBr_2$ and MABr were used as the precursors. Octylammonium bromide played the role of coordinating ligand. A solution of oleic acid and octy-lammonium bromide was magnetically stirred, and $PbBr_2$, MABr, and octylammonium bromide were added to it. The product mixture contained nanoparticles and nanoplatelets of $MAPbBr_3$. Upon centrifugation and sonication, crystalline $MAPbBr_3$ nanoplatelets having thickness of a single cell were obtained. When octylamine was used as the coordinating ligand nanoparticles in the size range 5–20 nm, nanoplatelets were obtained. Nanoplatelets showed a tendency to aggregate into polycrystalline nanoplatelets.

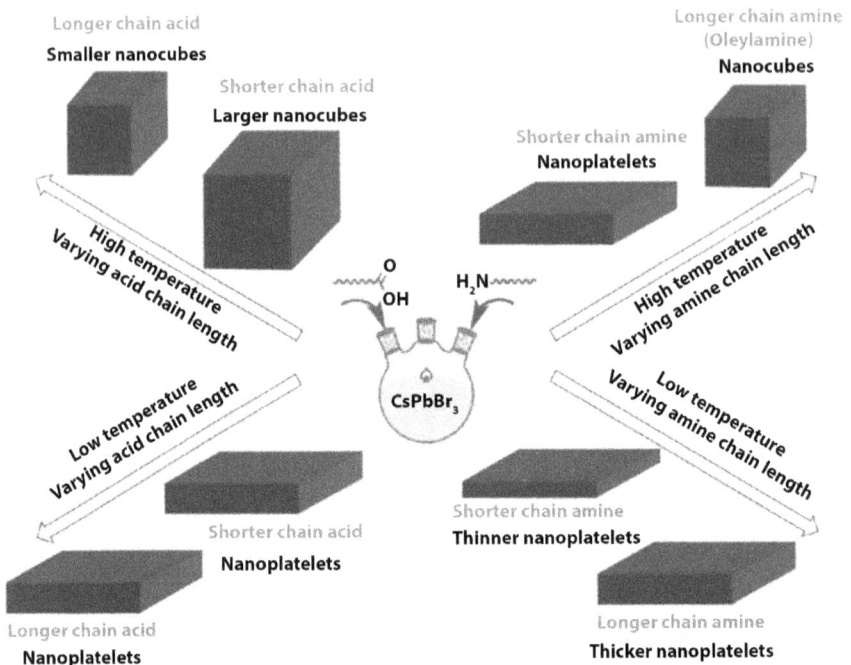

FIGURE 1.4   Summary of the shape and size dependence on the chain length of carboxylic acids and amines. (Reprinted with permission from [39]. Copyright 2016 American Chemical Society.)

Synthesis of MAPbI$_3$ nanocubes, nanowires, and nanoplatelets was demonstrated by Sun et al. [46] using oleylamine and oleic acid as the ligands. The nanocubes obtained had a mean particle size of 10 nm and absorption edge at about 750 nm. The nanoplatelets synthesized by this technique showed a tendency of agglomeration, and the shape of the obtained structures was not of satisfactory. Highly monodisperse and high-quality nanocubes were synthesized by Imran et al. [43] using oleyl ammonium bromide as the Br precursor and oleic acid as the ligand. They synthesized formamidinium lead bromide (FAPbBr$_3$), CsPbBr$_3$, and MAPbBr$_3$ NCs by the same method. FAPbBr$_3$ nanocubes were more robust and maintained bright PL emission even after two to three cycles of purification. CsPbBr$_3$ were nonluminescent and MAPbBr$_3$ crystals decomposed quickly. In the same procedure by using oleic acid and oleyl amine as the ligands and CH$_3$MgBr as the precursor, FAPbBr$_3$ NCs with board size distribution were obtained. The change of precursor may be the reason for this. The use of benzoyal halides as precursors lead to the formation of APbX$_3$ (A = Cs, MA, FA; X = Cl, Br, I) NCs having cubic shape [43]. Such NCs exhibited high phase stability, good size distribution, and excellent optical properties.

1.2.2   LIGAND-ASSISTED REPRECIPITATION

In this process [33], desired ion is dissolved in a solvent. This solution is converted to its nonequilibrium supersaturation stage by varying the temperature, by evaporating the solvent, or by adding a miscible co-solvent, in which the solubility of the ions is low. To regain the equilibrium state, spontaneous precipitation and crystallization reaction occurs. This process continues until the system reaches the equilibrium state again. The formation and growth of crystals can be controlled down to nanoscale if this process is carried out in the presence of ligands, hence the name LARP. This technique is used for the fabrication of colloidal NCs. It is an easy and fast method to obtain high-quality NCs by this strategy, but the yield is limited by the polar and nonpolar ratio.

In the LARP technique, the desired precursor salt is dissolved in a good polar solvent such as dimethylformamide (DMF) and dimethyl sulfoxide (DMSO) and is dropped into a poor solvent such as toluene or hexane in the presence of ligands. The salts that were used [21] in the LARP technique are PbX_2, CsX, MAX, and FAX (X = Cl, Br, and I). An instantaneous supersaturation is induced by the mixture of two solvents. This triggers the nucleation and growth of perovskite NCs [47]. This technique can be used for large-scale production of MHP NCs since it is carried out in air using simple chemical apparatus. But the nucleation and growth stages of LARP cannot be separated in time [48].

LARP is a room-temperature technique developed for the synthesis of colloidal ABX_3 NCs. By varying the solvent concentration of the capping ligands and the reaction time, one can control the morphology of the colloidal NCs. $CsPbX_3$ nanospheres, nanoplatelets, and nanobars are produced by using ethyl acetate as the antisolvent and oleylamine and oleic acid as the coordinating ligands [49]. When ethyl acetate was replaced by toluene nanocubes, nanorods and nanowires were obtained. NCs were of high crystalline quality and showed high PLQY. The morphology of $CsPbX_3$ NCs can be modified by choosing different combinations of organic acids and amine ligands. Figure 1.5 [46] gives the schematic representation.

Sichert et al. [18] reported the synthesis of $MAPbBr_3$ NCs through LARP using toluene as the antisolvent and octylammonium as the ligand. The thickness of the nanoplatelets thus obtained can be tuned by varying the concentration of octylammonium. Transmission electron microscope (TEM) characterization [32] showed the presence of quasi-spherical nanoparticles. This might be due to the degradation of nanoplatets on exposure with the electron beam. So the stability of the NCs needs to be enhanced.

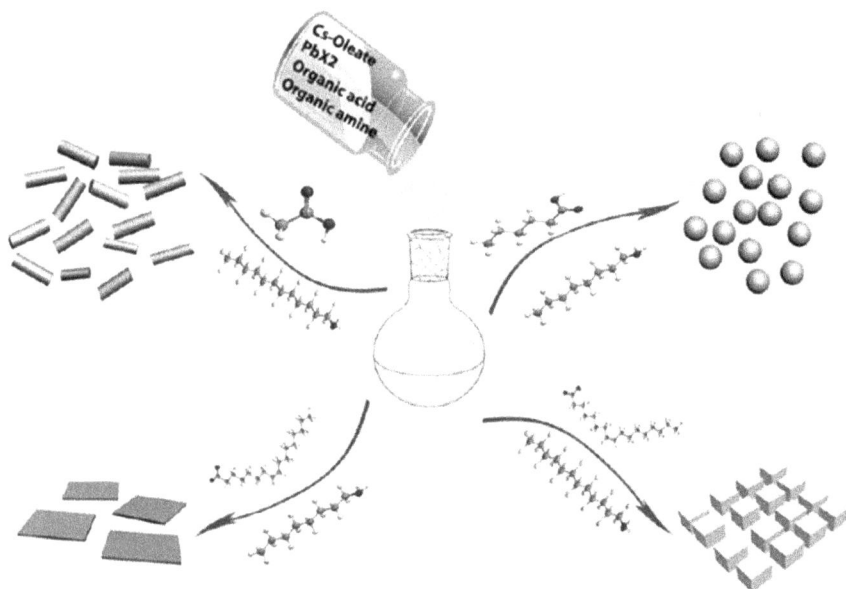

FIGURE 1.5   Schematic illustration of the formation process for different $CsPbX_3$ (X = Cl, Br, and I) NCs mediated by organic acid and amine ligands through LARP. Hexanoic acid and octylamine for nanospheres; oleic acid and dodecylamine for nanocubes; acetate acid and dodecylamine for nanorods; and oleic acid and octylamine for nanoplatelets. (Reprinted with permission from [46]. Copyright 2016 American Chemical Society.)

MAPbBr$_3$ nanoplatelets were also developed by Sichert et al. [18] using acetone as the antisolvent and introducing oleic acid and oleylamine as the ligand. The reverse LARP process was developed by Zhang et al. [50], in which the sequence of the solvent mixing was changed. Antisolvent was poured into the good solvent, which induced the nucleation and growth of NCs. They synthesized MAPbBr$_3$ NCs with nanocube and nanowire morphology by controlling the amount of ligands.

FAPbX$_3$ NCs were synthesized [51] with toluene (or cholroform) as the antisolvent. They had nanocube or nanoplatelet shapes. Moreover, they had better colloidal and chemical stability and high PLQY [20].

The advantage of the LARP method is that it facilitates the room-temperature synthesis of perovskite systems under air. But it has certain disadvantages that prevent its use in large-scale production. If the precursor solvent is added too much, the obtained NCs would be redissolved in the nonpolar solvent. The solvents such as DMF can easily degrade and even dissolve $CsPbX_3$ NCs [52]. The interaction between precursors and polar solvent leads to the formation of defective perovskite NCs. This is known as solvent effect. The effect of using different polar solvents on the crystallization of $MAPbI_3$ NCs was investigated by Zhang et al. [50]. They proved that in coordination solvents such as DMSO, DMF, and tetrahydrofuran PbI_2 forms stable intermediates. But this does not happen in noncoordinating solvents such as Υ-butyrolactone and acetonitrile [50]. So, these solvents have different impacts on the crystal structure of the synthesized NCs. Defective $MAPbI_3$ NCs are formed due to the strong bonding between PbI_2 and coordinating solvents. When noncoordinating solvents were used, defect-free $MAPbI_3$ NCs were obtained, as shown in Figure 1.6 [44]. Also, the polar solvents such as DMF and DMSO have high boiling point and are toxic. So, they are not suitable for large-scale production.

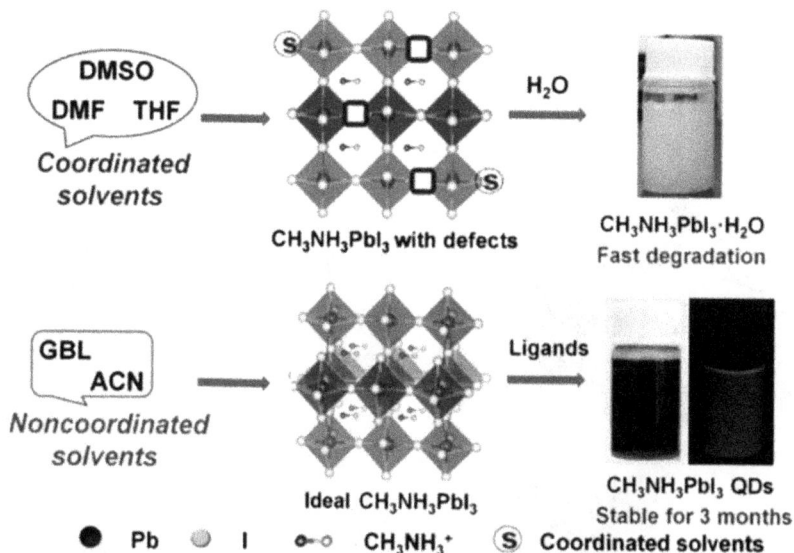

FIGURE 1.6 Effects of the solvents on the crystal structure of $CH_3NH_3PbI_3$ NCs: the use of coordinated solvents (top) leads to the formation NC defects, which are prone to degradation under humidity; noncoordinated solvents (bottom) allow for the formation of "defect-free" and stable NCs [50]. (Reprinted with permission from [50]. Copyright 2017 American Chemical Society.)

1.2.3 ULTRASONICATION-ASSISTED SYNTHESIS

In this method, nanostructures are produced through the effects of high-intensity ultrasonic irradiation of materials. Both physical and chemical effects of ultrasound are utilized in the production of nanostructured materials. In the ultrasonic method, the precursors are added to a nonpolar solvent. The precursor solution is then positioned in an ultrasonicator or treated by tip sonication. This will lead to NC formation [53]. This is an important way of obtaining MHP NCs with various shapes. The shape and size of the NCs depend on the concentration of the precursors and ligands and the reaction time. The advantage of this method is that no polar solvents are used.

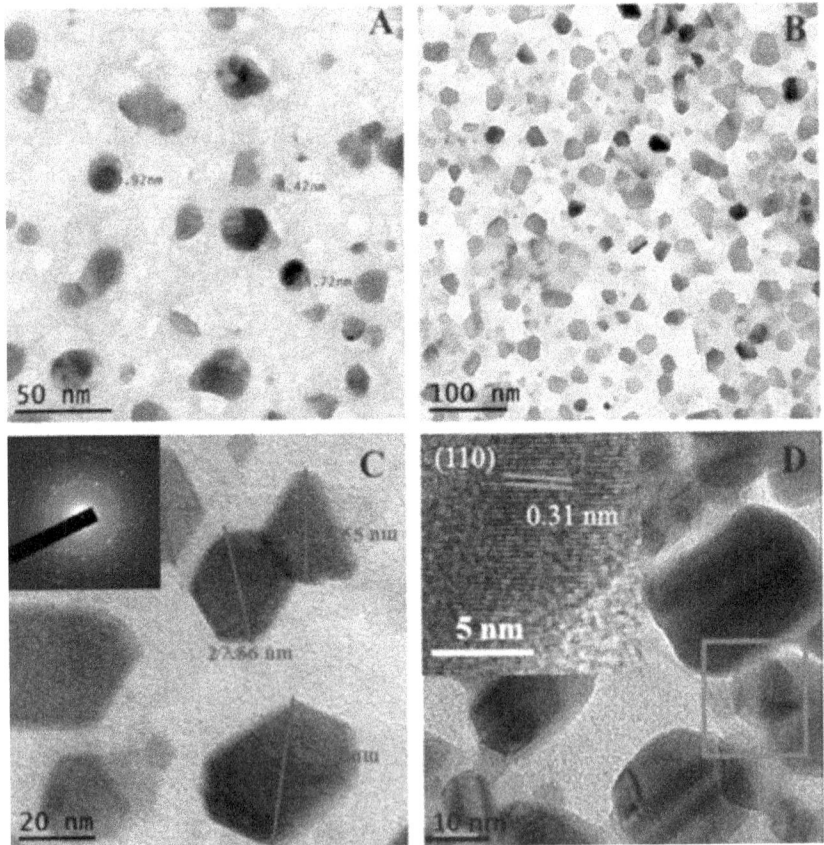

FIGURE 1.7   TEM images of the $CH_3NH_3PbI_3$ nanoparticles formed within 10 min (A) and 20 min (B) of ultrasonic irradiation. (C) Higher magnification image of the particles shown in (B). (D) A higher magnification image of several $CH_3NH_3PbI_3$ particles. Inset: lattice fringes of a single particle. (Reprinted with permission from [54]. Copyright 2016 Elsevier.)

Van Der Stam et al. [56] reported the synthesis of MAPbI$_3$ NCs by sonochemical synthesis. The solutions of methyl ammonium iodide and PbI$_2$ in isopropanol were tip sonicated together. NCs in the size range 10–40 nm were obtained. Figure 1.7 shows TEM images of the CH$_3$NH$_3$PbI$_3$ nanoparticles formed by this method. It can be seen that the size of the particles depends on the sonication time. Chen et al. [57] prepared MAPbX$_3$ quantum dots (QDs) with uniform particle size by the ultrasonic method. MAX and PbX$_2$ were dissolved in DMF. Prior to sonication, n-octylamine was added to the precursor solution to avoid agglomeration of the particles. The precursor solution was added to toluene placed in the bath sonicator (see Figure 1.8). The ultrasonic treatment limits the crystallization process, and NCs with controllable size properties are obtained. Furthermore, the particles obtained are of uniform size [55]. The PL and absorption spectra of the perovskites can be tuned by varying the halide compositions. The NCs obtained have good photodetection properties such as prolonged lifetime, improved photostability, increased external quantum yield, and faster response speed.

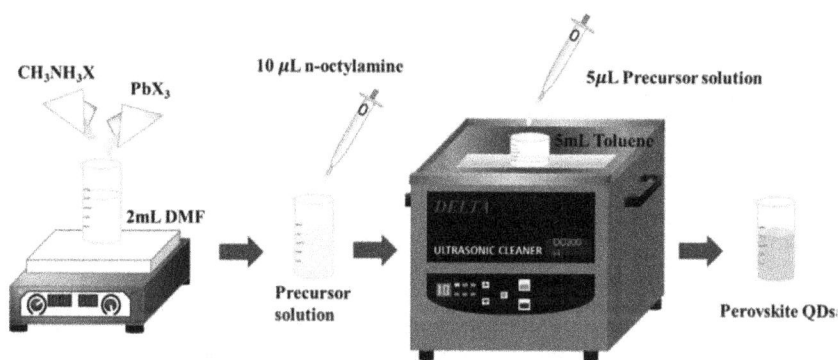

FIGURE 1.8 Schematic for the perovskite QD preparation in this study. (Reprinted with permission from [55]. Copyright 2017 Elsevier Ltd and Techna Group S.r.l.)

1.2.4 SOLVOTHERMAL METHOD

The solvothermal method is a commonly used method for the synthesis of NCs. The NCs are obtained due to the high temperature reduction of precursor salts in some suitable solvent. By varying the concentration of precursors, reaction time, and so on, we can tune the morphology of the NCs. Van Der Stam et al. [56] synthesized CsPbX$_3$NCs by this method. CsPbX$_3$ nanocubes were synthesized when the precursors were heated without predissociation, and nanowires were obtained when the precursors were heated after predissociation. The concentration of precursor ions was low without predissociation, so only small amount of $CsPbX_3$ NCs nucleated. Predissociation leads to an increase in the concentration of precursor of which self-assembled into nanowires through seed-mediated growth. Figure 1.9 shows [59] the synthesis path.

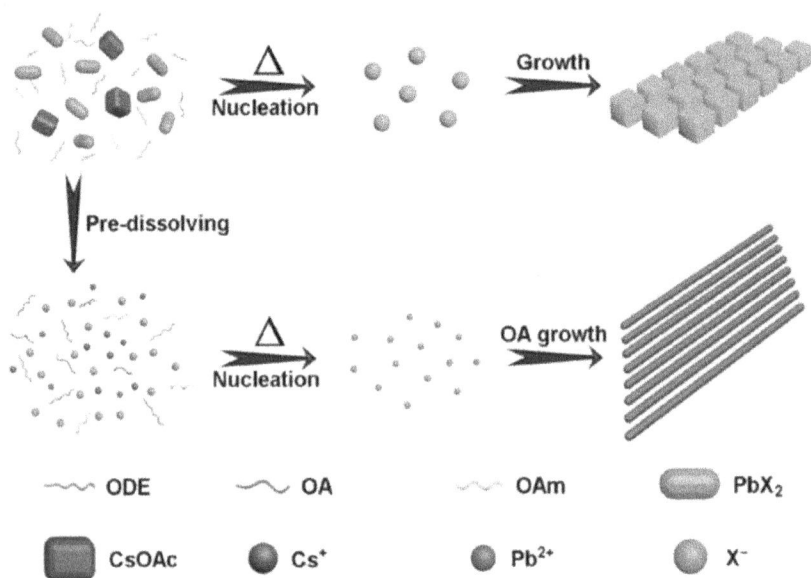

FIGURE 1.9 The proposed growth process of the CsPbX3 NCs obtained without predissolution and with predissolution of the precursors. ODE: 1-octadecene; OA: oleic acid; OAm: oleylamine; CsOAc: cesium acetate. (Reprinted with permission from [57]. Copyright 2017 the Royal Society of Chemistry.)

1.3 OPTICAL PROPERTIES HALIDE PEROVSKITE NCS

Colloidal MHP NCs have exceptional optical and optoelectronic properties. This is due [25] to wide-bandgap tunability, large absorption coefficient, high PLQY, narrow emission full width at half maximum, high carrier mobility, and long carrier diffusion length. HI and LARP approaches [21] provide the direct synthesis of nanostructured materials at room temperature. Nanostructures of $MAPbX_3$ (X = Cl, Br, I), $FAPbX_3$ (X = Cl, Br, I), and $CsPbX_3$ (X = Cl, Br, I) exhibit PL covering the entire visible spectrum.

One can tune the bandgap of the MHP NCs by [25] exchanging the individual components. $MAPbBr_3$ NCs had absorption onset and PL emission around 529 nm. The optical properties of NCs can be shifted throughout the entire visible spectrum by modifying the halide composition. In the case of $CsPbX_3$ NCs, the emission could be shifted from 410 nm (X = Cl) to 512 nm (X = Br) to 685 nm (X = I). This tunability is due [58] to the specific electronic structure of halide perovskites. As the size of A cation decreases (from formamidinium (FA) to methyl ammonium (MA) to Cs) cubic crystal structure is distorted due to the increase in the tilting angle of Pb–X–Pb bonds. This leads to a blue shift in the bandgap. The emission spectra of MA-based perovskite NCs vary with the halide content from 407 to 734 nm. Like A-site cation, B-site cation also has a role to play in dictating the final optical properties of MHP NCs. The bandgap and PL emission of the NCs strongly red shift when Pb^{2+} ion is replaced by Sn^{2+}. Higher electronegativity of Sn^{2+} may be the reason [21] for this. Upon quantum confinement, the PL of MHP NCs exhibits [33] a blue shift. This is due to the modification in the optical transition energy due to electron–hole pair interaction. Electron–hole pair interaction energy is also known as the exciton binding energy. In quantum-confined structures, the amount of binding energy [33] is much higher than that of bulk materials. Change in dimensionality and reduced screening due to the coulomb interaction due to surroundings are the reasons for this.

Exciton binding energy is an important parameter for an optoelectronic material [22]. When they are optically excited, the value of exciton binding energy determines the nature of their response. The exciton binding energy of LHPs is so small that they can be considered as Wannier–Mott excitons [22]. If the dimension of the NCs is significantly larger than the exciton Bohr radius, there is no quantum confinement, and the values are similar to the bulk counterparts. As the size of the NCs was reduced below Bohr radii, quantum-confinement-effect-induced variations in the PL emission spectrum are observed [25].

1.4 STABILITY

The resistance against atmosphere (moisture, oxygen, and temperature) and light are major challenges that hinder the development commercial PV devices using perovskite NCs. Understanding the degradation mechanism and further improvements in stability are the major issues in the current scenario. The stability of the perovskites can be improved by surface passivation, encapsulation, and doping or substitution.

Halide perovskite NCs can be easily synthesized due to [59] its electrovalent bond features. Hence, there are chances for easy breakdown during subsequent isolation and purification process. Surface passivation agents are used to maintain the perovskite structure intact. Amino capping ligands are used [59] for the passivation in MHP NCs. But they are disintegrated during the passivation process. Due to the synergic effect between carboxylic and amino groups, the capping efficiency was improved [60] in the presence of oleic acid and amine ligands. When 3-aminopropyl triethoxysilane and carboxylic acids were used [60] in tandem, NCs were well passivated, and they exhibited high PLQY and stability. When NCs were synthesized in halide-rich conditions, the halide was retained largely on the surface. It helped improve the self-passivation.

Coating is an efficient and effective method [60] to stabilize the moisture and solvent sensitive halide perovskite from harsh environment. Perovskite possesses poor thermal stability and intrinsic chemical instability due to lower crystal lattice energy. If the B-site cation is replaced by smaller cations such as Mn^{2+}, Sn^{2+}, Cd^{2+}, and Zn^{2+} ions, there was an enhancement [61] in the formation energies of the perovskite lattices and it lead to improved thermal stability [56].

The surface chemistry of the MHP NCs is also important. Ligands can influence [22] the surface termination and stability of the NCs. Some ligands passivate trap states and others introduce new trap states. There is huge variation in the surface chemistry and the nature of the energetic of trap states when one varies the halide ion from chloride to bromide and iodide. The surface energy of the NCs changes according to the ligands.

1.5 APPLICATIONS

1.5.1 SOLAR CELLS

Conventional solar cells use semiconducting materials such as silicon, cadmium telluride, organic thin films, and so on as the light-absorbing materials. High carrier mobility, long carrier diffusion length, higher optical absorption, excellent defect tolerance, and lower surface rate are the advantages of perovskite materials [61]. PSCs have two structures [62]: mesoscopic structure and planar heterojunction structure. Mesoscopic solar cells are based on dye-sensitized solar cells. Mesoporous oxide materials such as TiO_2 act as skeleton materials and transport electrons. NCs are attached to TiO_2, and hole transport material is deposited on its surface. These three together act as the hole transport layer (HTL). In plane heterostructure, the perovskite material is sandwiched between the electron transport layer (ETL) and the HTL (see Figure 1.10). Perovskite is an ambhipolar material, and it transmits electrons and holes at the same time [63].

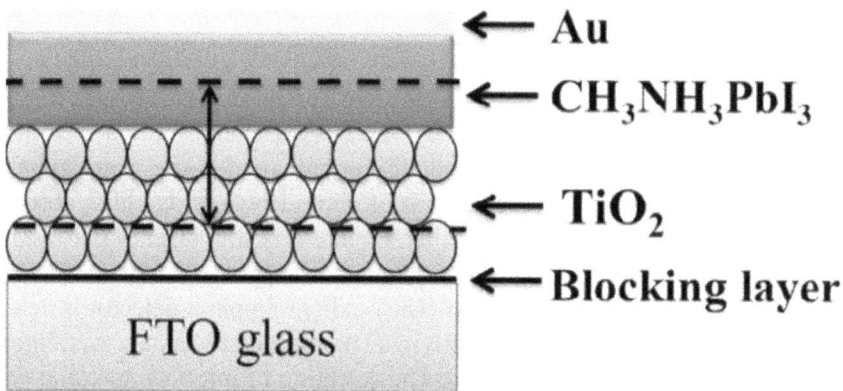

FIGURE 1.10 Structure of a PSC. (Reprinted with permission from [31]. Copyright received Physical Chemistry Chemical Physics 2014.)

As mentioned, colloidal MHP NCs have great potential [9] in solar cells due to their attractive optoelectronic properties. Appealing optical properties such as high quality, monodispersity, shape uniformity, and superlative self-assembly performance are the specialties [25] of colloidal MHP NCs. Swarnakar et al. [65] incorporated colloidal MHP NCs into PV applications. PV devices based on colloidal $CsPbI_3$ NCs as the light absorbent materials yielded [64] PCE as high as 10.7% and a V_{oc} of 1.23 V [25]. When $MAPbX_3$ (X = I, Br) was used [64], the PV performance improved by 13.4% [64]. In this case, they improved the carrier mobility of the NC films by enhancing the charge coupling between the NCs [64]. Fu et al. [25] used colloidal $MAPbBr_3$ NCs as the nucleation center to induce the growth of high-quality $MAPBI_3$ films. The performance of the device depends on the quality of the films. This method caused the heterogeneous nucleation of MHPs rather than homogenous nucleation due to the presence of similar lattice constants. The films obtained have good crystallinity and decreased grain size. Longer lifetime and lower trap-state density of these films improved [65] the PCE to 17.10% compared to the devices with homogenous nucleation. The PV performance of all-inorganic perovskite NCs is superior to that of organic–inorganic perovskites. Environmental stress causes [66] the dissociation of $MAPbI_3$ into PbI_2 and CH_3NH_3I. $CsPbX_3$ (X = Cl, Br) is more stable under various stresses [67]. Perovskite NCs exhibit multiple exciton generation effects [25]. This effect helps in extracting charges from the thicker absorber layer when used in solar cells. The PV performance of perovskite NCs can be tuned [21] by optical and energy bandgap manipulation [21].

1.5.2   PHOTODETECTORS

Photodetector is a device that converts an optical signal into an electrical signal. Organic–inorganic hybrid and all-inorganic perovskite material have great potential [68] in the development of photodetectors due high absorption efficiency and high carrier mobility. Even a very thin layer of perovskite absorbs [69] light strongly and gives quick photoresponse since the distance traveled by the carriers is very short. The quality of a photodetector is determined by its responsivity (R), detectivity (D), noise equivalent power, linear dynamic range, and response speed. The selection of interface material and design of device structure influences the performance of the photodetectors. Halide perovskite QDs/NCs show [71]more excellent detective performance than bulk counterparts due to its high exciton binding energy and quantum confinement. Photodetectors based on $CsPbI_3$ NCs were first demonstrated by Ramaswamy and group [66]. A schematic diagram of a photodetector based on $CsPbI_3$ is shown in Figure 1.11 [66]. These NCs are used in photodectorrs since it is defect-free single crystal [72], highly crystalline in nature [68], and have large specific surface area [73]. 1D nanostructure is suitable for high-performance devices [73] since its conductive channel could confine the active area and shorten the carrier transit time.

Hu and Zhu [74] used hybrid organic–inorganic $MAPbI_3$ NCs synthesized by the solution method as the active layer for photodetectors. They employed poly(3,4-ethylenedioxythiophene) polystyrene sulfonate as the HTL and [6,6]-phenyl-C61-butyric acid methy ester (PCBM) as the ETL. The photoresponse time is found [74] to vary with the interface material [75]. Sargent's group [69] fabricated the device by depositing TiO_2, Al_2O_3, and PCBM on the conductive glass of fluorine-doped tin oxide. This formed an additional hole-blocking layer, which passivated the trap state of the interface to create small dark current. This device had a responsivity of 0.4 W^{-1} at

600 nm and a detectivity of 10^{12} Jones at low bias. The stability of the device in air increased [69] due to the introduction of Al_2O_3 and PCBM [69]. One can tune [70] the spectrum response in the range 370–780 nm by controlling the halogen ratios [70, 71].

CsPbI3 NCs layer

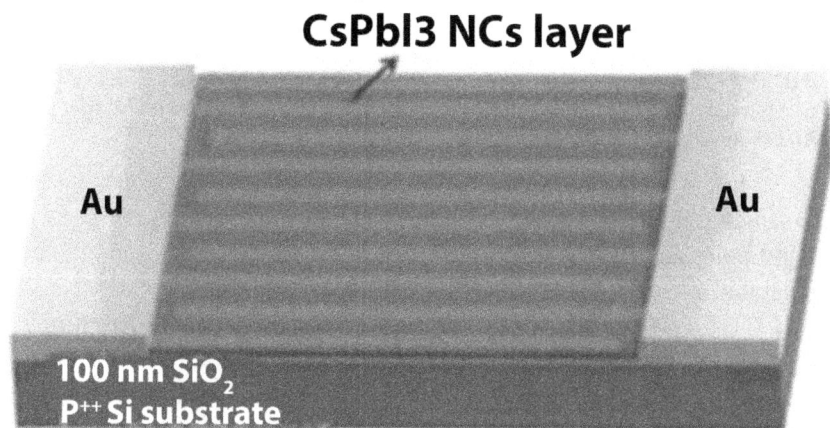

Au   **Au**

100 nm SiO$_2$
P^{++} Si substrate

FIGURE 1.11 Schematic diagram of the photodetectors based on CsPbI$_3$ NCs/QDs. (Reprinted with permission from [66]. Copyright 2016 Royal Chemical Society.)

Poor stability is the common problem associated with both NCs and devices. Perovskites are strongly affected [76] by environmental influences, including oxygen, light irradiation, humidity, and heat because of its ionic nature.

1.5.3 LIGHT-EMITTING DIODE

As already mentioned, MHPs provide high PLQY and highly saturated colors. The primary colors obtained from LHP NCs possess an impressive gamut. Because of its excellent optoelectronic properties and ease of fabrication, organic–inorganic LHP is suitable [77] for light-emitting diodes. Figure 1.12 describes a structure of an LED with perovskite as the emissive layer (EML). When voltage is applied, holes and electrons are injected from the anode and the cathode. They go through the ETL and HTL into the emitting layer, where they form excitons. Subsequently, radiative recombination takes place, and photons are emitted [74].

FIGURE 1.12 Cell configuration diagram of (a) the inverted LED structure of ETL/ EML/HTL (n-i-p), (b) normal LED structure of HTL/EML/ETL(p-i-n), and (c) operation mechanism diagram of the both structures. (Reprinted with permission from [78]. Copyright 2019American Chemical Society.)

The quantum well structure of perovskite QDs effectively confines electrons and holes, and this is useful for radiative recombination. Ionic defects are generated due to low interaction energy between metal cations and halide anions [77]. These defect sites can be passivated using amine-based materials. This treatment led to a reduction in undercoordinated Pb. This enhances the efficiency and device stability of LEDs [79]. Device performance can be improved [80] by using mixed ion perovskite and light extraction techniques.

1.6   COMPUTATIONAL MODELING IN PEROVSKITES

Atomistic modeling and simulation of materials is very important in material science since they provide valuable information about the known material processes and properties [2] and predict the shortest way to find new functional materials that meet requirements [32]. Lattice constants, electronic structure, linear response properties, and transport properties are the material characteristic of a crystalline solid. Using the first principle method based on quantum mechanics, we can determine these properties with an accuracy of 1% relative error to the experiment, in which atomic information and lattice structure are used as the input data. Computational material science is an interdisciplinary subject, which is useful both in software and hardware aspects. In hardware, it provides rapid and ever-increasing processor speed and memory capacity, whereas in software, it provides continuous progress in simulation algorithms and material theory. Density functional theory (DFT) [81] is used for atomistic modeling and simulation [81]. Similarly, many-body perturbation and pairwise interatomic potential molecular dynamics have been successfully employed in the case of halide perovskites.

We deal with the material properties such as crystal structures and electronic and optical properties in this session on atomistic modeling and simulation.

1.6.1   X-RAY DIFFRACTION STUDIES

Ancharova et al. [82] carried out computational modeling of X-ray scattering on perovskite-like oxide (ABO_3) to understand nanostructuring and mechanism of oxygen transport. From the viewpoint of structure investigation, especially interesting are the samples with oxygen composition $2.5 < (3 - \delta) < 2.7$, that is, strongly nonstoichiometric perovskites with high defect content [82]. The ceramic method was used for the synthesis of compounds for study. Various methods such as slow cooling in a furnace, annealing at 900 °C in dynamic vacuum, and annealing at 500 °C in 5% H_2/Ar atmosphere were employed to change oxygen stoichiometry. Diffraction studies were carried out by synchrotron radiation. Area Diffraction Machine Software was used for processing 2-D diffraction patterns.

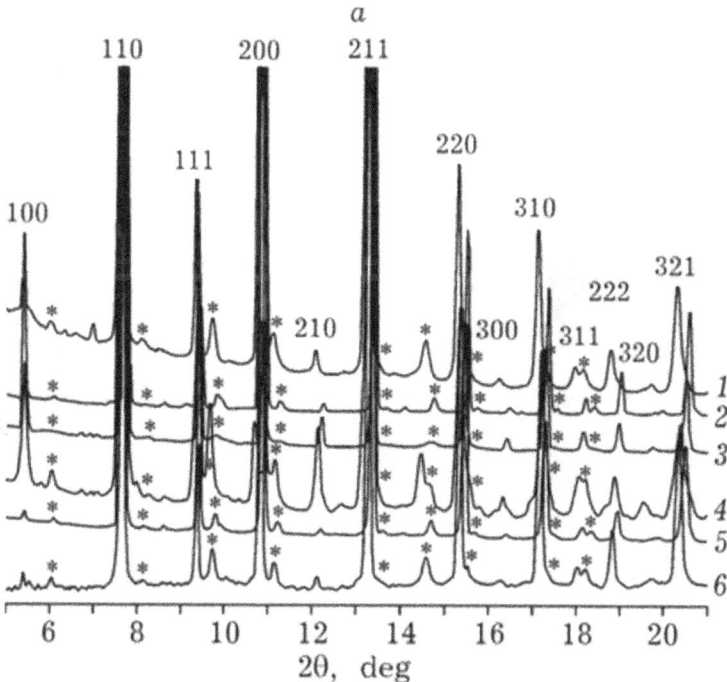

FIGURE 1.13   Diffraction patterns of nonstoichiometric perovskites.

A diffraction phenomenon is specific for samples with different cation composition and oxygen stoichiometry in the range $2.45 < (3 - \delta) < 2.66$. This is shown in Figure 1.13. The variation of oxygen stoichiometry leads to the formation of different stoichiometric phases of defect ordering. Homogeneous ordering of oxygen vacancies [82] proceed with the formation of a series of stoichiometric phases (brownmillerite, Grenier, etc.)

Homogenous and inhomogenous models were used to explain the vacancy ordering of oxygen. In the homogenous model, vacancy ordering occurs in a double cubic cell. This is a superstructure with heterogeneous nature. The diffraction pattern of compounds having a superstructure has oxygen composition in the range $2.45 < (3 - \delta) < 2.66$. The stoichiometric phase $ABO_{2.5}$ corresponds to brownmillerite phase. TEM data [83] corroborate the presence of the $ABO_3 + ABO_{2.5}$ heterogeneous system.

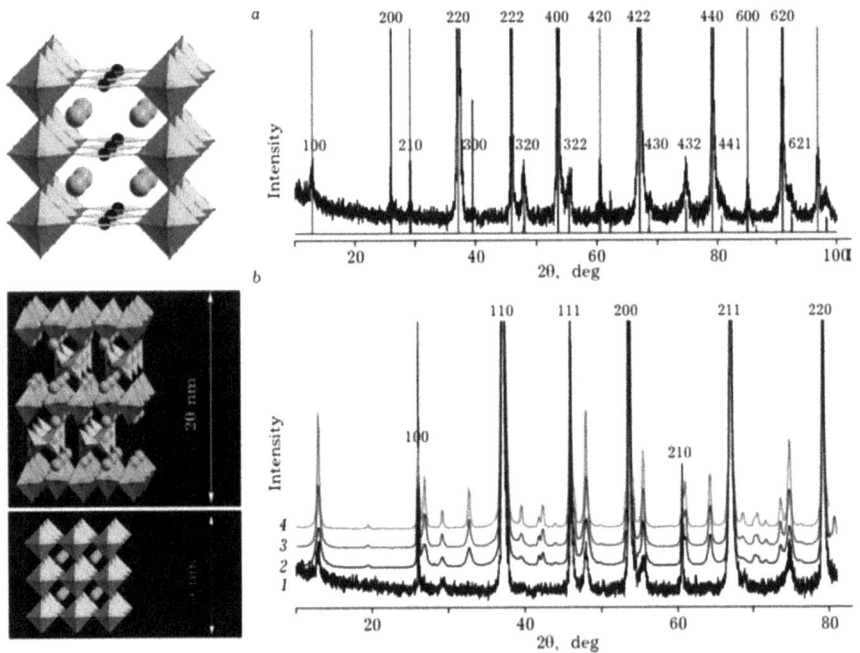

FIGURE 1.14   Model of homogeneous ordering of vacancies (a) and nonhomogeneous ordering of perovskite and brownmillerite structures $ABO_3 + ABO_{2.5}$ along one direction of the crystal (b): 1—Experimental diffraction pattern of sample $SrFe_{0.95}Mo_{0.05}O_{2.66}$ quenched in vacuum; 2–4—Probability of grouping of $(ABO_{2.5})$ cells.

The inhomogenous model has coherently joined lamellar components with perovskite and brownmillerite structures alternately. The distortion in order leads to diffuse scattering. Modeling of diffraction patterns showed that the diffused scattering is concentrated in regions [82] of superstructural peaks (see Figure 1.14). Also, brownmillerite layers are four times thicker than perovskite layers. This variation affects the intensity of peaks. The calculated diffraction pattern and experimental pattern are in good agreement when the thickness of the perovskite and brownillerite structures is in the range 5 and 20, respectively.

1.6.2 DIFFUSION

At ambient temperature, nonstoichiometric perovskite-related oxides exhibit fast oxygen transport. This is attributed to the microstructural texturing of nanoscale domains. The kinetics of oxygen incorporation in nanostructured oxides were explained by using the heterogeneous diffusion model (see [84]) for the description of diffusion processes in polycrystalline metals.

Nonstoichiometric perovskite samples for analysis were synthesized by a solid-state reaction between corresponding metal oxides and carbonates. Nanostructured materials were obtained by ball milling the so-obtained products. The powder obtained was calcined at 900 °C, pressed in pellets and annealed in air at 1400 °C for 6 h and cooled in the furnace. Oxygen stoichiometry was varied by slow cooling in the furnace and annealing at 900 °C under high vacuum and quenching.

The X-ray diffraction (XRD) data of $SrCo_{0.5}Fe_{0.2}Ta_{0.3}O_{3-y}$ have peaks corresponding to the cubic structure with lattice parameters 3.934 A° and 3.952 A° [84]. Oxygen stoichiometry of the samples was analyzed by iodometric titration, and the results showed that due to quenching, the oxygen content varied in the range $2.92 < (3 - y) < 2.70$. The oxidation of $SrCo_{0.5}Fe_{0.2}Ta_{0.3}O_{3-y}$ was analyzed by chronopotentiometric studies. The measurements of the chronopotentiometric proved that phase transition occurs with the change in oxygen stoichiometry. Phase reaction leads to the oxidation of samples, and the potential range within which the oxidation takes place was determined.

Kinetic studies of potentiometric oxidation of $SrCo_{0.5}Fe_{0.2}Ta_{0.3}O_{3-y}$ and analysis of current transients were carried out to find the optimum values of τ_1 (first diffusion time), τ_2 (second diffusion time), and α. The Laplace transform was used for it.

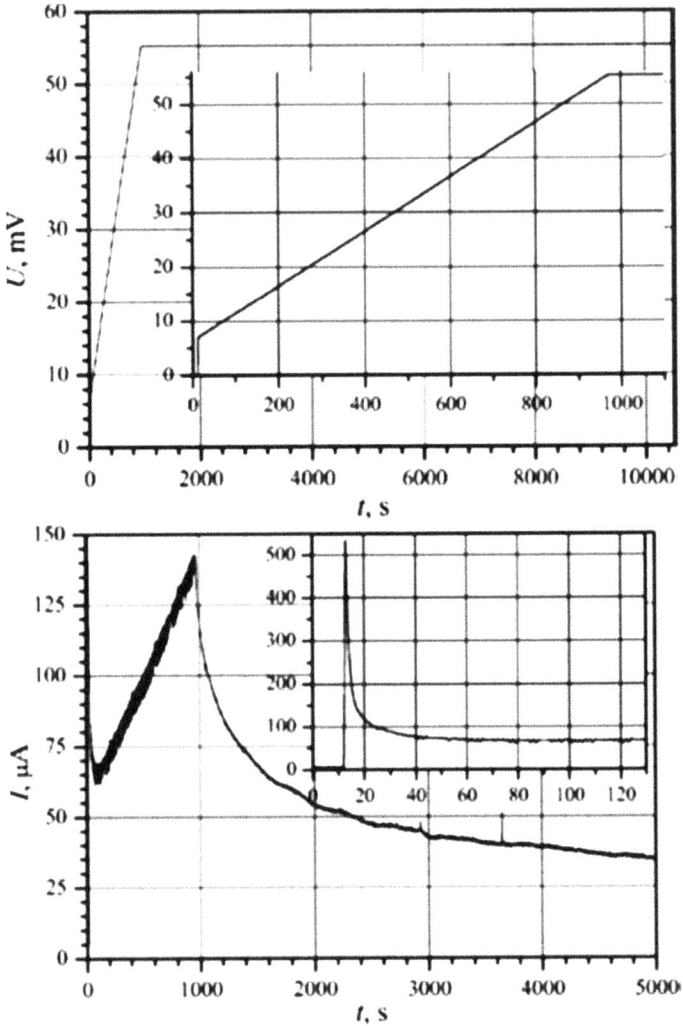

FIGURE 1.15 Applied potential (a) and current transient (b) versus time. (Reprinted with permission from [84]. Copyright 2006 Elsevier.)

Figure 1.15 gives the plot of potential and transient current versus time. By performing computational calculations using MATLAB, potential step parameters for Figure 1.15 were obtained. Using these parameters, Figure 1.16 was plotted, and τ_1, τ_2, and α were obtained as follows:

$$\tau_0 = \frac{r^2}{D_1}$$

$$\tau_2 = \frac{R^2}{D_2}$$

where D_1: coefficient of slow diffusion in nanodomain; D_2: effective fast diffusivity (due to interfaces); R: radius of sample particles; r: radius of nanodomains.

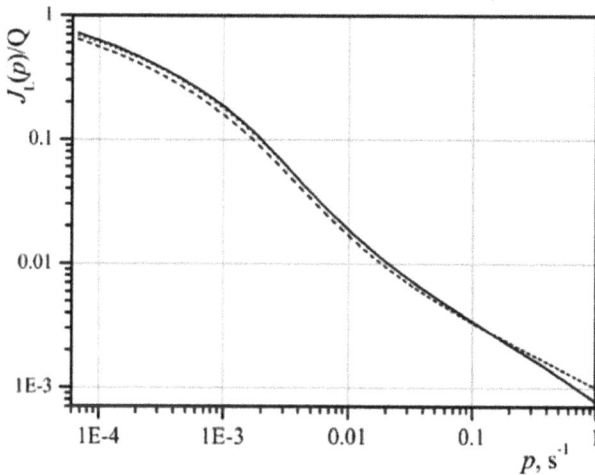

FIGURE 1.16 Normalized Laplace transform of experimental data (solid; $Q = 0.44$ C) and fitting model curves for inhomogeneous (dot) and homogeneous (dash; $a = 0$) model. (Reprinted with permission from [84]. Copyright 2006 Elsevier.)

By computational calculations, the diffusion coefficients obtained are $D_1 = 2 \times 10^{-13}$ cm^2/s and $D_2 = 5 \times 10^{-10}$ s. So, we have seen that nanostructuring leads to heterogeneous oxygen diffusion in domains and along the interfaces. The activation energy for ion migration along the domain boundaries is much lower than bulk in nanostructures. So, nanostructures provide higher magnitude of oxygen flux at working temperatures below order–disorder transition point.

1.6.3 HALIDE SUBSTITUTION

Using DFT, the effects of Br substitution for I on the structural, electronic, and optical properties of mixed iodide–bromide perovskite compounds $MAPb(_{1-x}Br_x)_3$ were studied. The virtual crystal approximation (VCA) method was used to compare experimental and theoretical values for lattice constants and bandgap when the Br content was varied in the range $0 < x < 1$. Thus, by varying the Br content, virtual atoms were constructed. The *Yu–Emmerich extended averaging approach* (YE^2A^2) averaging method was used to calculate the pseudopotential of these virtual atoms. XRD measurements confirmed the pseudo cubic structure of $MAPb(_{1-x}Br_x)_3$[85]. By changing the volume evenly, the lowest possible atomic force position was determined. The optimized lattice constants were calculated by feeding these values to the Birch–Murnaghan equation of state [86]. This process was repeated for each Br content x, $0 < x < 1$. The results show a linear relationship between lattice constant and Br; they are inversely proportional. This may be due to the fact that the ionic radius of bromine ion is smaller than that of iodine ion. Vegard's law was used to calculate lattice parameters of mixed perovskites experimentally [87]. The calculated constants were overestimated in comparison with the experimental values. But the linear coefficient in the fitting parameter was almost identical in both cases. So, we can conclude that Vegard's law is in agreement with the YE^2A^2.

The addition of Br influences the electronic structures as well. The energy band structure and the partial density of states (DOS) gradually change with the Br content. The VCA method is used to find out the change in the electronic band structure and the DOS. Another advantage of Br mixing is that the nature of bandgap changes to direct mode. This causes the generation electron–hole pair directly by the absorption of phonons. The calculated bandgaps of MHPs are slightly less than experimental values. This variation increases as we go from $MAPbI_3$ ($x = 0$) to $MAPbBr_3$ ($x = 1$). Br content x and bandgap (E_g) follows a quadratic relationship, and the bandgap increases with the increase in the Br content. The increase in bandgaps with increasing Br content is due to the stronger hybridization of the Br 4p states with the Pb s states than with the I 5p states, which leads to a downshift of valence band maximum, accompanied by a decrease in the lattice constant [87]

$$E_g(x) = E_g(0) + [E_g(1) - E_g(0) - b]x + bx^2$$

where $E_g(0)$ and $E_g(1)$ are bandgaps of MAPbI$_3$ ($x = 0$) and MAPbBr$_3$ ($x = 1$), respectively, and b is the so-called bowing parameter [35]. Since Br ion is smaller than I ion, substituting Br for iodine increases the interaction between Pb atoms and X atoms. This interaction causes modification in the band structure. Since the bandgap of MAPbBr$_3$ is more than that of MAPbI$_3$, it is not good for PSCs. The photoabsorption coefficients calculated using density functional perturbation theory (DFPT) indicate that low Br content is advisable for better light-harvesting properties.

Effective masses of electrons and holes give idea about the mobility of charge carriers. The calculated values of $= 0.18$, $= 0.19$, and $= 0.09$ by DFT are comparable with the experimental values 0.12, 0.15, and 0.09, respectively [88]. As the Br content increases from $x = 0$ to $x = 1$, reduced mass increases, which badly affects the charge carrier mobility. So, we can conclude that Br substitution is not much good for PSC application of MAPbI$_3$.

1.6.4 GOLDSCHMIDT TOLERANCE FACTOR

Perovskites have a general structure of ABX$_3$ where A is a monovalent cation, B is a divalent metal cation, and X is a halide anion. The Goldschmidt tolerance factor [89] is a geometric factor that restricts the formation of 3-D halide perovskite structure

$$t = \frac{r_A + r_X}{\sqrt{2}(r_B + r_X)}$$

where r_A, r_B, and r_X are the ionic radii of the A, B, and X ions, respectively. The stability of the perovskite structure depends on the value of t. For a perfect crystal, $t = 1$, and it ranges from 0.8 to 1.0 for perovskites with tetragonal, orthorhombic, and rhombohedral lattices. When $t > 1$, the hexagonal structure [90] is formed and for $t < 0.8$, different structures [91] are formed.

By changing halide ions (X= F-, Cl-, Br-, I-), B-site metal cations (B = Pb, Sn), and A-site cations, too many compounds are possible. Kieslich et al. [92] calculated the tolerance factors of 2352 amine–metal–anion compounds. They found out 180 stable halide perovskites with the tolerance factor in the range 0.8–1.0 (see Figure 1.17). Having a tolerance factor in a specific range can be a necessary condition for stable perovskite formation, but it is not sufficient [92]. We have to analyze so many factors in computational calculations.

The octahedral factor is defined as the ratio of the ionic radii of the B cation and the halide anion μ = [93]. For a stable perovskite, its value ranges between 0.44 and 0.9 [94]. A stable perovskite can be chosen from so many possible cases available by comparing the values of the tolerance factor and the octahedral factor.

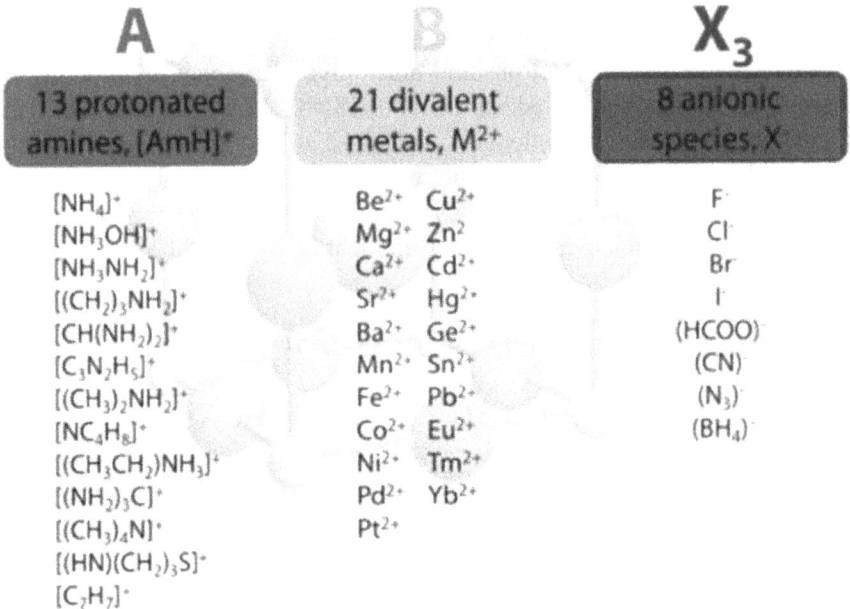

A

B

X₃

13 protonated amines, [AmH]⁺	21 divalent metals, M²⁺		8 anionic species, X⁻
[NH₄]⁺	Be²⁺	Cu²⁺	F⁻
[NH₃OH]⁺	Mg²⁺	Zn²	Cl⁻
[NH₃NH₂]⁺	Ca²⁺	Cd²⁻	Br⁻
[(CH₂)₃NH₂]⁺	Sr²⁺	Hg²⁺	I⁻
[CH(NH₂)₂]⁺	Ba²⁺	Ge²⁺	(HCOO)⁻
[C₃N₂H₅]⁺	Mn²⁺	Sn²⁺	(CN)⁻
[(CH₃)₂NH₂]⁺	Fe²⁺	Pb²⁻	(N₃)⁻
[NC₄H₈]⁺	Co²⁺	Eu²⁺	(BH₄)⁻
[(CH₃CH₂)NH₃]⁺	Ni²⁺	Tm²⁺	
[(NH₂)₃C]⁺	Pd²⁺	Yb²⁺	
[(CH₃)₄N]⁺	Pt²⁺		
[(HN)(CH₂)₃S]⁺			
[C₇H₇]⁺			

FIGURE 1.17 Based on the Goldschmidt tolerance factor calculation for 2352 amine–metal–anion compounds, 562 organic-anion-based and 180 halide-based perovskites are selected to have 0.8 < *t* < 1.0. (Reprinted with permission from [92]. Copyrights 2015 Royal Chemical Society.)

1.6.5 ELECTRONIC AND OPTICAL PROPERTIES

Electronic and optical properties of halide perovskites are of prime importance since they are used as light absorbers in PV applications. In atomistic modeling and simulation, we first refine a unit cell by applying structural optimization. Non-self-consistent field calculation is used for getting electronic energy band structure and DOS. The effective masses of the conductive electrons and holes are obtained by post-processing energy band. Exciton binding energy and photoabsorption coefficients are calculated from frequency-dependent dielectric constants. DFPT [98] is applied for calculating dielectric constants, and the Bethe–Salpeter equation [99] is used for obtaining the effect of electron–hole interaction.

The extended Hückel model was applied to study the electronic structure of hybrid perovskites [100]. $MAPbX_3$ consists of bands with antibonding character, so the extended Hückel model failed to predict [101] the electronic structure accurately. Even DFT cannot estimate the bandgaps of perovskites accurately. Recently, first-order scalar relativistic DFT (SR-DFT) and higher order spin-orbit coupling have been applied. But still they could not give accurate results for Pb and Sn perovskites. The offsetting of electron self-energy/many-body effects has a positive effect on the bandgap (E_g), and relativistic interactions have a negative impact on E_g [102]. The E_g values of $MAPbI_3$ were calculated using SR-DFT, and it is more precise, since the above-mentioned effects were taken into account.

But this approach failed in the case of $MASnI_3$ because of miner relativistic effect with a lighter metal center. For very complex systems, only quasi-particle self-consistent GW (QSGW) with spin–orbit corrections can precisely and accurately determine the electronic structure [101] of organic, inorganic, and hybrid MHPs [101].

In halide perovskites, the bandgap can be tuned [81] by static volume exchange, temperature change, and chemical substitution. Volume can be reduced by small perturbation or by increasing pressure. DFT calculation indicated a reduction in the bandgap as volume decreases. $MAPbI_3$ transforms from an indirect bandgap to direct bandgap [103] material as its volume reduces. Also, lattice expansion resulted in the stabilization [104] of out-of-phase band edge states. For $MAPbI_3$, the temperature-dependent PL drops from 1.61 eV at 300 K to 1.55 eV at 150 K [81]. Chemical substitution is most suitable for material design [105] in the case of halide perovskites.

QSGW calculation of $MAPbI_3$ gives fruitful results about the exchange of MA cations by small molecular cations such as NH_{4+} or H_+. The bandgap decreased by 0.3 eV in NH_4PbI_3 and less than 0.3 eV in $HPbI_3$ [106]. The calculations also indicated the contraction of lattice constants. Changing B-site ion has a direct effect on the conduction band (CB) without changing the crystal structure. Substitution of Sn in the place of Pb reduces the bandgap by 0.3eV. $MAPbI_3$ has a tetragonal 14 cm phase and $MASnI_3$ has pseudocubic P4 mm phase. So, the substitution induces a phase transition. But Sn_{2+} is not much chemically stable since it has a tendency to oxidize to Sn_{4+}.

The parabolic relationship between energy (E) and momentum (k) is used for calculating band extremum

$$E = \frac{h^2 k^2}{2m*}$$

where ℏ is the Planck's constant and $m*$ is the effective mass of electrons (holes) in the CB (valence band). The expression for effective mass is given by

$$m^* = \frac{h^2}{\frac{\partial^2 E}{\partial k^2}}$$

This is the well-known parabolic approximation for effective mass as derived from band dispersion. For MHPs, electrons and holes have comparable effective mass at the band edge due to similar dispersion. For Sn and Pb perovskites, relativistic effects cause noticeable deviation in the values of effective mass calculated by parabolic approximation. Theoretically, the estimated bandgap shows variation relying on the computational method adapted. For MAPbI$_3$, the average values of = 0.19 and = 0.25 are obtained by calculations using SOC-GW [102], where m_0 is the free electron mass. These results are in close agreement with the one obtained using high-field magnetoabsorption spectroscopy [100].

1.7 CONCLUSIONS

In spite of the popularity of MHPs as a highly promising solar cell absorber material, they have all the potential to be the next-generation semiconductors because its excellent optoelectronic properties coupled with good solution processability, high PLQY, and narrow emission width make them perfect for a number of optoelectronic applications. In the case of solar cells fabricated with perovskite QDs, better stability under environmental stress compared to its bulk forms points toward the additional potential of perovskites when used in nanoforms. The photocatalytic activity of perovskites in visible light is very useful in semiconductor systems. Perovskite NCs exhibit high PLQY and high saturated colors owing to their narrow emission bandwidth. Since the morphology and bandwidth of perovskite NCs can be tuned by changing composition, they possess excellent photodetection potential too. Atomistic modeling and simulation has been applied to study the material properties of perovskites. It gives an insight into the tolerance factor and chemical composition of stable perovskites.

KEYWORDS

- **nanocrystalline perovskites**
- **hot injection method**
- **ligand-assisted reprecipitation**
- **solvothermal method**
- **computational modeling**
- **Goldschmidt tolerance factor**

REFERENCES

1. Green, M.A., "The path to 25% silicon solar cell efficiency: History of silicon cell evolution," *Prog. Photovolt.: Res. Appl.*, vol. 17, pp. 183–189, 2009.
2. Kojima, A., Teshima, K., Shirai, Y., Miyasaka, T., "Organometal halide perovskites as visible-light sensitizers for photovoltaic," *J. Am. Chem. Soc.*, vol. 131, pp. 6050–6051, 2009.
3. Im, J., Lee, C., Lee, J., Park, S., Park, N., "6.5% efficient perovskite quantum-dot-sensitized solar cell," *Nanoscale*, vol. 2, pp. 4088–4093, 2011.
4. Tong, X., Lin, F., Wu, J., Wang, Z.M., "High performance perovskite solar cells," *Chem. Rev.*, vol. 120, pp. 1–18, 2016.
5. Yang, W.S., Park, B., Jung, E.H., Jeon, N.J., "Iodide management in formamidinium-lead-halide-based perovskite layers for efficient solar cells," *Science*, vol. 1379, pp. 1376–1379, 2017.
6. Prasanthkumar, S., Giribabu, L., "Recent advances in perovskite-based solar cells," *Curr. Sci.*, vol. 111, no. 7, pp. 1173–1181, 2016.
7. Murphy, J.P., "Novel hybrid perovskite composites and microstructures: Synthesis and characterization," Ph.D. dissertation, Dept. Mater. Sci., Montana Technol. Univ., Butte, MT, USA, 2018.
8. Weber, D., "$CH_3NH_3PbX_3$, ein Pb(II)-System mit kubischer perowskitstruktur," *Z. Naturforschung B*, vol. 33b, pp. 1443–1445, 1978.
9. Park, N.G., "Perovskite solar cells: An emerging photovoltaic technology," *Mater. Today*, vol. 18, no. 2, pp. 65–72, 2015.
10. Giorgi, G., Fujisawa, J., Segawa, H., Yamashita, K., "Small photocarrier effective masses featuring ambipolar transport in methylammonium lead iodide perovskite: A density functional analysis," *J. Phys. Chem. Lett.*, vol. 4, pp. 4213–4216, 2013.
11. Innocenzo, V.D., Ram, A., Kandada, S., De Bastiani, M., Gandini, M., Petrozza, A., "Tuning the light emission properties by band gap engineering in hybrid lead halide perovskite," *J. Am. Chem. Soc.*, vol. 136, pp. 17730–17733, 2014.

12. Haug, F., Yum, J., Ballif, C., "Organometallic halide perovskites: Sharp optical absorption edge and its relation to photovoltaic performance," *J. Phys. Chem. Lett.*, vol. 5, pp. 1035–1039, 2014.

13. Ponseca, C.S. et al., "Organometal halide perovskite solar cell materials rationalized: Ultrafast charge generation, high and microsecond-long balanced mobilities, and slow recombination," *J. Am. Chem. Soc.*, vol. 136, pp. 5189–5192, 2014.

14. Huang, H., Susha, A.S., Kershaw, S.V., Hung, T.F., Rogach, A.L., "Control of emission color of high quantum yield $CH_3NH_3PbBr_3$ perovskite quantum dots by precipitation temperature," *Adv. Sci.*, vol. 2, no. 9, pp. 1–5, 2015.

15. Jung, H.S., Park, N., "Perovskite solar cells: From materials to devices," *Small*, vol. 11, pp. 10–25, 2015.

16. Liu, Y., Yang, Z., Liu, S.F., "Recent progress in single-crystalline perovskite research including crystal preparation , property evaluation, and applications," *Adv. Sci.*, vol. 5, no. 1, 2018, Art. no. 1700471.

17. Liu, Y. et al., "Multi-inch single-crystalline perovskite membrane for high-detectivity flexible photosensors," *Nat. Commun.*, vol. 9, 2018, Art. no. 5302.

18. Sichert, J.A. et al., "Quantum size effect in organometal halide perovskite nanoplatelets," *Nano Lett.*, vol. 15, no. 10, pp. 6521–6527, 2015.

19. Kovalenko, M. V., Protesescu, L., Bodnarchuk, M.I., "Properties and potential optoelectronic applications of lead halide perovskite nanocrystals," *Science*, vol. 358, no. 6364, pp. 745–750, 2017.

20. Levchuk, I. et al., "Brightly luminescent and color-tunable formamidinium lead halide perovskite $FAPbX_3$ (X = Cl, Br, I) Colloidal nanocrystals," *Nano Lett.*, vol. 17, no. 5, pp. 2765–2770, 2017.

21. Dong, Y. et al., "Recent advances toward practical use of halide perovskite nanocrystals," *J. Mater. Chem. A*, vol. 6, pp. 21729–21746, 2018.

22. Shamsi, J., Urban, A.S., Imran, M., De Trizio, L., Manna, L., "Metal halide perovskite nanocrystals : Synthesis , post-synthesis modifications, and their optical properties," *Chem. Rev.*, vol. 119, pp. 3296–3348, 2019.

23. Dirin, D.N. et al., "Harnessing defect-tolerance at the nanoscale: Highly luminescent lead halide perovskite nanocrystals in mesoporous silica matrixes," *Nano Lett.*, vol. 16, no. 9, pp. 5866–5874, 2016.

24. Fu, P. et al., "Perovskite nanocrystals: Synthesis, properties and applications," *Sci. Bull.*, vol. 62, no. 5, pp. 369–380, 2017.

25. Fu, H., "Colloidal metal halide perovskite nanocrystals: A promising juggernaut in photovoltaic," *J. Mater. Chem. A*, vol. 7, no. 24, pp. 14357–14379, 2019.

26. Rivesta, J.B., Jain, P.K., "Cation exchange on the nanoscale: An emerging technique for new material synthesis, device fabrication, and chemical sensing," *Chem. Soc. Rev.*, vol. 42, pp. 89–96, 2013.

27. Nedelcu, G., Protesescu, L., Yakunin, S., Bodnarchuk, M.I., Grotevent, M.J., Kovalenko, M.V., "Fast anion-exchange in highly luminescent nanocrystals of cesium lead halide perovskites ($CsPbX_3$, X = Cl, Br, I)," Nano Lett., vol. 15, no. 8, pp. 5635–5640, 2015.

28. Akkerman, Q.A. et al., "Tuning the optical properties of cesium lead halide perovskite nanocrystals by anion exchange reactions," J. *Am.* Chem. Soc., vol. 137, no. 32, pp. 10276–10281, 2015.

29. Yantara, N. et al., "Loading of mesoporous titania films by $CH_3NH_3PbI_3$ perovskite, single step *vs.* sequential deposition," *Chem. Commun.*, vol. 51, no. 22, pp. 4603–4606, 2015.

30. Jang, D.M., Kim, D.H., Park, K., Park, J., Lee, J.W., Song, J.K., "Ultrasound synthesis of lead halide perovskite nanocrystals," *J. Mater. Chem. C*, vol. 4, pp. 10625–10629, 2016.

31. Aharon, S., Gamliel, S., El Cohen, B., Etgar, L., "Depletion region effect of highly efficient hole conductor free $CH_3NH_3PbI_3$ perovskite solar cells," *Phys. Chem. Chem. Phys.*, vol. 16, no. 22, pp. 10512–10518, 2014.

32. Bai, S., Yuan, Z., Gao, F., "Colloidal metal halide perovskite nanocrystals : Synthesis, characterization, and applications," *J. Mater. Chem. C*, vol. 4, pp. 3898–3904, 2016.

33. Huang, H., Polavarapu, L., Sichert, J.A., Susha, A.S., Urban, A.S., Rogach, A.L., "Colloidal lead halide perovskite nanocrystals: synthesis, optical properties and applications," *NPG Asia Mater.*, vol. 8, no. 11, 2016, Art. no. e328.

34. Liang, Z. et al., "Shape-controlled synthesis of all-inorganic $CsPbBr_3$ perovskite nanocrystals with bright blue emission," *ACS Appl. Mater. Interfaces*, vol. 8, no. 42, pp. 28824–28830, 2016.

35. Almeida, G. et al., "Role of acid–base equilibria in the size, shape, and phase control of cesium lead bromide nanocrystals," *ACS Nano*, vol. 12, no. 2, pp. 1704–1711, 2018.

36. Zi Long, Hong Ren, Jianghui Sun, Jin Ouyang, and Na Na., "High-throughput and tunable synthesis of colloidal CsPbX3 perovskite nanocrystals in a heterogeneous system by microwave irradiation," *Chem. Commun.*, vol. 71, pp. 9914–9917, 2017

37. Deka, S., Genovese, A., Zhang, Y., Miszta, K., Bertoni, G., "Phosphine-free synthesis of p-type copper (I) selenide nanocrystals in hot coordinating solvents," *J. Am. Chem. Soc.*, vol. 132, no. 26, pp. 8912–8914, 2010.

38. Park, Y., Guo, S., Makarov, N., Klimov, V.I., "Room temperature single-photon emission from individual perovskite quantum dots," *ACS Nano*, vol. 9, no. 10, pp. 10386–10393, 2015.

39. Pan, A. et al., "Insight into the ligand-mediated synthesis of colloidal $CsPbBr_3$ perovskite nanocrystals: The role of organic acid, base, and cesium precursors," *ACS Nano*, vol. 10, no. 8, pp. 7943–7954, 2016.

40. Zhu, F. et al., "Shape evolution and single particle luminescence of organometal halide perovskite nanocrystals," *ACS Nano*, vol. 9, no. 3, pp. 2948–2959, 2015.

41. Zhang, D., Eaton, S.W., Yu, Y., Dou, L., Yang, P., "Solution phase synthesis of cesium lead halide perovskite nanowires," *J. Am. Chem. Soc.*, vol. 137, no. 29, pp. 9230–9233, 2015.

42. Protesescu, L. et al., "Monodisperse formamidinium lead bromide nanocrystals with bright and stable green photoluminescence," *J. Am. Chem. Soc.*, vol. 138, no. 43, pp. 14202–14205, 2016.

43. Imran, M. et al., "Benzoyl halides as alternative precursors for the colloidal synthesis of lead based halide perovskite nanocrystals," *J. Am. Chem. Soc.*, vol. 140, no. 7, pp. 2656–2664, 2018.

44. Prato, M., Manna, L., "Colloidal synthesis of quantum confined single crystal $CsPbBr_3$ nanosheets with lateral size control up to the micrometer range," *J. Am. Chem. Soc.*, vol. 138, no. 23, pp. 7240–7243, 2016.

45. Tyagi, P., Arveson, S.M., Tisdale, W.A., "Colloidal organohalide perovskite nanoplatelets exhibiting quantum confinement," *J. Phys. Chem. Lett.*, vol. 6, no. 10, pp. 1911–1916, 2015.

46. Sun, S., Yuan, D., Xu, Y., Wang, A., Deng, Z., "Ligand-mediated synthesis of shape-controlled cesium lead halide perovskite nanocrystals via reprecipitation process at room temperature," *ACS Nano*, vol. 10, no. 3, pp. 3648–3657, 2016.

47. Ling, Y. et al., "Bright light-emitting diodes based on organometal halide perovskite nanoplatelets," *Adv. Mater.*, vol. 28, no. 2, pp. 305–311, 2016.

48. Lignos, I., Stavrakis, S., Nedelcu, G., Protesescu, L., deMello, A.J., Kovalenko, M.V., "Synthesis of cesium lead halide perovskite nanocrystals in a droplet-based microfluidic platform: Fast parametric space mapping," *Nano Lett.*, vol. 16, no. 3, pp. 1869–1877, 2016.

49. Seth, S., Samanta, A., "A facile methodology for engineering the morphology of $CsPbX_3$ perovskite nanocrystals under ambient condition," *Sci. Rep.*, vol. 6, 2016, Art. no. 37693.

50. Zhang, F., "Colloidal synthesis of air-stable $CH_3NH_3PbI_3$ quantum dots by gaining chemical insight into the solvent Effects," *Chem. Mater.*, vol. 29, no. 8, pp. 3793–3799, 2017.

51. Kumar, S. et al., "Ultrapure green light-emitting diodes using two-dimensional formamidinium perovskites: Achieving recommendation 2020 color coordinates," *Nano Lett.*, vol. 17, no. 9, pp. 5277–5284, 2017.

52. Li, M. et al., "Slow cooling and highly efficient extraction of hot carriers in colloidal perovskite nanocrystals," *Nat. Commun.*, vol. 8, 2017, Art. no. 14350..

53. Jiang, Q., Zeng, X., Wang, N., Xiao, Z., Guo, Z., Lu, J., "Electrochemical lithium doping induced property changes in halide perovskite $CsPbBr_3$ crystal," *ACS Energy Lett.*, vol. 3, no. 1, pp. 264–269, 2018.

54. Bhooshan, V., Gouda, L., Porat, Z., Gedanken, A., "Sonochemical synthesis of $CH_3NH_3PbI_3$ perovskite ultrafine nanocrystal sensitizers for solar energy applications," *Ultrason. Sonochem.*, vol. 32, pp. 54–59, 2016.

55. Chen, L., Chen, L., Tseng, Z., Chen, S., Yang, S., "An ultrasonic synthesis method for high-luminance perovskite quantum dots," *Ceram. Int.*, vol. 43, no. 17, pp. 16032–16035, 2017.

56. Van Der Stam, W., et al., "Highly emissive divalent-ion-doped colloidal $CsPb_{1-x}M_xBr_3$ perovskite nanocrystals through cation exchange," *J. Am. Chem. Soc.*, vol. 139, no. 11, pp. 4087–4097, 2017.

57. Xian-gang Wua, Jialun Tanga, Feng Jianga, Xiaoxiu Zhua, Yanliang Zhangb, Dengbao Hana, Lingxue Wangc, and Haizheng Zhong., "Highly luminescent red emissive perovskite quantum dots embedded composite films: ligands capping and caesium doping controlled crystallization process," *Nanoscale*, vol. 11, pp. 4942–4947, 2019.

58. Wojciechowski, K., Saliba, M., Leijtens, T., Abate, A., Snaith, H.J., "Sub-150 °C processed meso-superstructured perovskite solar cells with enhanced efficiency," *Energy Environ. Sci.*, vol. 7, no. 3, pp. 1142–1147, 2014.

59. Kojima, A., Ikegami, M., Teshima, K., Miyasaka, T., "Highly luminescent lead bromide perovskite nanoparticles synthesized with porous alumina media," *Chem. Lett.*, vol. 41, no. 4, pp. 397–399, 2012.

60. Wang, F., Yu, H., Xu, H., Zhao, N., "HPbI$_3$: A new precursor compound for highly efficient solution-processed perovskite solar cells," *Adv. Funct. Mater.*, vol. 25, no. 7, pp. 1120–1126, 2015.

61. Shi, Z., Jayatissa, A.H., "Perovskites-based solar cells : A review of recent progress , materials and processing methods," *Materials*, vol. 11, no. 5, 2018, Art. no. 729.

62. Kim, J.H. et al., "High-performance and environmentally stable planar heterojunction perovskite solar cells based on a solution-processed copper-doped nickel oxide hole-transporting layer," *Adv. Mater.*, vol. 27, no. 4, pp. 695–701, 2015.

63. Correa-Baena, J.-P., et al.,, "The rapid evolution of highly efficient perovskite solar cells,"*Energy Environ. Sci.*, vol. 10, no. 3, pp. 710–727, 2017.

64. Sanehira, E.M. et al., "Enhanced mobility CsPbI$_3$ quantum dot arrays for record-efficiency, high-voltage photovoltaic cells," *Sci. Adv.*, vol. 3, no. 10, 2017, Art. no. eaao4204.

65. Hak Beom Kim,a Young Jin Yoon,a Jaeki Jeong,a Jungwoo Heo,b Hyungsu Jang,a Junghwa Seo,c Bright Walkera, and Jin Young Kima, "Peroptronic devices: perovskite-based light-emitting solar cells," *Energy Environ. Sci.*, vol. 9, pp. 1950–1957, 2017.

66. Ramasamy, P., Lim, D., Kim, B., Lee, S.-H., Lee, M.-S., Lee, J.-S., "All-inorganic cesium lead halide perovskite nanocrystals for photodetector applications," *Chem. Commun.*, vol. 52, no. 10, pp. 2067–2070, 2016.

67. Dastidar, S., Li, S., Smolin, S.Y., Baxter, J.B., Fafarman, A.T., "Slow electron--Hole recombination in lead iodide perovskites does not require a molecular dipole," *ACS Energy Lett.*, vol. 2, no. 10, pp. 2239–2244, 2017.

68. Chen, W. et al., "Surface-passivated cesium lead halide perovskite quantum dots : toward efficient light-emitting diodes with an inverted sandwich structure," *Adv. Opt. Mater.*, vol. 6, no. 14, 2018, Art. no. 1800007.

69. Peng, L., Hu, L., Fang, X., "Low-dimensional nanostructure ultraviolet photodetectors," *Adv. Mater.*, vol. 25, no. 37, pp. 5321–5328, 2013.

70. Sutherland, B.R. et al., "Sensitive, fast, and stable perovskite photodetectors exploiting interface engineering," *ACS Photon.*, vol. 2, no. 8, pp. 1117–1123, 2015.

71. Deng, W. et al., "Ultrahigh-responsivity photodetectors from perovskite nanowire arrays for sequentially tunable spectral measurement," *Nano Lett.*, vol. 17, no. 4, pp. 2482–2489, 2017.

72. Wu, H. et al., "Fine-tuned multilayered transparent electrode for highly transparent perovskite light-emitting devices," *Adv. Electron. Mater.*, vol. 4, no. 1, 2017, Art. no. 1700285.

73. Chen, H., Liu, H., Zhang, Z., Hu, K., Fang, X., "Nanostructured photodetectors : from ultraviolet to terahertz," *Adv. Mater.*, vol. 28, no. 3, pp. 403–433, 2016.

74. Hu, M.Z., Zhu, T., "Semiconductor nanocrystal quantum dot synthesis approaches towards large-scale industrial production for energy applications," *Nanoscale Res. Lett.*, vol. 10., 2015, Art. no. 469.

75. Dou, L. et al., "Solution-processed hybrid perovskite photodetectors with high detectivity," *Nat. Commun.*, vol. 5, 2014, Art. no. 5404.

76. Schmidt, L.C. et al., "Nontemplate synthesis of CH$_3$NH$_3$PbBr$_3$ perovskite nanoparticles," *J. Am. Chem. Soc.*, vol. 136, no. 3, pp. 850–853, 2014.

77. Xiao, Z. et al., "Giant switchable photovoltaic e fect in organometal trihalide perovskite devices," *Nature Mater*, vol. 14, pp. 193–198, 2014.

78. Wei, Z., Xing, J., "The rise of perovskite light-emitting diodes," *J. Phys. Chem. Lett.*, vol. 10, pp. 3035–3042, 2019.

79. Zhang, L. et al., "Ultra-bright and highly efficient inorganic based perovskite light-emitting diodes," *Nat. Commun.*, vol. 8, pp. 1–8, 2017.

80. Chen, L.-J., Lee, C.-R., Chuang, Y.-J., Wu, Z.-H., Chen, C., "Synthesis and optical properties of lead-free cesium tin halide perovskite quantum rods with high-performance solar cell application," *J. Phys. Chem. Lett.*, vol. 7, no. 24, pp. 5028–5035, 2016.

81. Yu, C.-J., "Advances in modelling and simulation of halide perovskites for solar cell applications," *J. Phys. Energy,* vol. 1, no. 2, 2019, Art. no. 022001.

82. Ancharova, U.V., Cherepanova, S.V., Lyakhov, N.Z., "Modelling of X-ray diffraction patterns from nanostructured perovskites Sr (Fe,Co) $O_{3-\delta}$," Chem. Sustain. Develop., vol. 20, pp. 351–359, 2012.

83. Hiroi, Z., Ikeda, Y., Takano, M., Bando, Y., "Electron microscopy study of oxide superconductors," *Phys. B: Condens. Matter*, vol. 165–166, pp. 1693–1694, 1990.

84. Zhogin, I.L., Nemudry, A.P., Glyanenko, P. V., Kamenetsky, Y.M., "Oxygen diffusion in nanostructured perovskites," *Catalysis Today*, vol. 118, nos. 1/2, pp. 151–157, 2006.

85. Baikie, T. et al., "Synthesis and crystal chemistry of the hybrid perovskite (CH 3 NH 3) PbI 3 for solid-state sensitised solar cell applications," *J. Mater. Chem. A*, vol. 1, pp. 5628–5641, 2013.

86. Born, M., "finite elastic strain of cubic crystals," *Phys. Rev.*, vol. 71, no. 11, pp. 809–824, 1947.

87. Jong, U.-G., Yu, C.-J., Ri, J.-S., Kim, N.-H., Ri, G.-C., "Influence of halide composition on the structural , electronic , and optical properties of mixed CH_3NH_3Pb $(I_{1-x}$ $Brx)_3$ perovskites calculated using the virtual crystal approximation method," *Phys. Rev. B*, vol. 94, 2016, Art. no. 125139.

88. Tanaka, K., Kondo, T., Technol, S., Mater, A., Tanaka, K., Kondo, T., "Bandgap and exciton binding energies in lead-iodide-based natural quantum-well crystals Bandgap and exciton binding energies in lead-iodide-based natural quantum-well crystals," *Sci. Technol. Adv. Mater.*, vol. 4, no. 6, pp. 599–604, 2003.

89. Goldschmidt, V.M., "Die Gesetze der Krystallochemie.," *Naturwissenschaften,* vol. 14, no. 21, pp. 477–485, 1926.

90. Stoumpos, C.C., Kanatzidis, M.G., "The renaissance of halide perovskites and their evolution as emerging semiconductors," *Accounts. Chem. Res.*, vol. 48, no. 10, 2791–2802, 2015.

91. Travis, W., Glover, E. N. K., Bronstein, H., Scanlon, D. O., and Palgrave, R. G. "On the application of the tolerance factor to inorganic and hybrid halide perovskites: a revised system," *Chem. Sci.*, vol. 7, pp. 4548–4556, 2016.

92. Kieslich, G., Sun, S., Cheetham, A.K., "An extended Tolerance Factor approach for organic–inorganic perovskites," *Chem. Sci.*, vol. 6, no. 6, pp. 3430–3433, 2015.

93. Li, C., Lu, X., "Formability of ABX_3 (X = F, Cl, Br, I) halide perovskites," *Acta. Crystallogr. Sec. B.*, vol. 64, no. 6, pp. 702–707, 2008.

94. Yin, W., Yan, Y., Wei, S., "Anomalous alloy properties in mixed halide perovskites," *J. Phys. Chem. Lett.*, vol. 5, no. 21, pp. 3625–3631, 2014.

95. Hautier, G., Miglio, A., Ceder, G., Rignanese, G., Gonze, X., "Identification and design principles of low hole effective mass p-type transparent conducting oxides," *Nat. Commun.*, vol. 4, 2013, Art. no. 2292.

96. Isayev, O. et al., "Materials cartography: Representing and mining materials space using structural and electronic fingerprints," *Chem. Mater.*, vol. 27, no. 3, pp. 735—743, 2015.

97. Oganov, A.R., Lyakhov, A.O., Valle, M., "How evolutionary crystal structure prediction works—and why," *Acc. Chem. Res.*, vol. 44, no. 3, pp. 227–237, 2011.
98. Íñiguez, J., Vanderbilt, D., Bellaiche, L., "First-principles study of (BiScO3) – (PbTiO$_3$) piezoelectric alloys," *Phys. Rev. B*, vol. 67, 2003, Art. no. 224107.
99. Onida, G., Nazionale, I., Vergata, R.T., Scientifica, R., Roma, I., "Electronic excitations : density-functional versus many-body Green's-function approaches," *Rev. Mod. Phys.*, vol. 74, no. 2, pp. 601–659, 2002.
100. Manser, J.S., Christians, A., Kamat, P. V., "Intriguing Optoelectronic Properties of Metal Halide Perovskites," *Chem. Rev.*, vol. 116, no. 21, pp. 12956–13008, 2016.
101. Huang, L., Lambrecht, W.R.L., "Electronic band structure, phonons, and exciton binding energies of halide perovskites," *Phys. Rev. B*, vol. 88, 2013, Art. no. 165203.
102. Umari, P., Mosconi, E., De Angelis, F., "Relativistic GW calculations on CH$_3$NH$_3$PbI$_3$ and CH$_3$NH$_3$SnI$_3$ perovskites for solar cell applications," *Sci. Rep.*, vol. 4, no. 1, 2014, Art. no. 4467.
103. Bokdam, M., Lahnsteiner, J., Ramberger, B., Schäfer, T., Kresse, G., "Assessing density functionals using many body theory for hybrid perovskites," *Sci. Rep.*, vol. 4, 2017, Art. no. 4467.
104. Frost, J.M., Butler, K.T., Brivio, F., Hendon, C.H., Van Schilfgaarde, M., Walsh, A., "Atomistic origins of high-performance in hybrid halide perovskite solar cells," *Nano Lett.*, vol. 14, no. 5, pp. 2584–2590, 2014.
105. Amat, A. et al., "Cation-induced band-gap tuning in organohalide perovskites: Interplay of spin-orbit coupling and octahedra tilting," *Nano Lett.*, vol. 14 no. 6, pp. 3608—3616, Art. no. 2014.
106. Brivio, F., Butler, K.T., Walsh, A., "Relativistic quasiparticle self-consistent electronic structure of hybrid halide perovskite photovoltaic absorbers," *Phys. Rev. B*, vol. 89, 2004, Art. no. 155204.
107. Lindblad, R. et al., "Electronic structure of CH$_3$NH$_3$PbX$_3$ perovskites: Dependence on the halide moiety," *J. Phys. Chem. C*, vol. 119, no. 4, pp. 1818–1825, 2015.
108. Jong, U.-G., Yu, C.-J., Jang, Y.-M., Ri, G.-C., Hong, S.-N., Pae, Y.-H., "Revealing the stability and efficiency enhancement in mixed halide perovskites MAPb (I$_{1-x}$Cl$_x$)$_3$ with ab initio calculations," *J. Power Sources*, vol. 350, pp. 65–72, 2017.
109. Baroni, S., de Gironcoli, S., Dal Corso, A., Giannozzi, P., "Phonons and related crystal properties from density-functional perturbation theory," *Rev. Mod. Phys.*, vol. 73, no. 2, pp. 515–562, Apr. 2001.
110. Tenuta, E., Zheng, C., Rubel, O., "Thermodynamic origin of instability in hybrid halide perovskites," *Sci. Rep.*, vol. 6, 2016, Art. no. 37654.

CHAPTER 2

INVESTIGATION ON THE EFFECT OF METHOD OF SYNTHESIS ON THE THERMAL DECOMPOSITION OF CERIA NANOSTRUCTURES

K. NUSRATH and K. MURALEEDHARAN*

Department of Chemistry, University of Calicut, Calicut, Kerala 673635, India

*Corresponding author. E-mail: kmuralika@gmail.com

ABSTRACT

Nanoflower petals, nanoplates, nanodiscs, and array of nanohexagonal Ce-Ox were synthesized *via* simple precipitation as well as hydrothermal methods. Executing the thermal decomposition strategy for the oxalate precursors, surface-modified ceria nanostructures such as branched hexagonal nanorod, multibranched, 2D nanoplate, and nanosheet were synthesized and characterized. Surface modification of Ce-Ox has a pivotal role on the thermal decomposition strategy, particle size and shape, and formation of oxygen ion vacancies. Kinetic characteristics revealed the significance of Ce-Ox surfaces upon the thermal decomposition strategy to the formation of ceria nanostructures. It was explored that for the formation of nanosheet-like ceria, activation energy values are lowered. But nanodiscs/flower of Ce-Ox has higher thermal prevention for the decomposition process to produce *nanoplates* of ceria. The thermal decomposition process does not retain the surface morphology of oxalate precursors. Upon thermal decomposition, ceria nanostructures with *branched hexagonal nanorod, multibranched, 2-D nanoplates,* and *sheets* were synthesized. Activation energy was found to be the lowest for producing *2-D nanosheets* of ceria.

2.1   INTRODUCTION

The design and synthesis of versatile ceria nanostructures have substantial interest in material synthesis field. The morphological and dimensional aspects of ceria differentiate its performance. Ceria is one of the major components of three-way catalysts for the removal of toxic automobile exhaust gases (CO, NO, etc.) [1–3]. It can act as the oxygen sensors [4, 5], humidity sensors [6], etc. Ceria performs as excellent ultraviolet (UV) absorbent and filter [7]. It can be utilized as the good absorbent for the removal of fluoride-ion- and arsenic-based compounds [8]. Ceria with controlled morphology exposes different crystal planes on the solid crystallites, which exhibits interesting chemical and physical properties. Ceria having different morphologies is synthesized and studied, which involves nanorods, nanocube, octahedron or polyhedron, etc. [9]. Surfaces of ceria can be activated by different factors such as surface area, elemental composition, defects, and reactive facets [10–13]. Ceria nanosystems such as nanowire, nanorod, and nanoparticle have different redox behavior toward CO oxidation. This occurs due to the distinguishing exposed crystal plane on the surface of ceria nanostructures [14]. Ceria nanorods and nanowires expose (1 0 0) and (1 1 0) as the performing crystal plane, whereas nanocubes have (1 0 0) and nanoparticle and polyhedron have (1 1 1) [9]. It was explored that ceria nanorods selectively exhibit higher activity for CO oxidation and NO reduction [15, 16], whereas nanocubes show superior properties in soot combustion [17], hydrogen oxidation [18], and preferential oxidation of CO [19]. But the existence of large proportion of reactive planes on the surface of ceria nanowire made it as potential redox catalyst for CO oxidation. Presence of oxygen vacancies and mobility of oxygen in the lattices are significantly altered with morphological parameters [14]. For water gas shift reaction processes [20], gold-supported ceria nanorod performed as the best catalyst, while Cu-based ceria polyhedral nanoparticles contributed to the best structural support [21, 22]. 3D flower-like ceria has owned enhanced catalytic activity toward oxidation of CO for the removal of As(V) and Cr(VI) [23].

Nanoceria can be prepared through variety of methods such as hydrothermal, solvothermal [24, 25], sol–gel [26], microemulsion [27], thermal decomposition [28], etc. The design and synthesis of rare-earth oxide ceria with chelating ligand oxalate have found rare works in the literature. Thermal decomposition of cerium oxalate by the microwave heating method produced ceria nanoparticle [29]. Upon thermal decomposition of oxalate, it followed the stepwise thermal decomposition strategy [30–32]. It is more important to have a fair knowledge on the thermal decomposition behavior of surface-modified cerium oxalate up to the formation of CeO_2/Ce_2O_3. Morphological dependency on activation energy can be derived in terms of conversion fraction α. Up to now, no exploratory work has been done to deal with the effect of morphological aspects of cerium oxalate on the creation of ceria with unique surface textures. Hence, it has substantial interest to synthesis ceria in different designs with appropriately choosing input reactant and surfactant. Besides, it is equally important to understand the influence of structural features as the thermal stability or rate of decomposition of oxalate samples. It was investigated that surface modification of a polymeric material is an important tool to tune its degradation rate [33]. This happens because thermal conductivity of the surface functionalizing agents has a pivotal role in reducing thermal stability of the polymeric material, thereby increasing rate of its degradation process [34]. It was noted that the size of oxalate nanoparticle of calcium has a noticeable effect on the kinetics of the thermal decomposition process. On reducing the size of oxalates, the amount of activation energy needed for the decomposition process was decreased [35]. Bogatyreva et al. [36] studied the effect of surface modification of the nanodiamond on its thermal stability. High-temperature activation followed by chemical treatment with mineral acids reduced the impurities on the top of the surface of diamond, thereby reducing the chance of oxidation. Hence, it was suggested that up to 773 K, nanodiamond withstands the oxidation process at its surfaces.

2.2 EXPERIMENTAL DETAILS

2.2.1 MATERIALS AND METHODS

All the chemicals used for the synthesis process of ceria are of analytical grade and used without further purification. The chemicals used were $Ce(NO_3)_3 \cdot 6H_2O$ (Himedia), $Na_2C_2O_4$ (Sigma Aldrich), and surfactants CTAB (Merck) and PEG-800 (Merck). In this chapter, cerium oxalate is prepared through simple precipitation and hydrothermal methods. Each sample is prepared as shown in the following:

Sample H_1:

3.07-mmol cerium nitrate is dissolved in 60 mL of distilled water. Add 1.02-mmol CTAB and stirred the solution for 2 h. The required amount of sodium oxalate is added and stirred well for 30 min. The synthesized precipitate is filtered and washed with ethanol and water and kept at 80 °C for 12 h.

Sample H$_2$:

3.07-mmol cerium nitrate is dissolved in 60 mL of distilled water. Add 1.02-mmol CTAB and stirred the solution for 2 h. Add the required amount of sodium oxalate and stirred well. The resulting supernatant solution is transferred into a sealed autoclave and kept at 120 °C for 24 h. The formed precipitate is filtered and washed with ethanol and kept at 80 °C for 6 h.

Sample H$_3$:

3.07-mmol cerium nitrate is dissolved in 60 mL of distilled water. Add 1.02-mmol PEG and stirred the solution for 2 h. The required amount of sodium oxalate is added and stirred well for 30 min. The synthesized precipitate is filtered and washed with ethanol and water and kept at 80 °C for 12 h.

Sample H$_4$:

3.07-mmol cerium nitrate is dissolved in 60 mL of distilled water. Add 1.02-mmol PEG and stirred the solution for 2 h. The required amount of sodium oxalate is added and stirred well for 30 min. The resulting solution is transferred into a sealed autoclave and kept at 120 °C for 24 h. The formed precipitate is filtered and washed with ethanol and kept at 80 °C for 6 h.

Each cerium oxalate sample is identified by the Fourier transform infrared (FTIR) spectrum recorded by the transmittance method using a spectrometer (Jasco, FT-IR-4100). The prepared cerium oxalate precursor was calcined at 450 °C in muffle furnace for 5 h. The chemical composition and crystal structures of cerium oxide (prepared *via* thermal decomposition) were studied by X-ray diffraction (XRD) (Rigaku D/Max, Miniflex 600) with Cu-Kα (0.15418 nm) radiation (40 kV, 15 mA). The morphology and topographical studies of the oxide samples were brought with transmission electron microscope (TEM) (Jeol, JEM 2100) with an accelerating voltage of 200 kV and field-emission scanning electron microscope (FE-SEM) (Carl Zeiss, Gemini SEM 300). Samples for the FE-SEM image are prepared by placing dry and cleaned sample on a carbon tape with tweezers for mounting on the stub. UV–visible spectra of the oxide samples were taken with Jasco V-550 spectrophotometer. Photoluminescence (PL) properties were well characterized by using a Perkin Elmer LS 55 fluorescence spectrometer at room temperature. Electronic and crystalline properties are well characterized by using the HRTEM and selected area electron diffraction (SAED) pattern. Thermal decomposition properties of oxalates were analyzed by differential scanning calorimetry (DSC) (TA instruments, Q20) and *thermal gravimetric analysis* (Perkin Elmer Thermal Analyzer, STA 8000) analysis in N_2 (50 mL) atmosphere. Raman spectra of the samples were collected using the confocal Raman microscope (WITec GmbH, alpha 300 A) with the excitation of a 532-nm laser (Nd: YAG dye laser, 40 mW).

2.3 RESULTS AND DISCUSSION

2.3.1 MATERIAL CHARACTERIZATION

Figure 2.1A represents the FTIR spectrum of surface-modified Ce-Ox. Figure 2.1Aa, b, c, and d, respectively, shows the FTIR spectrum of samples H_1, H_2, H_3, and H_4. The broadband observed at 3080–3433 cm^{-1} (see Figure 2.1A) associated with water molecules, which are removed only at higher temperature [37].

Depending upon the bond formation of surfactants, the characteristic peaks show slight difference in the stretching and bending mode of vibration. The very strong peak observed at 1624.87 (see Figure 2.1Aa), 1337.46 (see Figure 2.1Ab and c), and 1613.41 cm^{-1} (see Figure 2.1Ad) corresponds to the combined effect of asymmetric bending and stretching of the water molecule. The peak observed at 1316.4 cm^{-1} (see Figure 2.1Aa and d), 1039.41 cm^{-1} (see Figure 2.1Ab), and 1028.76 cm^{-1} (see Figure 2.1Ac) represent the asymmetric stretching of CO_2 molecule associated with the oxalate ligand group [38]. Consequently, thermal decomposition behavior of cerium oxalate depends on the nature of surface modifier, its chemical bond with surface of the oxalate sample, mechanism, and kinetics of the thermal reaction. From the TG curve (see Figure 2.1B), it is understood that the method of preparation of the sample (history) also depends on the rate of thermal decomposition. Effect of surface textures of cerium oxalate upon the rate of formation of ceria *via* thermal decomposition route can be realized. The mass loss (%) corresponding to each surface-modified Ce-Ox up to the formation of ceria nanostructures by the thermal decomposition process is shown in Table 2.1. It was found that mass loss (%) for the sample H_4 (42.05%) is lower than others indicating lower-thermal stability of array of nanohexagonal Ce-Ox. Highest mass loss was found for nanodisc Ce-Ox (H_3), 54.76%, and nanoflower petals Ce-Ox (H_1), 53.06%, prepared by the simple precipitation method. Ce-Ox synthesized through hydrothermal methods (H_2 and H_4) has lower percentage of mass loss. Theoretical mass loss for the formation of ceria from cerium oxalate was found to be 55.06%.

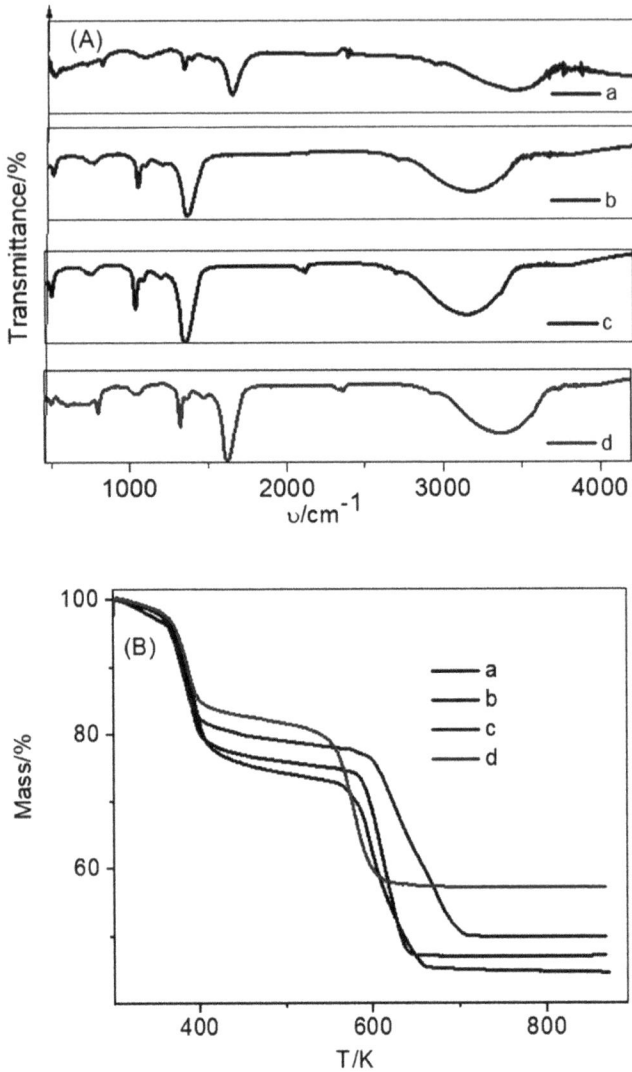

FIGURE 2.1 FTIR spectrum (A) and TG curves (B) [at 5 K/min in N$_2$ atmosphere (50 mL)] of surface-modified Ce-Ox: H$_1$ (a), H$_2$ (b), H$_3$ (c), and H$_4$ (d).

Figure 2.2A represents FE-SEM images of surface-modified Ce-Ox (sample H$_1$, H$_2$, H$_3$ and H$_4$). Figure 2.2Aa shows the Ce-Ox flower petals (H$_1$). Figure 2.2Ab shows like plates of Ce-Ox (H$_2$) while Figure 2.2Ac that disc like of Ce-Ox (H$_3$). Array of hexagonal Ce-Ox (H$_4$) is represented in Figure 2.2Ad. Figure 2.2B displays the DSC curves of Ce-Ox at 2 K/min. DSC curves show low-temperature dehydration at <500 K and the high-temperature decomposition of oxalates forming ceria nanostructures at >500 K.

TABLE 2.1 Mass Loss (%) for Each Ce-Ox Up To the Formation of Ceria Nanostructures Obtained from the TG Curve

Sample	Mass loss (%)
H_1	53.06
H_2	49.68
H_3	54.76
H_4	42.05

It is shown that changing the surface morphology of Ce-Ox causes the change in dehydration as well as decomposition temperature (see Figure 2.2B). This happens due to the difference in the diffusion-controlled reaction mechanisms of solid-state decomposition of each Ce-Ox. H_1 (see Figure 2.2Aa) undergoes dehydration at lower temperature (380.13 K) (see Figure 2.2Ba) than others. The samples H_2, H_3, and H_4 lose water at 394.22, 395.94, and 387.4 K (see Figure 2Bb, c, and d). On moving to the decomposition part at 2 K/min, H_4 has lower decomposition temperature 669.5 K (see Figure 2.2Bd) than H_1, H_2 and H_3. Hence, it shows the significant influence of morphology of precursor on the formation of ceria nanostructures.

Figure 2.3 shows the FE-SEM images of formed ceria nanomaterials. Each surface-modified ceria P_1, P_2, P_3, and P_4 was prepared by the calcination of Ce-Ox (H_1, H_2, H_3, and H_4) at 723 K in muffle furnace for 5 h. Nanoceria P_1 (see Figure 2.3a) has surfaces of *hexagonal rod* shape with branching in one direction having width in the range of 2.31–4.12 nm. But P_2 shows *nanorod*-like morphology with branches in many directions (see Figure 2.3b) having size in the range of 4.00–9.02 nm. These branches to the surfaces were occurred due to the surface coverage of PEG. Nanomaterial ceria P_3 (see Figure 2.3c) and P_4 (see Figure 2.3d) display like array of *2-D nanoplates* with size of 5.02–8.5 nm and array of *2-D nanosheets* with size of 3.04–9.5 nm, respectively. Nanoplate and sheet-like ceria were prepared by the directive influence of surfactant CTAB. From the surface morphological analysis of ceria, it can be seen that the thermal decomposition process does not allow retaining the surface morphology of Ce-Ox. Both the synthetic methods and surfactants followed affect the morphology of Ce-Ox, which has influence on the creation of varieties of ceria nanostructures.

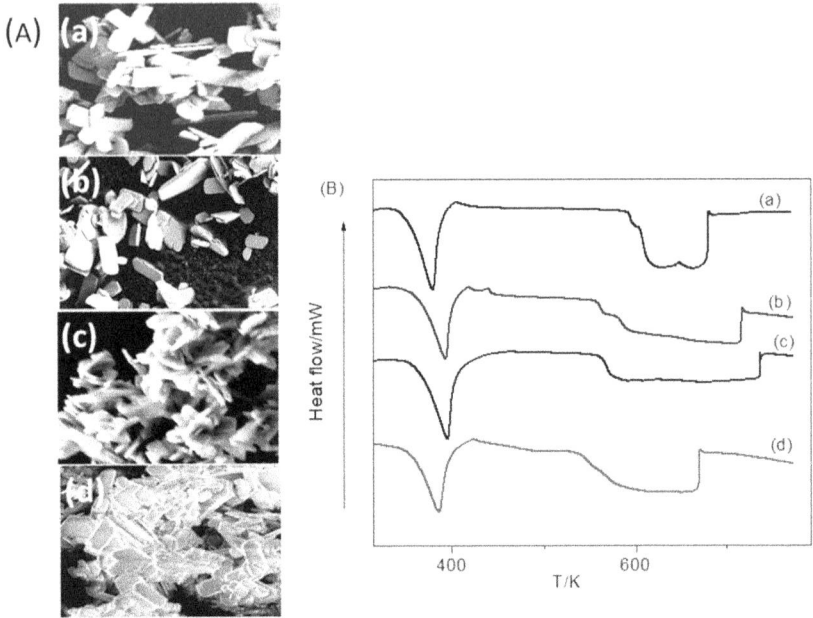

FIGURE 2.2 FE-SEM image (A) and DSC curves (B) of oxalate samples at 2 K/min: H_1 (a), H_2 (b), H_3 (c), and H_4 (d).

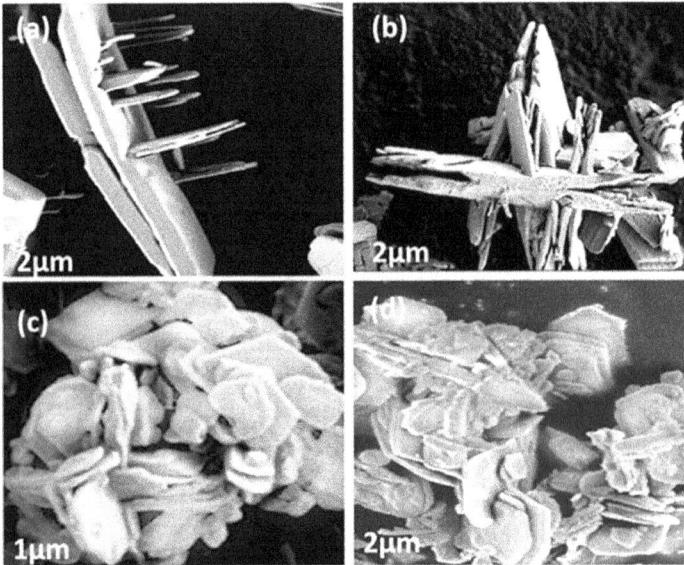

FIGURE 2.3 FE-SEM images of ceria nanostructures prepared from Ce-Ox: P_1 (a), P_2 (b), P_3 (c), and P_4 (d).

Figure 2.4A and B shows the FTIR spectrum and XRD pattern (normalized) of prepared ceria by thermal decomposition of Ce-Ox (*calcination at 723 K for 5 h*), where a, b, c, and d represent, respectively, P_1, P_2, P_3, and P_4. The band corresponds to 3011–3686.5 cm^{-1} is due to O–H stretching of adsorbed water molecules on the surface of the oxides. This hydroxyl group is eliminated only at higher temperature [39]. Surface of metal oxides is a good adsorbent of water from humidity even they are calcined at higher temperature. The band at 520–550 cm^{-1} corresponds to M–O bond [37, 38]. Figure 2.4B represents the diffraction peaks corresponding to cubic fluorite type structure of ceria (JCPDS Card No. 34–0394). Since they belong to the one type structure, their XRD patterns are similar with some deviations in the diffracting angle and peak area of intense peaks. Most intense peak was observed for crystalline P_3 (see Figure 2.4Bc), but lower intensities for peaks were observed for P_4 (see Figure 2.4Bd).

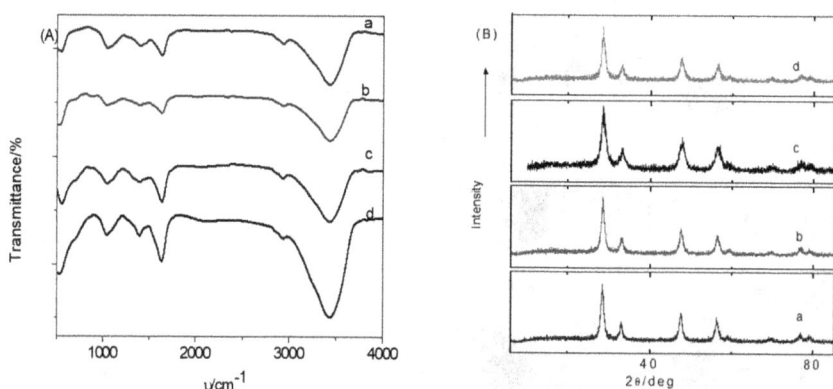

FIGURE 2.4   FTIR and XRD spectrum of synthesized ceria: P_1 (a), P_2 (b), P_3(c), and P_4 (d).

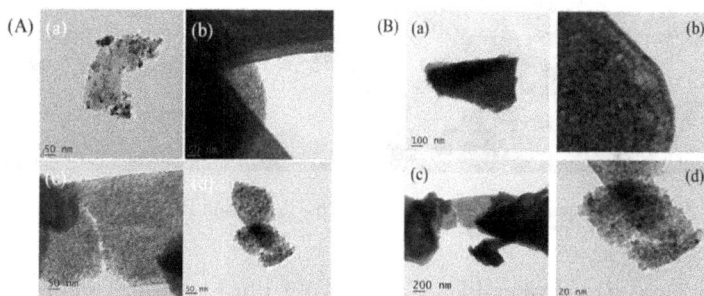

FIGURE 2.5   TEM images of ceria nanostructures at 50 nm (A): P_1 (a), P_2 (b), P_3 (c), and P_4(d) and at various nanometer scale (B): P_1 (a) at 100 nm, P_2 (b) at 20 nm, P_3 (c) at 200 nm, and P_4(d) at 20 nm.

Figure 2.5A shows the TEM images of ceria nanomaterials at uniform scale 50 nm. Thermal decomposition of Ce-Ox nanoflower petals H_1 yields nanohexagonal particles having size 2.11–3.98 nm (see Figure 2.5Aa and Ba). The interplanar distance was found to be 0.3 nm (see Figure S3a). This indicates the presence of branched nanohexagons with lattice fringes similar to interplanar distance of cubic fluorite-type structure of ceria. Thermal reactions of branched Ce-Ox (H_2) resulted in the formation of aggregates of nanoparticle with size in the range of 3.8–8.39 nm (see Figure 2.5Ab and Bb). The distance between two successive planes of this branched ceria was found to be 0.32 nm (see Figure S3b).

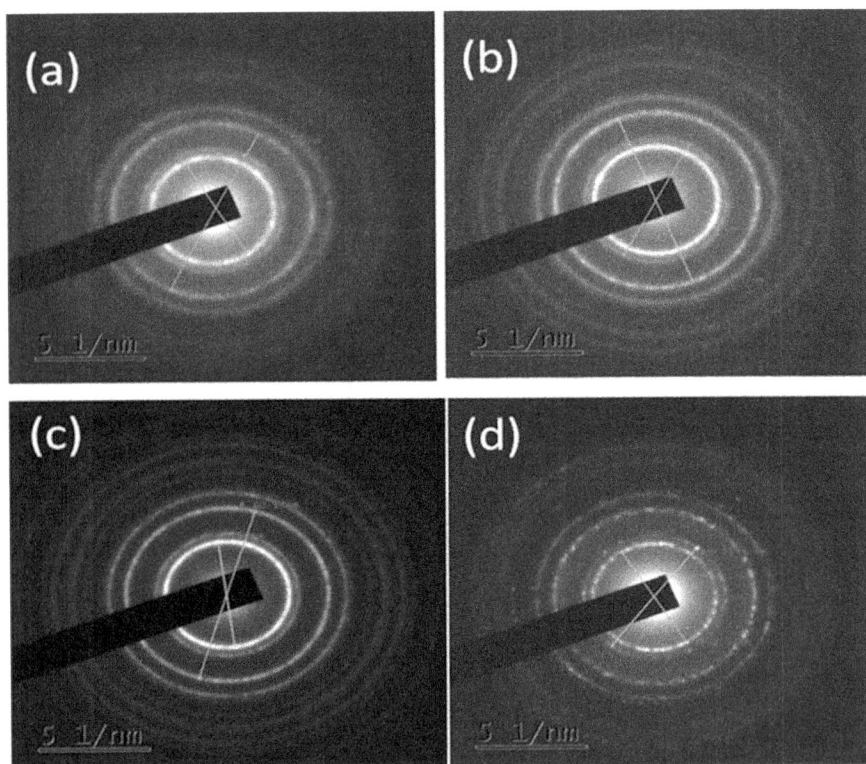

FIGURE 2.6　SAED pattern of ceria: P_1 (a), P_2 (b), P_3 (c), and P_4 (d).

The nanodisc/flower-like Ce-Ox (H_3) upon thermal decomposition destructed its morphology and formed aggregation of *2-D nanoplates*

(P$_3$). These nanoplates are composed of ceria nanohexagons of size 4.6–8.8 nm (see Figure 2.5Ac and Bc). The interplanar distance was observed to be 0.33 nm (see Figure S3c). Array of nanohexagonal Ce-Ox (H$_4$) altered its surface textures upon the thermal decomposition to yield *2-D nanosheets* (P$_4$) (Figure 2.3d), these sheets are composed of ceria nanospheres having size in the range of 2.92–9.32 nm (see Figure 2.5Ad and Bd). The interplanar distance of P$_4$ was observed to be 0.28 nm (see Figure S3d), indicating the cubic fluorite-type structure. The calculated values of lattice constant for each ceria nanostructures are listed in Table 2.2.

It is observed that lattice constant of each ceria nanostructure is altered. Changes occurred for each ceria lattice constant and interplanar distance are ascribed to the existence of difference in the experiencing lattice strain, which arise due to the difference in particle size and oxygen ion vacancies. Decrease in particle size and interplanar distance for the samples causes the lattices to be relaxed by creating new oxygen ion vacancies. Sample P$_1$ acquired lower particle size distribution compared to others. Hence, this sample reduces its lattice strain by creating more number of oxygen ion vacancies by converting more Ce^{4+} to Ce^{3+}.

TABLE 2.2 The Values of Lattice Constant, Particle Size, and Interplanar Distance of Ceria Nanostructures

Sample	Lattice Constant (nm)	Particle Size (nm)	Interplanar Distance (nm)
P$_1$	0.519	2.11–3.98	0.30
P$_2$	0.554	3.80–8.39	0.32
P$_3$	0.572	4.60–8.80	0.33
P$_4$	0.485	2.92–9.32	0.28

Figure 2.6 shows SAED pattern of each synthesized ceria nanostructures. P$_1$, P$_2$, and P$_3$ show highly crystalline ceria nanoparticle. But P4, *nanosheets*, composed of nanosphere-like particles exhibited semicrystalline nature as evident from the SAED pattern (see Figure 2.6d).

2.4   PHOTOPHYSICAL PROPERTIES

2.4.1   UV ABSORBANCE

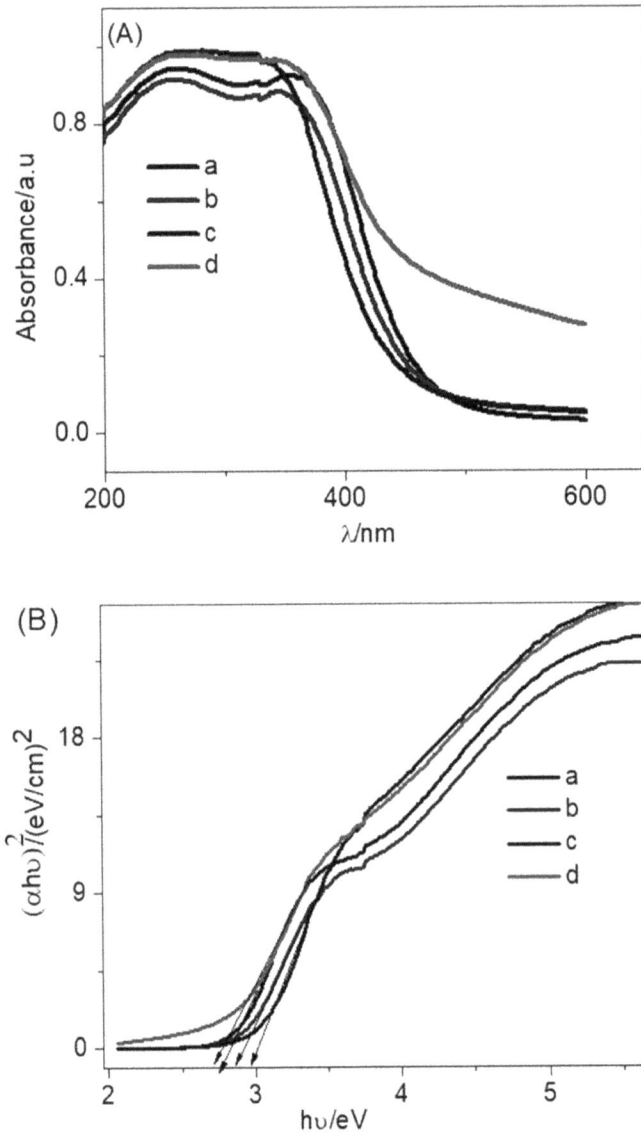

FIGURE 2.7   UV–visible absorbance spectra of ceria nanostructures (A): P_1 (a), P_2 (b), P_3 (c), and P_4 (d); Tauc plot (B): P_1 (a), P_2 (b), P_3 (c), and P_4 (d).

Figure 2.7A represents UV–visible absorbance spectra of the entire ceria nanostructures, which shows broad strong absorption below 400 nm. The absorbance spectrum shows two absorption maxima. The peaks about 256 nm correspond to charge transfer from $O^{2-}(2p)$ to Ce^{3+} (4f) orbitals in CeO_2, while UV absorption at 340 nm are responsible for CT from $O^{2-}(2p)$ to Ce^{4+} (4f) orbitals [40]. Figure 2.7Aa corresponds to ceria P_1, while Figure 2.7Ab, c, and d, respectively, represent P_2, P_3, and P_4. On comparing the UV absorbance, *nanoplate-like ceria* (P_3) show the highest UV absorbance. An estimate of the optical bandgap, E_g, can be determined by the Tauc equation $(\alpha h\upsilon)^n = B(h\upsilon - E_g)$, where $h\upsilon$ is the photon energy, α is the absorption coefficient, B is a constant relative to the material, and n is either 2 for direct transition or 1/2 for an indirect transition [30]. The $(\alpha h\upsilon)^2$ versus $h\upsilon$ curve (Tauc plot) is shown in Figure 2.7B. Figure 2.7Ba, b, c, and d, respectively, represent Tauc plots for P_1, P_2, P_3 and P_4. Lower E_g value was observed for 2D *nanosheets* of ceria (see Figure 2.7Bd). It is found that E_g value in the order of $P_4 < P_1 < P_2 < P_3$. E_g values calculated from the Tauc plot are listed in Table 2.3. This confirms that UV absorbance and optical bandgap significantly altered with respect to morphology of ceria nanostructure.

TABLE 2.3 The Values of Optical Bandgap (E_g) Calculated From the Tauc Plot

Sample	E_g (eV)
P_1	2.77
P_2	2.87
P_3	2.97
P_4	2.73

2.4.2 LUMINESCENT PROPERTIES

Figure 2.8 shows the significance of PL spectra of ceria nanostructures at an excitation wavelength of 340 nm. It was observed that depending upon the surface morphology of ceria, distinct level of intensity of emission band was occurred. Hence, surface morphology of nanostructures plays significant influence in the emission band intensity. Figure 2.8a represents PL of P_1. Figure 2.8b, c, and d represent PL of P_2, P_3, and P_4, respectively. Emission band consists of good violet bands at 419 nm (2.96 eV), blue emission band at 483 nm (2.57 eV), and green emission band at 529 nm (2.35 eV). It was reported that PL emission in the range of 400–550 nm (<3 eV) is highly associated with oxygen vacancies with trapped electrons, which localized between Ce (4f) and O (2p) band in CeO_2. Violet, blue, and green emission band intensities are high for P_3. This is attributed due to the variation in the concentration of oxygen ion vacancies. The relative contribution of emission bands strongly depends on irradiation temperature, electronic energy loss, and irradiation influence.

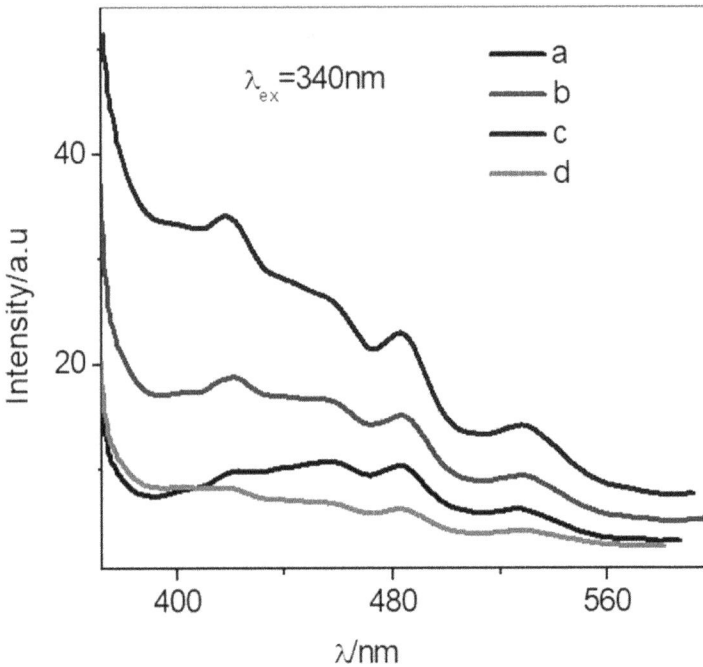

FIGURE 2.8 PL spectra of ceria nanostructures: P_1 (a), P_2 (b), P_3 (c), and P_4 (d).

2.5 THERMAL PROPERTIES

2.5.1 ESTIMATION OF EA VALUES FOR THE DECOMPOSITION PROCESS OF CE-OX UP TO CERIA NANOSTRUCTURES

Thermal decomposition of cerium oxalate to ceria is pass off via through low-temperature dehydration (<500 K) and high-temperature decomposition, which occurs through the breaking of oxalate bonds and formation of oxide bonds (>500 K). Even though there occur mainly two stages below and above 500 K, each stage is conducting through partially overlapped reaction stages [30]. Hence, the kinetic analysis regarding separately taking each stage is kinetically difficult. So, the method of kinetic deconvolution is the satisfactory method for studying partially overlapped multistep reactions. Each cerium oxalate removes its water molecules through two partially overlapping reactions [see Equations (2.1) and (2.2)], whereas the decomposition stage follows through three partially overlapped reactions. At the initiation of high-temperature reaction, removal of surfactants (PEG and CTAB) from the corresponding oxalates occurs followed by the formation of intermediate cerium oxycarbonate $Ce_2O_2 \cdot CO_3$ [see Equation (2.3)] [32]. At the final stage, ceria is formed from the intermediate by releasing CO [see Equation (2.4)]

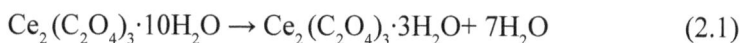

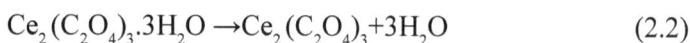

$$Ce_2(C_2O_4)_3 \cdot 10H_2O \rightarrow Ce_2(C_2O_4)_3 \cdot 3H_2O + 7H_2O \tag{2.1}$$

$$Ce_2(C_2O_4)_3 \cdot 3H_2O \rightarrow Ce_2(C_2O_4)_3 + 3H_2O \tag{2.2}$$

$$Ce_2(C_2O_4)_3 \rightarrow Ce_2O_2 \cdot CO_3 + 4CO_2 + CO \tag{2.3}$$

$$Ce_2O_2 \cdot CO_3 \rightarrow 2CeO_2 + CO \tag{2.4}$$

Figure 2.9 shows DSC curves for each cerium oxalate H_1, H_2, H_3, and H_4 at a β value of 2, 4, 6, and 8 K/min in the atmosphere of N_2. During the thermal decomposition process, the self-generated reaction condition has a predominant role on controlling the reaction path ways [41]. Change in the internal gaseous pressure can alter the smoothness of the heat transfer process, which affects kinetics of the reaction. Cerium oxalate decomposes to ceria through multistage decomposition stages. Due to the experimental inconvenience of separately tracking the component process, deconvolution of overall kinetic information into the reaction component is the only possible method for interpreting the multistage reaction scheme [42, 43]. Each kinetically resolved stage is studied by formal kinetic analysis. Figure 2.10 shows the typical deconvoluted DSC curve of sample H_4. Apparent activation energy for each independent process was measured by the isoconversional Friedman method. The determination of E_a values for each stage was based on (5). Plots of ln $(d\alpha/dt)$ versus T^{-1} (Friedman plot) [44–46] were carried out for the series of kinetic data recorded at different α values under linear nonisothermal condition. By this method, the coefficient of linearity was found to be near to unity. From the slope of the plot, the values of E_a can be evaluated.

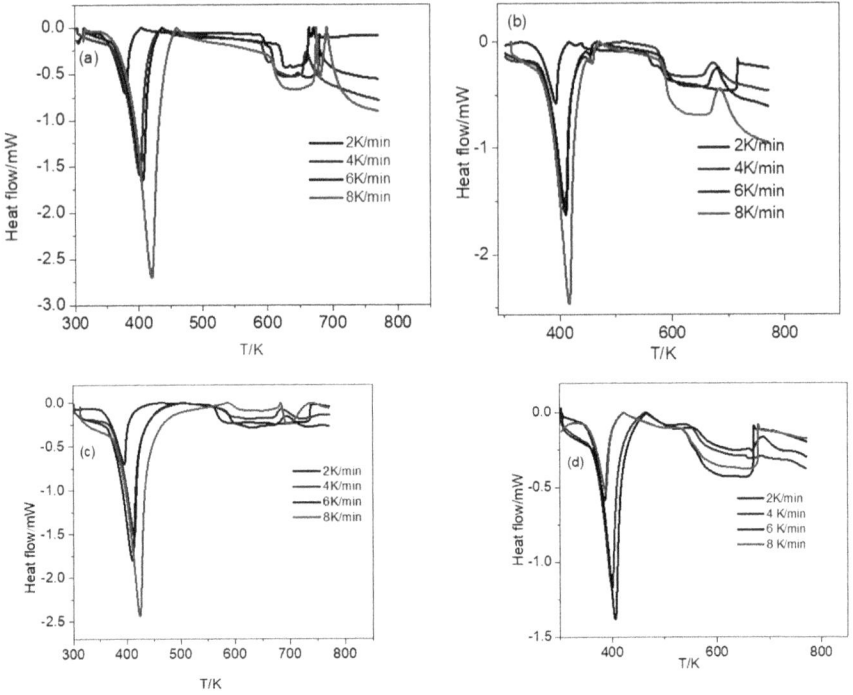

FIGURE 2.9   DSC curves for the thermal decomposition of each cerium oxalate at a heating rate of 2, 4, 6, and 8 K/min in the atmosphere of N_2 (50 mL): H_1 (a), H_2 (b), H_3(c), and H_4 (d).

$$\frac{d\alpha}{dt} = A \exp\left(-\frac{E_a}{RT}\right) f(\alpha)$$

(2.5)

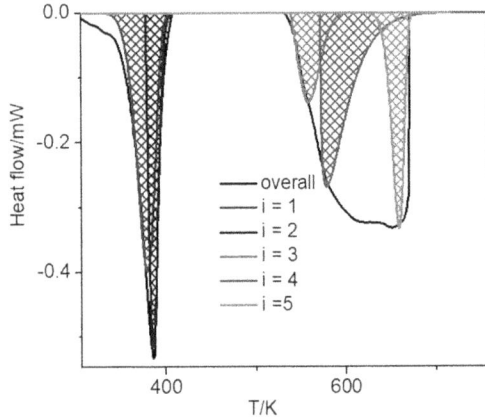

FIGURE 2.10   The deconvoluted DSC curve of sample H_4 at 2 K min^{-1}.

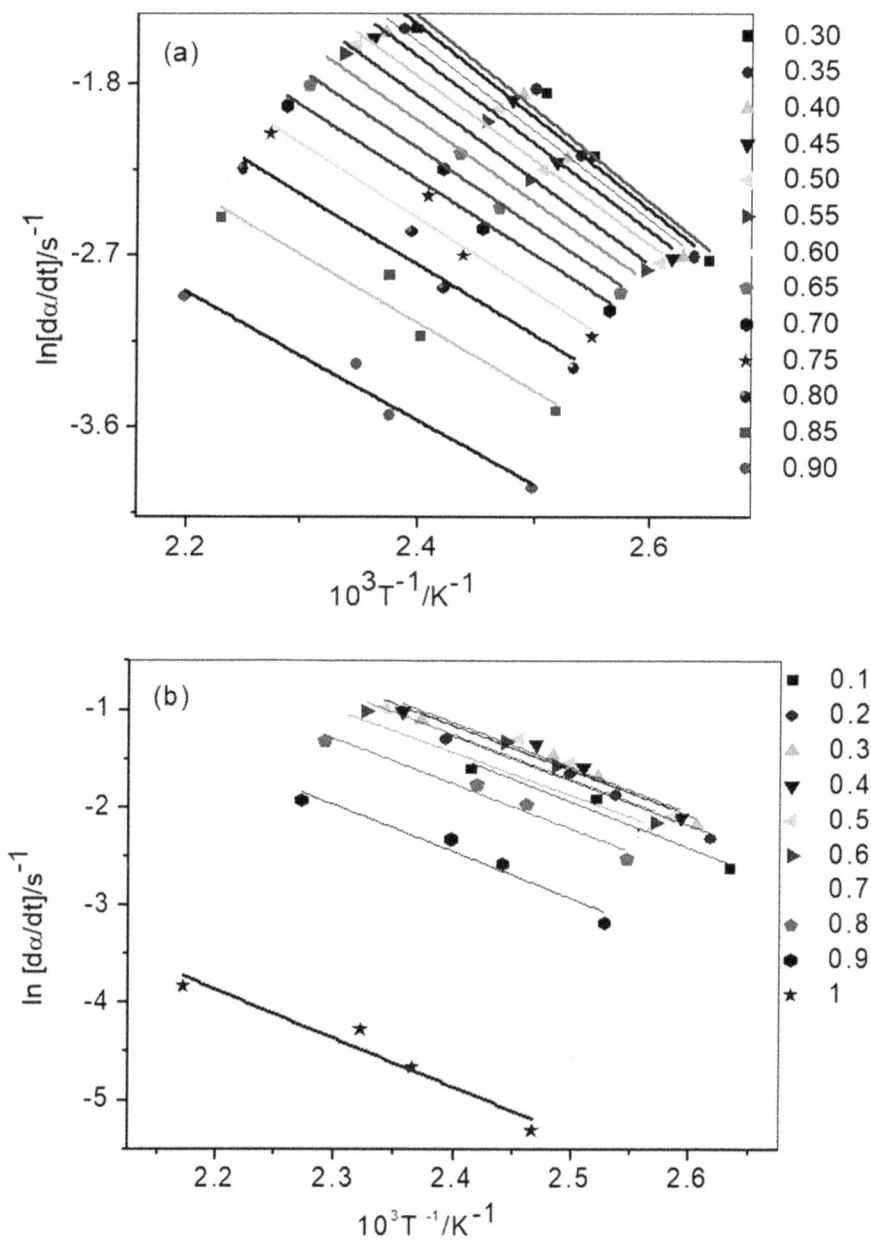

FIGURE 2.11 Friedman plot at different α values of first and second steps of dehydration reaction of array of nanohexagonal Ce-Ox (H₄).

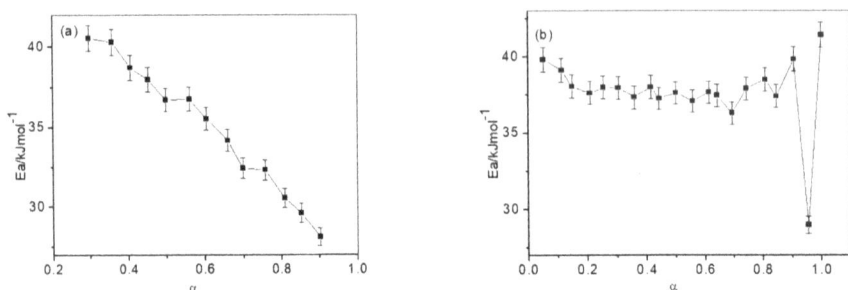

FIGURE 2.12   Dependence of E_a versus α for first stage (a) and second stage (b) for the sample H_4.

Figure 2.11 shows the Friedman plot at different fraction of conversion for the first stage (see Figure 2.11a) and the second stage (see Figure 2.11b) of thermal dehydration reaction of the sample H_4. For the third, fourth, and fifth stages, the plots are represented in Figure S1. Figure 2.12 depicts how thermal event occurs in the initial stages of the sample H_4 (see Figure 2.12a and b), reminding that E_a value decreases with reaction (for first) and E_a value almost constant toward the end of the stage (for second). Change in E_a value occurs due to the happening of change in reaction pathway as the reaction progresses. Figure S2 shows the dependence of E_a with respect to α for third, fourth, and fifth stages. The comparison of E_a value for each nanostructure is listed in Table 2.4. Significance of surface characteristics of each Ce-Ox upon the thermal reaction condition can be depicted from finding the E_a value throughout each stage.

From Table 2.4, it is known that array of nanohexagonal Ce-Ox (H_4) takes the easiest reaction path for removing crystallized water and for forming *2-D nanosheets* as the end product. Comparatively higher amount of activation energy was needed by Ce-Ox nanodisc/flower (H_3) to form *2-D nanoplates* as the reaction product. This means that Ce-Ox nanodisc/flower proceeds through the diffusion controlled reaction mechanism, which tends to higher thermal prevention for the decomposition. Change in the E_a value was observed throughout the reaction at different extents for samples H_1 and H_2. This informs that synthetic route and substrate surface characteristics control the thermal reaction programs for forming ceria nanostructures.

TABLE 2.4   The Average Values of E_a for Each Stage of Thermal Decomposition of Ce-Ox

Sample	E_a (kJ/mol)				
	1	2	3	4	5
H_1	47.15	48.13	130.37	146.24	68. 40
H_2	42.98	79.59	123.26	134.07	245.03
H_3	89.04	62.47	53.96	162.77	266.26
H_4	36.67	37.67	43.76	69.75	75. 53

2.6   CONCLUSION

Nanoflower petals, nanoplates, nanodiscs, and array of nanohexagonal Ce-Ox were synthesized *via* simple precipitation and hydrothermal methods. Executing the thermal decomposition strategy for the oxalate precursors, surface-modified ceria nanostructures such as branched hexagonal nanorod, multibranched, 2-D nanoplate, and nanosheet were synthesized and characterized. Surface modification of Ce-Ox has a pivotal role on the thermal decomposition strategy, particle size, and shape. The lowering of E_a values for the formation of nanosheet-like ceria has been observed. But nanodiscs/flower of Ce-Ox has higher thermal prevention for the decomposition process to produce *nanoplates* of ceria. Enhanced luminescent and UV-absorbing properties were observed for *nanoplate*-like ceria.

KEYWORDS

- nanostructured
- thermal decomposition
- nanoplates
- nanosheets
- oxygen ion vacancies

REFERENCES

1.   Sun C, Li H, Zhang H, Wang Z. Controlled synthesis of CeO$_2$ nanorods by a solvothermal method. *Nanotechnology*. 2005; **16**: 1454–63.

2. Trovarelli A. Catalytic properties of ceria and CeO_2-containing materials. *Catal Rev Sci Eng*. 1996; **38**: 439–520.

3. Laha SC, Ryoo R. Synthesis of thermally stable mesoporous cerium oxide with nano crystalline frameworks using mesoporous silica templates. *Chem Commun*. 2003; **17**: 2138–9.

4. Beie HJ, Gnorich A. Oxygen gas sensors based on CeO_2 thick and thin-films. *Sens Actuators B*. 1991; **4**: 393–9.

5. Jasinski P, Suzuki T, Anderson HU. Nano crystalline undoped ceria oxygen sensor. *Sens Actuators B*. 2003; **95**: 73–7.

6. Khadse VR, Sharada T, Patil KR, Pradip P. Humidity-sensing studies of cerium oxide nanoparticles synthesized by non-isothermal precipitation. *Sens Actuators B Chem*. 2014; **203**: 229–38.

7. Hue C, Zhang Z, Liu H, Gao P, Wang ZL. Direct synthesis and structure characterization of ultrafine CeO_2 nano particle. *Nanotechnology*. 2006; **17**: 5983–7.

8. Dong-En Z, Xiao-Jun Z, Xiao-Min N, Ji-Mei S, Hua-Gui Z. Fabrication of novel threefold shape CeO_2 dendrites: optical and electrochemical properties. *Chem Phys Lett*. 2006; **430**: 326–9.

9. Ren Z, Peng F, Li J, Liang X, Chen B. Morphology-dependent properties of Cu/CeO_2 catalysts for the water-gas shift reaction. *Catalysts*. 2017; 7: 48.

10. Yang F, Wei J, Liu W, Guo J, Yang Y. Copper doped ceria nanospheres: surface defects promoted catalytic activity and a versatile approach. *J Mater Chem A*. 2014; **2**: 5662–7.

11. Carrettin S, Concepcion P, Corma A, Nieto JML, Puntes VF. Nano crystalline CeO_2 increases the activity of Au for CO oxidation by two orders of magnitude. *Angew Chem Int Ed*. 2004; **43**: 2538–40.

12. Mai HX, Sun LD, Zhang YW, Si R, Feng W, Zhang HP, Liu HC, Yan CH. Shape-selective synthesis and oxygen storage behavior of ceria nanopolyhedra, nanorods and nanocubes. *J Phys Chem B*. 2005; **109**: 24380–5.

13. Nolan M, Parker SC, Watson GW. The electronic structure of oxygen vacancy defects at the low index surfaces of ceria. *Surf Sci*. 2005; **595**: 223–32.

14. Du X, Zhang D, Shi L, Gao R, Zhang J. Morphology dependence of catalytic properties of Ni/CeO_2 nanostructures for carbon dioxide reforming of methane. *J Phys Chem C*. 2012; **116**: 10009–16.

15. Huang XS, Sun H, Wang LC, Liu YM, Fan KN, Cao Y. Morphology effects of nano scale ceria on the activity of Au/CeO_2 catalysts for low-temperature CO oxidation. *Appl Catal B*. 2009; **90**: 224–32.

16. Zhang ML, Li J, Li HJ, Li Y, Shen WJ. Morphology-dependent redox and catalytic properties of CeO_2 nanostructures: nanowires, nanorods and nanoparticles. *Catal Today*. 2009; **148**: 179–83.

17. Aneggi E, Wiater D, de Leitenburg C, Llorca J, Trovarelli A. Shape-dependent activity of ceria in soot combustion. *ACS Catal*. 2013; **4**: 172–81.

18. Désaunay T, Bonura G, Chiodo V, Freni S, Couzinié JP, Bourgon J, Ringuedé A, Labat F, Adamo C, Cassir M. Surface-dependent oxidation of H_2 on CeO_2 surfaces. *J Catal*. 2013; **297**: 193–201.

19. Monte M, Gamarra D, Cámara AL, Rasmussen SB, Gyorffy N, Schay Z, Martínez-Arias A, Conesa J. Preferential oxidation of CO in excess H_2 over CuO/CeO_2 catalysts: performance as a function of the copper coverage and exposed face present in the CeO_2 support. *Catal Today*. 2014; **229**: 104–13.

20. Si R, Flytzani-Stephanopoulos M. Shape and crystal-plane effects of nano scale ceria on the activity of Au-CeO$_2$ catalysts for the water-gas shift reaction. *Angew Chem.* 2008; **120**: 2926–9.

21. Yao SY, Xu WQ, Johnston-Peck AC, Zhao FZ, Liu ZY, Luo S, Senanayake SD, Martínez-Arias A, Liu WJ, Rodriguez JA. Morphological effects of the nanostructured ceria support on the activity and stability of CuO/CeO$_2$ catalysts for the water-gas shift reaction. *Phys Chem Chem Phys.* 2014; **16**: 17183–95.

22. Rao KN, Bharali P, Thrimurthulu G, Reddy BM. Supported copper-ceria catalysts

23. for low temperature CO oxidation. *Catal Commun.* 2010; **11**: 863–6.

24. Wei L, Lijun F, Cong Z, Hongxiao Y, Jinxin G, Xiufang L, Xueying Z, Yanzhao Y. A facile hydrothermal synthesis of 3D flowerlike CeO$_2$ via a cerium oxalate precursor. *J Mater Chem A.* 2013; **1**: 6942–8.

25. Wright CS, Fisher J, Thompsett D, Walton RI. Hydrothermal synthesis of a cerium (IV) pyrochlore with low-temperature redox properties. *Angew Chem Int Ed.* 2006; **45**: 2442–6.

26. Corradi AB, Bondioli FB, Ferrari AM, Manfredini T. Synthesis and characterization of nano sized ceria powders by microwave-hydrothermal method. *Mater Res Bull.* 2006; **41**: 38–44.

27. Christel LR, Jeffrey WL, Erik ML, Katherine AP, Rhonda MS, Michael SD, Debra RR. Sol−gel-derived ceria nano architectures: synthesis, characterization, and electrical properties. *Chem Mater.* 2006; **18**: 50–8.

28. Ali B, Mohamed IZ, Julian E, Lata P. Micro emulsion-based synthesis of CeO$_2$ powders with high surface area and high-temperature stabilities. *Langmuir.* 2004; **20**: 11223–33.

29. Gabal MA, Shabaan AKE, Obaid AY. Synthesis and characterization of nano-sized ceria powder via oxalate decomposition route. *Powder Technol.* 2012; **229**: 112–8.

30. Miyazaki H, Kato JI, Sakamoto N, Wakiya N, Ota T, Suzuki H. Synthesis of CeO$_2$ nanoparticles by rapid thermal decomposition using microwave heating. *Adv Appl Ceram.* 2010; 109: 123–7.

31. Nusrath K, Muraleedharan K. Effect of Ca (II) on the multistep kinetic behavior of thermally induced oxidative decomposition of cerium(III) oxalate to CeO$_2$ (IV). *J Anal Appl Pyrol.* 2016; **120**: 379–88.

32. Nusrath K, Muraleedharan K. Effect of Ca(II) additive on the thermal dehydration kinetics of cerium oxalate rods. *J Therm Anal Calorim.* 2017; **128**: 541–52.

33. Rao VVS, Rao RVG, Biswas AB. Thermo gravimetric analysis of La, Ce, Pr, and Nd oxalates in air and in carbon dioxide atmosphere. *J Inorg Nucl Chem.* 1965; **27**: 2525–31.

34. Hoglund A, Hakkarainen M, Edlund U, Albertsson A. Surface modification changes the degradation process and degradation product pattern of polylactide. *Langmuir.* 2010; **26**: 378–83.

35. Moraczewski K. Effect of surface layer modification method on thermal stability of electroless metallized polylactide. *Polimery.* 2017; **62**: 750–6.

36. Fu Q, Cui Z, Xuea Y. Size dependence of the thermal decomposition kinetics of nano-CaC$_2$O$_4$: a theoretical and experimental study. *Eur Phys J Plus.* 2015; **130**: 1–14.

37. Bogatyreva GP, Marinich MA, Ya V, Zabuga, Tsapyuk GG, Panova AN, Bazalii GA. The effect of surface modification on thermal stability of nano diamonds. *J Superhard Mater.* 2008; **30**: 305–10.

38. Miguel G, Juan H, Leticia B, Joaquin N, Mario ERG. Characterization of calcium carbonate calcium oxide and calcium hydroxide as starting point to the improvement of lime for their use in construction. *J Mater Civil Eng*. 2009; **2**: 625–708.
39. Oman Z, Haznan A, Widayanti W. Synthesis and characterization of nanostructured CeO_2 with dyes adsorption property. *Process Appl Ceram*. 2014; **8**: 39–46.
40. Pavel J, Tomas HK, Martin K, Jakub E, Martin S. Thermal treatment of cerium oxide and its properties: adsorption ability versus degradation efficiency. *Adv Mater Sci Eng*. 2014; **2014**: 1–12.
41. Aškrabić S, Dohčević-Mitrović ZD, Araújo VD, Ionita G, de Lima Jr MM, Cantarero A. F-centre luminescence in nano crystalline CeO_2. *J Phys D: Appl Phys*. 2013; **46**: 495306.
42. Wada T, Nakano M, Koga N. Multistep kinetic behavior of the thermal decomposition of granular sodium per carbonate: hindrance effect of the outer surface layer. *J Phys Chem A*. 2015; **119**: 9749–60.
43. Yoshikawa M, Yamada S, Koga N. Phenomenological interpretation of them thermal decomposition of silver carbonate to form silver metal. *J Phys Chem C*. 2014; **118**: 8059–70.
44. Koga N, Goshi Y, Yamada S, Perez-Maqueda LA. Kinetic approach to partially overlapped thermal decomposition processes: co-precipitated zinc carbonates. *J Therm Anal Calorim*. 2013; **111**: 1463–74.
45. Ozawa T. Applicability of Friedman plot. *J Therm Anal*. 1986; **31**: 547–51.
46. Koga N. Kinetic analysis of thermo analytical data by extrapolating to infinite temperature. *Thermochimica Acta*. 1995; **258**: 145–59.
47. Gotor FJ, Criado JM, Malek J, Koga N. Kinetic analysis of solid-state reactions: the universality of master plots for analyzing isothermal and non-isothermal experiments. *J Phys Chem A*. 2000; **104**: 10777–82.

CHAPTER 3

KINETIC ANALYSIS OF THE FORMATION OF BARIUM-ZINC OXIDE NANOPARTICLES FROM THEIR OXALATE PRECURSORS

K. SABIRA, TINU LOWERENCE, and K. MURALEEDHARAN*

Department of Chemistry, University of Calicut, Malappuram, India

Corresponding author. E-mail: kmuralika@gmail.com

ABSTRACT

Zinc oxalate dihydrate and barium containing zinc oxalate of barium concentrations 2, 3, 4, and 5 mol% have been synthesized and characterized by Fourier transform-infrared spectroscopy (FTIR). The thermal events were studied by thermogravimetry in air and differential scanning calorimetry in a nitrogen atmosphere at different linear heating rates 2, 4, 6, 8, and 10 K/min. Zinc oxide and barium containing zinc oxide nanoparticles were characterized by FTIR, ultraviolet-diffuse reflectance spectrum, and X-ray diffraction. The activation energy for the formation of zinc oxide and barium containing zinc oxide nanoparticles was calculated by using different isoconversional methods and found that the activation energy varies with the addition of barium.

3.1 INTRODUCTION

The thurst for making small-sized materials having novel properties with atoms or molecules has led to the emergence of a new field of research, nanochemistry, or nanotechnology. In the past few years, research in the field of nanomaterials had grown, as the most exciting and vibrant field of endeavor. The nanotechnology revolutionized the world in many fields like nanomedicine. Nanomaterials are among the most challenging areas of current scientific and technological research because of their tremendous possibilities in generating novel shapes, structures, and the unusual phenomena associated with these materials [1]. Nanomaterials are exceptionally strong, hard, and ductile at high temperatures, wear resistant, and chemically very active. They exhibit unique electronic, magnetic, optical, photonic, and catalytic properties and their size are ideal for use as building blocks, which include metals, semiconductors, core–shell nanostructures, and organic polymeric materials [2]. Mixed metal oxide (MMO) nanoparticles (also called heterometal oxide nanoparticles) can play an appreciable role in many areas of chemistry and physics. The unique electronic and magnetic properties obtained when combining two metals in an oxide matrix have been well studied. However, the most common use for MMOs has been in the area of catalysis [3].

Among the nanostructured metal oxides, ZnO is considered to be one of the best metal oxides that can be used at a nanoscale level. ZnO normally has a hexagonal or wurtzite structure and it is an n-type II–VI semiconductor with a wide direct bandgap of about 3.37 eV and a large exciton binding energy of 60 meV [4]. From this point of view, nanostructured ZnO powders display great power in many applications such as gas sensors, solar cells, ultraviolet (UV) photonic devices, blue-green laser diodes, nonlinear optical devices, varistors, and photocatalyst with high chemical activity [5]. They display good photoconductivity and high transparency in the visible region and have been used as transparent electrodes for solar cells [6]. It also shows promising applications in piezoelectric and wireless devices. The optical characteristics of zinc oxide (ZnO) materials seem to depend on their microstructures, morphologies, and particle sizes. ZnO is known to be used as a highly efficient desulfurizer of coal-derived fuel gas and chemically synthesized gases since it can reduce the concentration of H_2S to a few parts per million [7]. Amongst many metal oxides, ZnO has remained as a very prominent low-cost wide bandgap semiconducting material.

Doping ZnO nanostructures with metal ions is a strategy to modify their electronic and optical performance and improve their applications [8]. Doping ZnO with transition metal enhances the ferromagnetic and piezoelectric coefficients with predicted Curie temperature above room temperature [9]. The literature review revealed that there are plenty of reports showing the optical and magnetic properties of transition and nontransition metals doped ZnO. Gupta et al. [10] reported ferroelectric transition around 343 K in K-doped ZnO nanorods, Yang et al. observed and explained the multiferroic behavior of Cr-doped ZnO [11]. Effects of Ba doping on structural, optical, and ferroelectric properties of ZnO nanoparticles prepared by a low-cost thermal decomposition method are presented by Srinet et al. [12] and they observed that some structural transformation in the morphology of nanostructure has occurred with Ba doping. The doping of Ba also causes redshift in the UV-visible spectra, a high value of dielectric constant, transition temperature, a high value of remnant polarization, and low value of the coercive field which can be useful for potential applications. Bukkitgar et al. studied the mefenamic acid-sensing properties of 5% barium doped ZnO nanoparticles using glassy carbon electrodes and the developed method was used in the in-vitro analysis of mefenamic acid in pharmaceutical formulations and spiked human urine samples [13]. Ba-doped ZnO nanospheres show superior photo-catalytic efficiency compared to pure ZnO and commercial TiO_2 [3]. The doping of Ba on ZnO nanoparticles causes changes in the optical, mechanical, photocatalytic, and electric properties of ZnO, so it is very important to study the effect of Ba doping on the kinetics of the formation of ZnO nanoparticles.

Transition metal oxalates act as the precursors for the synthesis of oxide nanoparticles. Transition metal oxalates, where oxalate ($C_2O_4^{2-}$) is the simplest dicarboxylate, are representative of transition metal oxide precursors because of the advantages of low cost, various preparation methods, and easy transformation at relatively low temperatures [14]. Ultrafine transition metal particles are a base for developing technologies such as metal injection molding, ceramics, and thick/thin-film applications.

The thermal decomposition processes of transition metal oxalates are relatively complicated because of the reduction property of $C_2O_4^{2-}$ the thermal decomposition of transition metal oxalates involves the cleavage of the C–C bond since the products are CO and CO_2 which contain only one carbon atom each. In many cases, the C–C bond cleavage is the rate-determining step. The cleavage may be heterolytic to produce CO_2 and CO_2^{2-} or homolytic to produce two CO_2^- anions [15]. Thermal decomposition study of mixed metal oxalates has been found useful for the preparation of MMOs possessing pores and lattice imperfections; therefore they serve as reactive solids. The MMOs may result in the modification of their thermal behavior, geometry, and electronic properties which lead to changes in their catalytic functions. The thermal decomposition of mixed metal oxalates, prepared by different techniques has been studied by Barbara et al. [16].

In recent years, the study of kinetics and the mechanism of solid-state thermal dehydration and decomposition reactions has become one of the important topics in solid-state chemistry. Kinetics study is important due to industry needs measurements of kinetic parameters for the accurate design of installations and treatment conditions because augmentation of temperature or elongation of reaction time means more cost. Using an appropriate mathematical expression, the thermoanalytical experiments can be applied for the modeling of industrial thermal processes. The results of the kinetic investigation can also be applied to problems such as useful lifetime of certain components, oxidative, and thermal stability and quality control.

There is plenty of reports available on the thermal decomposition of zinc oxalate [16–21] but there is no report on the effect of Ba addition on the kinetics of the formation of Ba-ZnO from their oxalate precursors. The present investigation aims to the synthesis and characterization of Ba doped ZnO nanoparticles by a simple, low cost, coprecipitation method followed by decomposition and the examination of the kinetics of the thermal decomposition of the formation of ZnO and Ba-ZnO nanoparticles from their oxalates precursors by nonisothermal thermogravimetry (TG) in the air with a view of obtaining the kinetic parameters and also to study the effect of barium doping on the kinetics of dehydration and decomposition of zinc oxalate dihydrate, the knowledge of which helps in the modification of the decomposition products by suitably altering the oxalate precursor.

3.2 EXPERIMENTAL

3.2.1 MATERIALS

AnalaR grade zinc nitrate dihydrate, oxalic acid ($H_2C_2O_4 \cdot 2H_2O$), barium nitrate, all of Merck, India, assay $\geq 99.9\%$ were used.

3.2.2 METHODS

3.2.2.1 PREPARATION OF ZNOX AND BA-ZNOX

The solutions of zinc nitrate with four different concentrations of barium, 2, 3, 4, and 5 mol% which are designated as $BZOx_2$, $BZOx_3$, $BZOx_4$, and $BZOx_5$ are prepared in distilled water. The solutions were coprecipitated by adding 0.1 M oxalic acid solution with warming and stirring. The resultant solution was stirred for another 2 h. The reacted solution was kept for some time to settle down the precipitate, then filtered off and washed several times with distilled water and air-dried in an oven kept at 343K and was used for characterization and thermal analysis. The precipitates of Ba-ZnOx were powdered. Pure zinc oxalates (ZnOx) were also prepared as per the abovementioned method as a reference sample without adding barium. The ZnO and barium-containing ZnO nanoparticles were designated as ZO, BZO_2, BZO_3, BZO_4, and BZO_5 of barium concentrations 2, 3, 4, and 5 mol%, respectively.

3.2.2.2   CHARACTERIZATION OF THE SAMPLE

Fourier transform infrared spectra (FTIR) of the samples was recorded by the transmittance method using a spectrometer (Model: Jasco FTIR-4100) after diluting the samples with KBr powder. For taking the FTIR spectra of the samples, the samples were well-grounded and the powdered samples were pressed using the hydraulic pellet press (KP, SR. No. 1718) under a pressure of 50 kg/cm². The optical bandgap (E_g) of the decomposed product was calculated from UV–visible reflectance which was measured using UV-visible diffuse reflectance spectrum (Model: Jasco V-550 spectrophotometer). The powder X-ray diffraction (XRD) patterns of the samples were recorded using a diffractometer (Miniflex 600, Model: RigakuD/Max) with Cu-Kα ($\lambda = 1.5418$ Å) radiation (40 kV, 15 mA) with a scan rate of 2θ/min in the region of 20°–90°.

3.2.2.3   MEASUREMENT OF THERMAL BEHAVIOR

TG measurements were made on a Perkin Elmer Pyris Thermal Analyser STA8000 at five different heating rates 2, 4, 6, 8, and 10 K/min. The operational characteristics of the thermal analysis system are as follows: atmosphere: air, sample mass: 5mg, and sample pan: silica. Duplicate runs were made under similar conditions and found that the data overlap with each other, indicating satisfactory reproducibility. The differential scanning calorimetric (DSC) measurements of the samples were done in the temperature range 303–773 K at five different heating rates, 2, 4, 6, 8, and 10 K/min on a Mettle Toledo DSC822e. The operational characteristics of the DSC system are as follows: atmosphere: flowing N_2 at a flow rate of 50 mL/min; sample mass 5 mg; and sample holder: silica.

3.3   RESULTS AND DISCUSSION

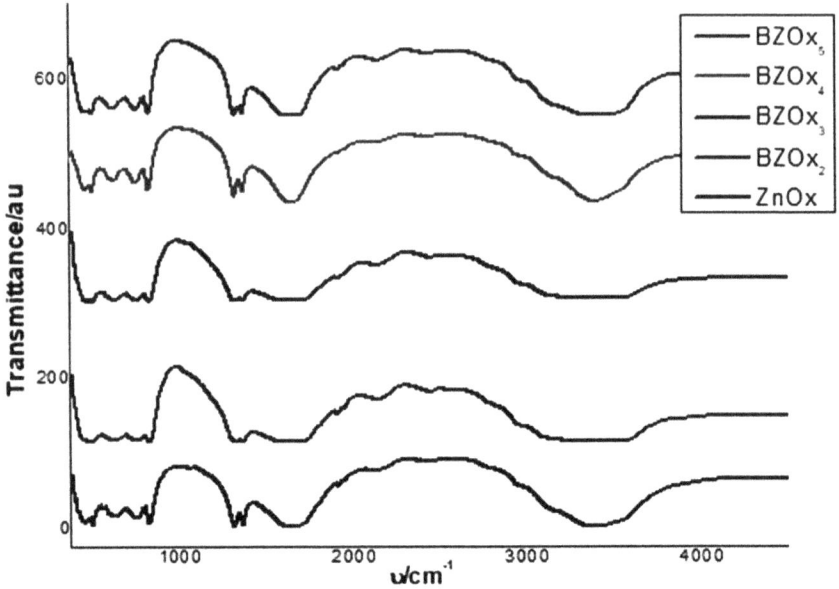

FIGURE 3.1   FTIR spectra of ZnOx and Ba-ZnOx.

Figure 3.1 shows the FTIR spectra of zinc oxalate and barium added zinc oxalate of barium concentrations 2, 3, 4, and 5 mol%. The strong broadband centered at about 3410 cm^{-1} is assigned to stretching modes of hydrated water, (O–H); this band arises due to the presence of water present in the sample (Figure 3.1). The strong band at 1640 cm^{-1} is characteristic of C=O antisymmetric stretching modes $_{as}$(C=O). The small peaks at 1349 and 1323 cm^{-1} is due to the O–C–O stretching modes and the small bands at 821 and 468 cm^{-1} are due to the O–C=O bending modes (O–C=O) and Zn–O stretching modes (Zn–O). Figure 3.2 shows the FTIR spectra of ZnO and Ba-ZnO nanoparticles. The FTIR spectra of ZnO and Ba-ZnO nanoparticles show only the stretching mode of Zn–O at 465 cm^{-1} indicating that ZnO only is the final product. The FTIR spectra obtained for $ZnC_2O_4 \cdot 2H_2O$ and ZnO were compared with the reported value [18] and found that the bands obtained were perfectly matching.

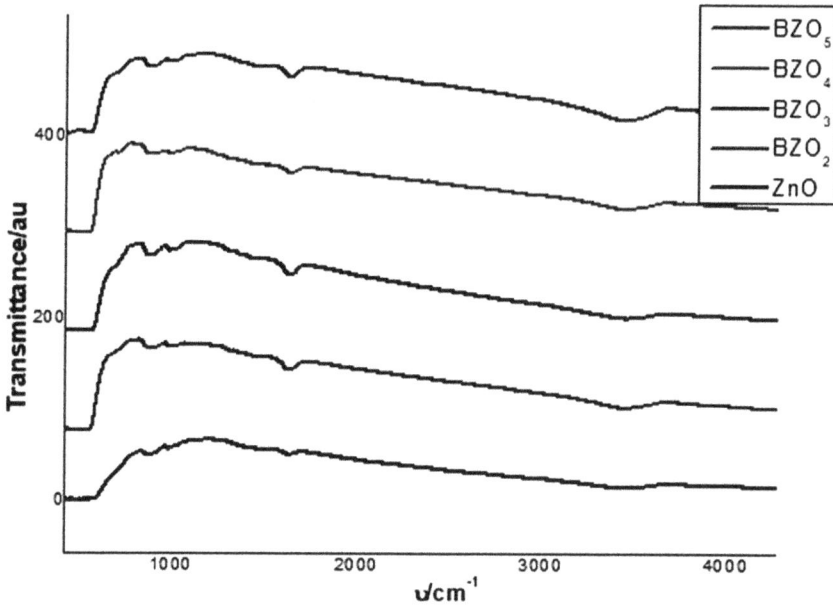

FIGURE 3.2   FTIR spectra of ZnO and Ba-ZnO nanoparticles.

Figure 3.3 shows the structural characteristics of ZnO nanoparticles and its mixed oxides with barium in different ratios (2, 3, 4, and 5 mol%) investigated by powder XRD. These XRD patterns reveal that both the zinc oxalate and its mixed oxalates with barium are highly crystalline in nature. The peaks centered at 31.75°, 34.69°, 36.58°, 47.70°, 56.74°, 63.05°, 66.62°, 68.31°, 69.36°, 72.71°, 77.12°, and 81.17° can be assigned to the (100), (022), (101), (102), (110), (103), 200), (112), and (201) planes of wurtzite ZnO. The absence of an additional peak of barium in the XRD pattern indicates the successful doping of barium in the ZnO lattices.

The position of the peaks and crystallinity did not change by the incorporation of barium in the ZnO lattice. The crystallite size is determined from the XRD line broadening using the Debye–Scherrer equation [22]:

$$d = 0.9\lambda/\beta\cos\theta \qquad (3.1)$$

where d is the crystallite size, λ is the wavelength used in XRD (0.15418 nm), θ is the Bragg angle in radians, β is the full width at half maximum intensity in radians. The crystallite sizes of ZnO and Ba-ZnO nanoparticles are different in different planes. The crystallite sizes along (100) planes for

ZnO and BZO$_2$, BZO$_3$, BZO$_4$, and BZO$_5$ are 35.6, 46.9, 47.7, 45.8, and 45.6 nm, respectively. The crystallite size first increased with Ba addition up to 3 mol% then it is decreasing on further addition of Ba loading. The decrease in size is not much visible by the addition of barium greater than 4 mol%. The average crystallite size in all the barium added ZnO nanoparticles were higher than that of pure ZnO nanoparticles.

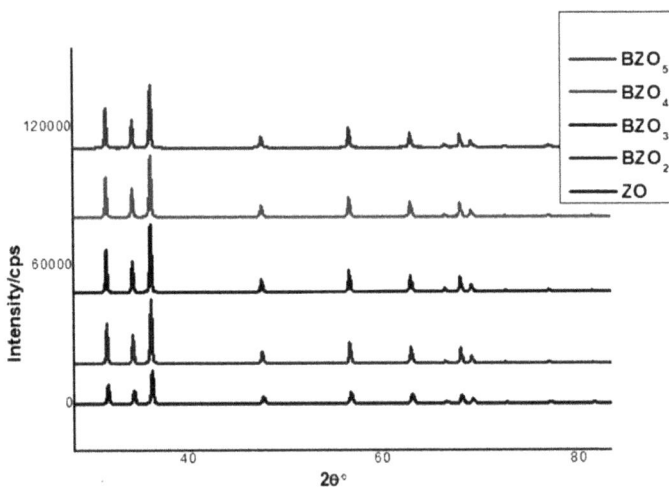

FIGURE 3.3   Powder XRD pattern of ZnO and Ba-ZnO nanoparticles.

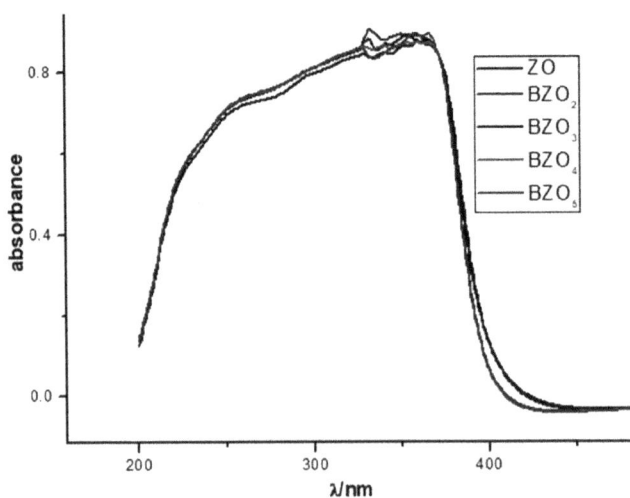

FIGURE 3.4   UV–visible spectra of ZnO and Ba-ZnO nanoparticles.

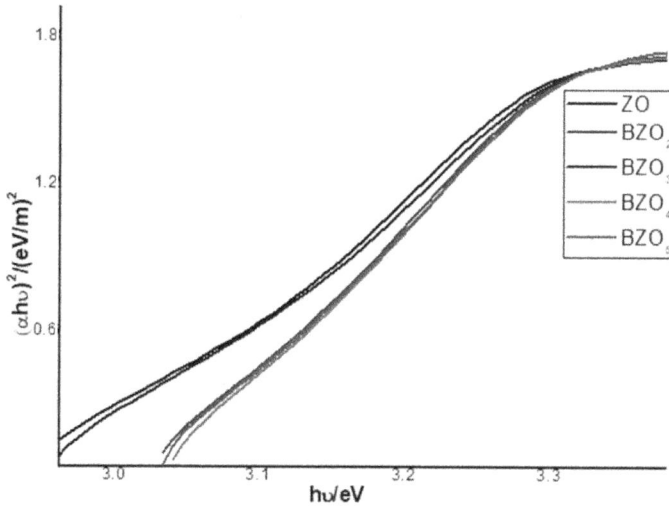

FIGURE 3.5   Tauc-plot of ZnO and Ba-ZnO nanoparticles.

Figure 3.4 shows the UV–visible spectra of ZnO and Ba-ZnO nanoparticles and from the spectra it is observed that all oxide nanoparticles have an adsorption maximum at about 360 nm. Figure 3.5 shows the Tauc-plot of ZnO and Ba-ZnO nanoparticles. The optical bandgap (E_g) of the decomposed product was calculated from UV–vis reflectance which was measured using the UV–vis diffuse reflectance spectrum. The optical bandgap is determined by using the Tauc equation

$$\left(\alpha h\upsilon\right)^{n} = B\left(h\upsilon - E_{g}\right) \tag{3.1}$$

where $h\upsilon$ is the photon energy, α is the absorption coefficient, B is a constant relative to the material, and n is either 2 for direct transition or 1/2 for an indirect transition. The bandgap obtained for ZnO nanoparticles is 2.95 eV. The bandgap obtained for barium added ZnO nanoparticles are higher than that of pure ZnO. The bandgap obtained for BZO_2, BZO_3, BZO_4, and BZO_5 is 3.02, 2.96, 3.02, and 3.01 eV, respectively.

The DSC curves obtained for the thermal decomposition of zinc oxalate and the barium-zinc oxalates are shown in Figure 3.6. The DSC curves of zinc oxalate and barium-zinc oxalate show two endothermic peaks which correspond to the dehydration reaction of oxalate dihydrate (temperature 364–442 K) and decomposition of oxalate into the oxides (temperature 608–687 K). The temperature for the dehydration and decomposition reaction reactions are shifted to higher values with the increase of heating rates indicating that the reactions are kinetically controlled. The peak temperature for the dehydration and decomposition reactions is not much affected by the addition of barium.

Figure 3.7a–e shows the TG plot of ZnOx and Ba-ZnOx samples of different barium concentrations at different heating rates 2, 4, 6, 8, and 10 K/min, and Figure 7.7f shows the TG curve of pure and Ba-ZnOx samples at the heating rate of 2 K min^{-1}. The thermogravimetric analysis shows two mass loss stages, the first one corresponding to the dehydration reaction (380–440 K) and the second stage corresponding to the decomposition of oxalate into oxide (597–684 K).

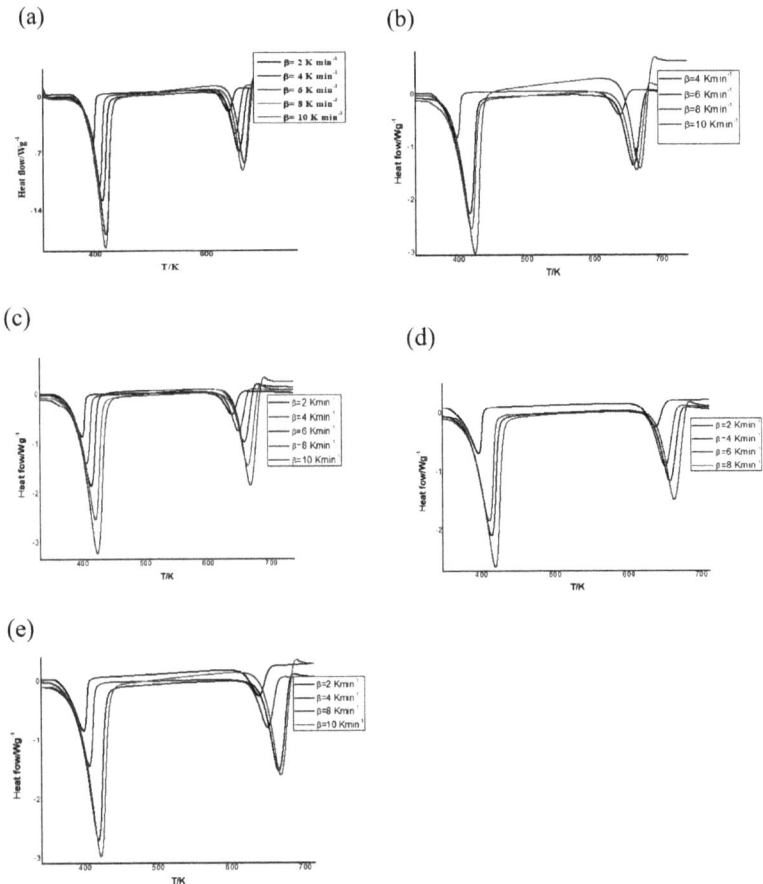

FIGURE 3.6 DSC plots of ZnOX (a), BZOx$_2$ (b), BZOx$_3$ (c), BZOx$_4$ (d), and BZOx$_5$ (e).

From Figure 3.7.a–e, it is observed that the temperature for the dehydration and decomposition reaction is shifted to higher values with the increase of heating rates as observed in the case of DSC. It is observed from Figure 3.7f that the temperature for dehydration and decomposition reaction is not much affected by barium addition. Thermogravimetric analysis indicates that the decomposition reaction occurs in a single step in the temperature range 597–684 K. The first mass loss of 17% corresponding to dehydration of two molecules of water and the second mass loss of 39.25% corresponding to the decomposition of zinc oxalate into ZnO nanoparticles.

The conversion function, α in the range 0.05–1.00 with an interval of 0.05 and their corresponding temperatures are determined from the experimental data obtained from the TG for the dehydration and decomposition reactions. From the values of α and the corresponding temperature, activation energy for the dehydration and decomposition reactions was calculated by utilizing the model-free isoconversional methods.

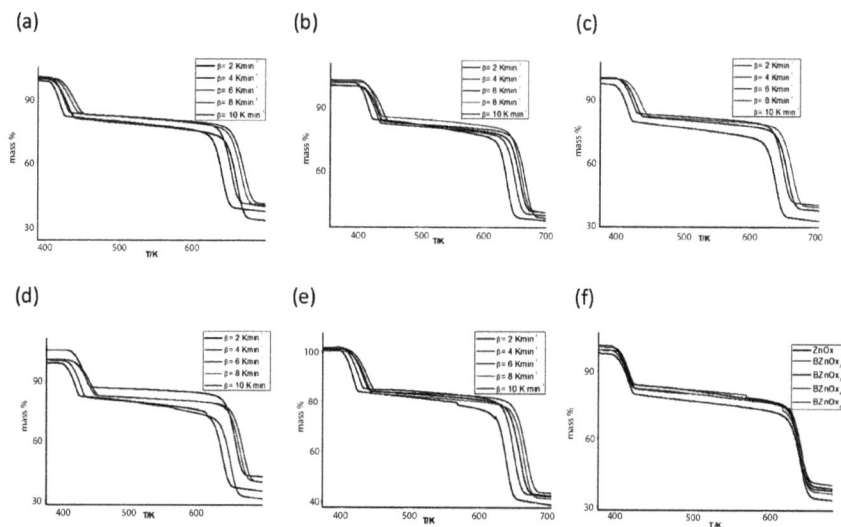

FIGURE 3.7   TG of ZnOx (a), BZOx$_2$ (b), BZOx$_3$ (c), BZOx$_4$ (d), and BZOx$_5$ (e) at different heating rates and pure and Ba-ZnOx samples at the heating rate 2 K/min (f).

3.3.1   ISOCONVERSIONAL METHODS USED FOR THE CALCULATION OF EA

(1) Flynn–Wall–Ozawa (FWO) method [23,24]

$$\ln \beta = \ln \left(AE / g(\alpha)R \right) - 5.331 - 1.052 E_a / RT_{ai} \qquad (3.2)$$

where α is the conversion function. From the slope of the graph $\ln\beta$ against $1/T_{ai}$, E_a can be calculated.

(2) Kissinger–Akahira–Sunose (KAS) method [25]

KAS is also an integer isoconversional method similar to FWO.

$$\ln \beta / T^2 = \ln \left(AR / E_a g(\alpha) \right) - E_a / RT \qquad (3.3)$$

E_a can be obtained from the slope of the straight-line graph $\ln\beta/T^2$ against $1/T$.

(3) Tang method [26]

$$\ln \left(\beta / T^{1.894661} \right) = \ln \left(AE_a / Rg(\alpha) \right) + 3.635041 - 1.894661 \ln E_a \\ -1.00145033 E_a / RT \qquad (3.4)$$

E_a can be obtained from the slope of the straight-line graph $\ln \left(\beta/T^{1.894661} \right)$ against $1/T$.

(4) Starink method [27]

$$\ln \beta / T^{1.95} = C^{1.95}(\alpha) - E_a / RT \qquad (3.5)$$

$$\ln \beta / T^{1.92} = C^{1.92}(\alpha) - 1.0008 E_a / RT \qquad (3.6)$$

(5) Boswell method

$$\ln \beta / T = C(\alpha) - E_a / RT_{ai}$$

The obtained α-T data were subjected to linear least-squares analysis (in the range $\alpha = 0.05$–1.00) using the isoconversional methods of FWO, KAS, Tang, Starink$^{1.92}$, Starink$^{1.95}$, and Boswell to obtain the values of activation energy (E_a). The obtained values of activation energies from the isoconversional methods are given in Tables 3.1–3.10. The average values obtained for the dehydration reaction of ZnOx and Ba-ZnOx and their decomposition reaction are given in Tables 3.11 and 3.12. The average value of activation energies for the dehydration reaction of pure zinc oxalate dihydrate in air obtained by the isoconversional methods FWO, KAS, Tang, Starink$^{1.92}$,

Starink[1.95], and Boswell are 115.05, 112.39, 112.60, 11.73, 112.54, and 117.45 kJ/mol, respectively. This value gets changed by the addition of barium. The activation energy value for all the samples for the dehydration reaction keeps on decreasing with the increase of conversion function indicating the complexity of the reaction. The average value of activation energies for the decomposition reaction of pure zinc oxalate dihydrate in air obtained by the isoconversional methods FWO, KAS, Tang, Starink[1.92], Starink[1.95], and Boswell are 189.56, 188.55, 155.85, 188.84, 188.83, and 193.99 kJ/mol, respectively. The activation energy value is also changed with the addition of barium. The activation energy value for all the samples for the decomposition reaction keeps on changing with the increase of conversion function indicating the complexity of the reaction; however, this change has no regular pattern.

TABLE 3.1 Activation Energy Obtained for the Dehydration Reaction of ZnOx Dihydrate

α	FWO	KAS	Tang	Starink[1.92]	Starink[1.95]	Boswell
0.3	133.63					136.97
0.35	130.34	130.10	130.28	129.59	130.07	133.50
0.4	127.18	126.75	126.94	126.29	126.72	130.17
0.45	124.20	123.60	123.79	123.19	123.60	127.04
0.5	121.57	120.81	121.00	120.42	120.78	124.24
0.55	118.51	117.57	117.78	117.26	117.62	121.06
0.6	116.89	115.86	116.06	115.46	115.78	119.28
0.65	113.03	111.77	111.99	111.47	111.82	115.27
0.7	111.25	109.88	110.10	109.57	109.87	113.36
0.75	108.71	107.19	107.41	106.82	107.15	110.66
0.8	105.65	103.96	104.19	103.59	103.94	107.45
0.85	103.08	101.23	101.46	100.77	101.09	104.67
0.9	98.93	96.84	97.08	96.31	96.92	100.41
0.95	97.71	95.52	95.77	91.76	97.63	100.14

TABLE 3.2 Activation Energy Obtained for the Dehydration Reaction of BZOx$_2$

α	FWO	KAS	Tang	Starink[1.92]	Starink[1.95]	Boswell
0.1	154.06	155.30	155.43	154.34	155.47	158.69
0.15	146.69	147.51	147.66	146.61	147.68	150.92
0.2	140.07	140.51	140.67	139.67	140.68	143.93

TABLE 3.2 *(Continued)*

α	FWO	KAS	Tang	Starink$^{1.92}$	Starink$^{1.95}$	Boswell
0.25	133.13	133.17	133.34	132.38	133.34	136.61
0.3	127.74	127.47	127.65	126.74	127.65	130.93
0.35	123.19	122.66	122.85	121.96	122.83	126.13
0.4	119.17	118.40	118.60	117.74	118.58	121.89
0.45	116.76	115.85	116.05	115.20	116.02	119.34
0.5	113.64	112.54	112.75	111.93	112.72	116.05
0.55	111.31	110.07	110.28	109.47	110.24	113.58
0.6	108.93	107.55	107.76	106.97	107.72	111.07
0.65	107.39	105.90	106.12	105.34	106.08	109.44
0.7	105.97	104.39	104.61	103.84	104.57	107.94
0.75	104.59	102.93	103.15	102.39	103.10	106.48
0.8	102.16	100.34	100.57	99.83	100.52	103.91
0.85	100.95	99.05	99.28	98.54	99.23	102.62
0.9	98.43	96.38	96.62	95.90	96.56	99.97
0.95	94.63	92.35	92.60	91.90	92.53	95.95

TABLE 3.3 Activation Energy Obtained for the Dehydration Reaction of BZOx$_3$

α	FWO	KAS	Tang	Starink$^{1.92}$	Starink$^{1.95}$	Boswell
0.05	130.64	130.68	130.85	129.91	130.85	134.06
0.1	121.15	120.64	120.82	119.95	120.81	124.04
0.15	115.99	115.16	115.35	114.52	115.33	118.59
0.2	111.82	110.74	110.94	110.13	110.91	114.18
0.25	108.37	107.08	107.29	106.51	107.25	110.54
0.3	105.65	104.19	104.41	103.64	104.37	107.67
0.35	103.40	101.80	102.02	101.26	101.97	105.28
0.4	101.49	99.76	99.99	99.25	99.94	103.26
0.45	101.08	99.31	99.54	98.81	99.49	102.83
0.5	99.35	97.47	97.70	96.98	97.65	100.99
0.55	98.32	96.37	96.60	95.89	96.55	99.90
0.6	97.19	95.17	95.40	94.69	95.34	98.71
0.65	95.55	93.42	93.66	92.96	93.60	96.97
0.7	94.75	92.56	92.80	92.11	92.74	96.12
0.75	93.42	91.14	91.39	90.70	91.32	94.71

TABLE 3.3   *(Continued)*

α	FWO	KAS	Tang	Starink$^{1.92}$	Starink$^{1.95}$	Boswell
0.8	91.87	89.49	89.74	89.07	89.67	93.07
0.85	90.67	88.21	88.46	87.79	88.39	91.79
0.9	89.21	86.65	86.90	86.25	86.83	90.25
0.95	86.68	83.96	84.22	83.58	84.14	87.57
1.0	77.76	74.47	74.75	74.17	74.66	78.14

TABLE 3.4   Activation Energy Obtained for the Dehydration Reaction of BZOx$_4$

α	FWO	KAS	Tang	Starink$^{1.92}$	Starink$^{1.95}$	Boswell
0.05	143.86	144.58	144.72	143.70	144.75	147.96
0.1	131.49	131.51	131.68	130.73	131.68	134.92
0.15	123.82	123.39	123.57	122.68	123.56	126.82
0.2	118.41	117.66	117.86	117.00	117.84	121.11
0.25	114.52	113.54	113.74	112.92	113.72	117.01
0.3	110.72	109.52	109.73	108.93	109.70	113.00
0.35	108.45	107.10	107.31	106.53	107.28	110.59
0.4	106.10	104.61	104.83	104.06	104.79	108.12
0.45	104.29	102.69	102.91	102.15	102.86	106.20
0.5	101.94	100.19	100.42	99.67	100.37	103.72
0.55	100.82	99.00	99.22	98.49	99.17	102.53
0.6	98.96	97.01	97.25	96.53	97.19	100.56
0.65	97.42	95.37	95.61	94.90	95.55	98.93
0.7	95.97	93.83	94.07	93.37	94.01	97.39
0.75	94.28	92.03	92.28	91.59	92.21	95.61
0.8	92.39	90.02	90.27	89.59	90.20	93.61
0.85	90.11	87.61	87.86	87.20	87.79	91.20
0.9	88.14	85.52	85.77	85.12	85.70	89.12
0.95	85.04	82.22	82.48	81.85	82.40	85.84
1.0	78.71	75.47	75.75	75.16	75.66	79.14

TABLE 3.5   Activation Energy Obtained for the Dehydration Reaction of BZOx$_5$

α	FWO	KAS	Tang	Starink[1.92]	Starink[1.95]	Boswell
0.05	160.71	162.29	162.41	161.27	162.46	165.68
0.1	147.21	148.03	148.17	147.12	148.20	151.45
0.15	139.80	140.19	140.35	139.35	140.36	143.63
0.2	132.75	132.74	132.92	131.96	132.92	136.20
0.25	128.42	128.16	128.34	127.42	128.33	131.63
0.3	123.95	123.42	123.61	122.72	123.60	126.91
0.35	121.13	120.44	120.63	119.76	120.61	123.93
0.4	118.04	117.16	117.36	116.51	117.34	120.67
0.45	116.40	115.41	115.62	114.78	115.59	118.93
0.5	114.67	113.58	113.78	112.95	113.75	117.10
0.55	112.85	111.64	111.85	111.03	111.82	115.18
0.6	111.06	109.74	109.96	109.15	109.92	113.29
0.65	109.97	108.57	108.79	107.99	108.75	112.13
0.7	108.05	106.54	106.76	105.97	106.71	110.10
0.75	106.43	104.82	105.05	104.27	105.00	108.40
0.8	104.47	102.74	102.96	102.21	102.92	106.32
0.85	102.99	101.16	101.39	100.64	101.34	104.75
0.9	100.32	98.33	98.57	97.84	98.51	101.93
0.95	98.03	95.90	96.14	95.42	96.08	99.51
1.0	86.42	83.59	83.85	83.81	83.77	87.25

TABLE 3.6   Activation Energy Obtained for the Decomposition Reaction of ZnOx

α	FWO	KAS	Tang	Starink[1.92]	Starink[1.95]	Boswell
0.05	181.06	180.01	180.30	180.28	180.27	185.24
0.1	193.53	193.00	193.27	193.27	193.26	198.29
0.15	195.17	194.66	194.94	194.93	194.93	199.99
0.2	193.86	193.24	193.52	193.51	193.51	198.59
0.25	192.84	192.13	192.42	192.41	192.40	197.50
0.3	191.06	190.23	190.52	190.50	190.49	195.61
0.35	190.45	189.56	189.86	189.84	189.83	194.96
0.4	189.17	188.20	188.49	188.48	188.47	193.60
0.45	187.88	186.81	187.11	187.09	187.08	192.23
0.5	187.32	186.21	186.51	186.49	186.48	191.63

TABLE 3.6 *(Continued)*

α	FWO	KAS	Tang	Starink$^{1.92}$	Starink$^{1.95}$	Boswell
0.55	185.91	184.71	185.01	184.99	184.98	190.14
0.6	185.17	183.90	184.21	184.19	184.17	189.35
0.65	184.23	182.89	183.20	183.18	183.17	188.35
0.7	184.31	182.96	183.27	183.25	183.23	188.42
0.75	183.31	181.89	182.20	182.18	182.16	187.37
0.8	182.70	181.22	181.53	181.51	181.49	186.71
0.85	182.64	181.14	181.45	181.43	181.41	186.64
0.9	182.35	180.79	181.11	181.08	181.06	186.31
0.95	183.85	182.32	182.64	182.62	182.60	187.86
1	234.41	235.23	235.49	235.50	235.51	240.92

TABLE 3.7 Activation Energy Obtained for the Decomposition Reaction of BZOx$_2$

α	FWO	KAS	Tang	Starink$^{1.92}$	Starink$^{1.95}$	Boswell
0.05	228.31	229.75	229.96	177.66	230.01	234.97
0.1	213.64	214.19	214.44	161.04	214.46	219.47
0.15	206.60	206.73	206.99	153.02	206.99	212.04
0.2	202.32	202.19	202.45	145.73	202.45	207.52
0.25	199.71	199.40	199.68	138.08	199.67	204.75
0.3	197.62	197.17	197.45	132.06	197.44	202.53
0.35	196.19	195.64	195.92	126.97	195.91	201.01
0.4	194.71	194.06	194.34	122.53	194.33	199.44
0.45	193.61	192.88	193.17	119.79	193.15	198.28
0.5	192.73	191.93	192.23	116.38	192.20	197.34
0.55	191.44	190.56	190.85	113.78	190.83	195.98
0.6	191.25	190.34	190.63	111.05	190.61	195.77
0.65	189.45	188.43	188.72	109.24	188.70	193.86
0.7	188.99	187.92	188.22	107.72	188.19	193.37
0.75	188.90	187.81	188.11	106.23	188.08	193.27
0.8	188.01	186.84	187.15	103.55	187.11	192.31
0.85	186.72	185.46	185.77	102.15	185.73	190.95
0.9	185.43	184.07	184.38	99.54	184.34	189.57
0.95	170.72	168.52	168.86	95.92	168.79	174.06

TABLE 3.8 Activation Energy Obtained for the Decomposition Reaction of $BZOx_3$

α	FWO	KAS	Tang	Starink$^{1.92}$	Starink$^{1.95}$	Boswell
0.05	175.21	173.85	174.15	174.02	174.12	179.09
0.1	175.54	174.12	174.42	174.29	174.38	179.39
0.15	173.38	171.79	172.09	171.96	172.05	177.09
0.2	172.06	170.35	170.67	170.53	170.62	175.68
0.25	170.90	169.10	169.42	169.28	169.37	174.45
0.3	169.48	167.58	167.90	167.77	167.85	172.94
0.35	167.56	165.53	165.86	165.72	165.80	170.91
0.4	166.52	164.41	164.74	164.60	164.68	169.79
0.45	164.84	162.62	162.95	162.82	162.89	168.01
0.5	164.02	161.74	162.07	161.94	162.01	167.14
0.55	163.04	160.69	161.03	160.89	160.96	166.11
0.6	162.77	160.38	160.72	160.58	160.65	165.81
0.65	162.54	160.12	160.46	160.32	160.39	165.55
0.7	162.23	159.78	160.12	159.98	160.05	165.22
0.75	162.21	159.73	160.08	159.94	160.01	165.19
0.8	161.64	159.11	159.45	159.31	159.38	164.57
0.85	161.58	159.02	159.37	159.23	159.29	164.50
0.9	162.21	159.65	159.99	159.86	159.92	165.14
0.95	164.33	161.83	162.18	162.04	162.11	167.35

TABLE 3.9 Activation Energy Obtained for the Decomposition Reaction of $BZOx_4$

α	FWO	KAS	Tang	Starink$^{1.92}$	Starink$^{1.95}$	Boswell
0.05	167.11	165.33	165.64	164.43	165.59	170.56
0.1	172.04	170.42	170.73	169.48	170.68	175.70
0.15	178.58	177.24	177.54	176.26	177.51	182.56
0.2	180.46	179.17	179.47	178.17	179.43	184.50
0.25	181.23	179.94	180.24	178.94	180.21	185.30
0.3	181.18	179.86	180.17	178.86	180.13	185.23
0.35	181.09	179.74	180.04	178.74	180.01	185.12
0.4	180.36	178.95	179.25	177.95	179.22	184.34
0.45	180.35	178.92	179.23	177.93	179.19	184.33
0.5	180.23	178.77	179.08	177.78	179.04	184.18
0.55	179.72	178.21	178.52	177.23	178.48	183.64
0.6	179.32	177.77	178.09	176.79	178.05	183.21

TABLE 3.9   *(Continued)*

α	FWO	KAS	Tang	Starink$^{1.92}$	Starink$^{1.95}$	Boswell
0.65	179.42	177.85	178.17	176.87	178.12	183.30
0.7	179.10	177.50	177.82	176.53	177.78	182.96
0.75	178.60	176.95	177.27	175.98	177.23	182.42
0.8	178.87	177.21	177.53	176.24	177.49	182.69
0.85	178.61	176.92	177.24	175.95	177.19	182.41
0.9	177.90	176.14	176.46	175.17	176.41	181.64
0.95	179.06	177.31	177.63	176.34	177.59	182.84
1.0	168.15	165.66	166.01	164.79	165.94	171.27

TABLE 3.10   Activation Energy Obtained for the Decomposition Reaction of BZOx$_5$

α	FWO	KAS	Tang	Starink$^{1.92}$	Starink$^{1.95}$	Boswell
0.05	111.34	106.80	107.19	106.21	107.06	111.96
0.1	143.67	140.64	140.99	139.78	140.90	145.89
0.15	151.35	148.64	148.98	147.72	148.91	153.93
0.2	164.20	162.09	162.42	161.07	162.36	167.42
0.25	169.88	168.02	168.34	166.96	168.29	173.37
0.3	172.65	170.91	171.22	169.82	171.18	176.27
0.35	174.65	172.98	173.29	171.87	173.25	178.35
0.4	175.45	173.79	174.11	172.68	174.06	179.18
0.45	175.60	173.93	174.25	172.82	174.20	179.33
0.5	176.33	174.68	175.00	173.56	174.95	180.09
0.55	176.72	175.06	175.38	173.94	175.33	180.48
0.6	176.50	174.81	175.13	173.69	175.08	180.24
0.65	176.99	175.31	175.63	174.19	175.58	180.75
0.7	177.02	175.32	175.64	174.20	175.59	180.77
0.75	177.82	176.14	176.46	175.01	176.41	181.60
0.8	177.47	175.75	176.07	174.62	176.02	181.22
0.85	177.30	175.55	175.87	174.42	175.82	181.04
0.9	176.90	175.09	175.41	173.97	175.36	180.59
0.95	176.36	174.46	174.79	173.35	174.74	179.99
1.0	180.92	179.05	179.39	177.91	179.33	184.69

TABLE 3.11 Average Values Obtained for the Dehydration Reaction of ZnOx and Ba-ZnOx Samples from Different Isoconversional Methods

Sample	FWO	KAS	Tang	Starink[1.92]	Starink[1.95]	Boswell
ZnOx	115.05	112.39	112.60	111.73	112.54	117.45
BZOx$_2$	117.16	116.24	116.44	115.60	116.42	119.75
BZOx$_3$	100.72	100.20	99.14	98.41	99.09	102.43
BZOx$_4$	104.27	102.64	102.87	102.11	102.82	106.17
BZOx$_5$	117.18	116.22	116.42	116.41	116.40	119.75

TABLE 3.12 Average Values Obtained for the Decomposition Reaction of ZnOx and Ba-ZnOx Samples from Different Isoconversional Methods

Sample	FWO	KAS	Tang	Starink[1.92]	Starink[1.95]	Boswell
ZnOx	189.56	188.55	188.85	188.84	188.83	193.99
BZOx$_2$	195.07	194.41	192.74	194.68	192.72	197.86
BZOx$_3$	166.42	164.28	164.61	164.48	164.55	169.68
BZOx$_4$	178.07	176.49	176.81	175.52	176.76	181.91
BZOx$_5$	169.46	167.45	167.78	166.39	167.72	172.86

3.4 CONCLUSION

In this study, we have succeeded in the synthesis of barium-containing ZnO nanoparticles through a simple, low cost, oxalate coprecipitation route followed by the thermal decomposition. The synthesized oxalate samples were characterized by FTIR. The formed barium ZnO particles were characterized by FTIR, ultraviolet-diffuse reflectance spectrum, and XRD and found that the oxide particles are in the nanometer range. The bandgap is increased by the addition of barium. The activation energy for the formation of ZnO and barium containing ZnO nanoparticles was calculated by using different isoconversional methods and found that the activation energy varies with the addition of barium. The average value of activation energies for the dehydration reaction of pure zinc oxalate dihydrate in air obtained by the isoconversional methods FWO, KAS, Tang, Starink[1.92], Starink[1.95], and Boswell are 115.05, 112.39, 112.60, 11.73, 112.54, and 117.45 kJ/mol, respectively. The average value of activation energies for the decomposition reaction of pure zinc oxalate in air obtained by the isoconversional methods FWO, KAS, Tang, Starink[1.92], Starink[1.95], and Boswell are 189.56, 188.55, 155.85, 188.84, 188.83, and 193.99 kJ/mol, respectively. The activation energy value for all the samples studied for the dehydration and decomposition reactions is changed with the addition of barium. The activation energy value for all the samples for the dehydration and decomposition reactions keeps on changing with the increase of conversion function indicating the complexity of the reaction.

KEYWORDS

- **kinetics**
- **isoconversional methods**
- **zinc oxalate**
- **barium-zinc oxalate**
- **barium-zinc oxide nanoparticles**

REFERENCES

1. Ozin, G.A., Arsenault, A., Nanochemistry: a chemical approach to nanomaterials, Royal Society of Chemistry, 2015.
2. Verelst, M., Ely, T.O., Amiens, C., Snoeck, E., Lecante, P., Mosset, A., Respaud, M., Broto, J.M., Chaudret, B., Synthesis and characterization of CoO, Co_3O_4 and mixed Co/CoO nanoparticules, Chem. Mater. 11 (1999) 2702–2708. https://doi.org/10.1021/cm991003h.
3. Jesudoss, S.K., Vijaya, J.J., Selvam, N.C.S., Kombaiah, K., Sivachidambaram, M., Adinaveen, T., Kennedy, L.J., Effects of Ba doping on structural, morphological, optical and photocatalytic properties of self-assembled ZnO nanospheres, Clean Technol. Environ. Policy. 18 (2016) 729–741. https://doi.org/10.1007/s10098–015-1047–1.
4. Kansal, S.K., Singh, M., Sud, D., Studies on photodegradation of two commercial dyes in aqueous phase using different photocatalysts, J. Haz. Mater. 141 (2007) 581–590. https://doi.org/10.1016/j.jhazmat.2006.07.035.
5. Suwanboon, S., Amornpitoksuk, P., Sukolrat, A., Dependence of optical properties on doping metal, crystallite size and defect concentration of M-doped ZnO nanopowders (M = Al, Mg, Ti), Ceram. Int. 37 (2011) 1359–1365. https://doi.org/https://doi.org/10.1016/j.ceramint.2010.12.010.
6. Shankar, R., Srivastava, R.K., Prakash, S.G., Study of dark-conductivity and photoconductivity of ZnO nano structures synthesized by thermal decomposition of zinc oxalate, Electron. Mater. Lett. 9 (2013) 555–559. https://doi.org/10.1007/s13391–013-2166–7.
7. Dar, G.N., Umar, A., Zaidi, S.A., Ibrahim, A.A., Abaker, M., Baskoutas, S., Al-Assiri, M.S., Ce-doped ZnO nanorods for the detection of hazardous chemical, Sens. Actuators B Chem. 173 (2012) 72–78. https://doi.org/https://doi.org/10.1016/j.snb.2012.06.001.
8. Water, W., Yan, Y.S., Characteristics of strontium-doped ZnO films on love wave filter applications, Thin Solid Films. 515 (2007) 6992–6996. https://doi.org/10.1016/j.tsf.2007.02.028.

9. Dietl, T., Ohno, H., Matsukura, F., Cibert, J., Ferrand, D., Zener model description of ferromagnetism in zinc-blende magnetic semiconductors, Science. 287 (2000) 1019–1022.

10. Gupta, M.K., Sinha, N., Singh, B.K., Kumar, B., Synthesis of K-doped p-type ZnO nanorods along (100) for ferroelectric and dielectric applications, Mater. Lett. 64 (2010) 1825–1828. https://doi.org/https://doi.org/10.1016/j.matlet.2010.05.044.

11. Yang, Y.C., Zhong, C.F., Wang, X.H., He, B., Wei, S.Q., Zeng, F., Pan, F., Room temperature multiferroic behavior of Cr-doped ZnO films, J. Appl. Phys. 104 (2008) 64102. https://doi.org/10.1063/1.2978221.

12. Srinet, G., Kumar, R., Sajal, V., High Tc ferroelectricity in Ba-doped ZnO nanoparticles, Mater. Lett. 126 (2014) 274–277. https://doi.org/https://doi.org/10.1016/j.matlet.2014.04.054.

13. Bukkitgar, S.D., Shetti, N.P., Kulkarni, R.M., Nandibewoor, S.T., Electro-sensing base for mefenamic acid on a 5% barium-doped zinc oxide nanoparticle modified electrode and its analytical application, RSC Adv. 5 (2015) 104891–104899. https://doi.org/10.1039/C5RA22581G.

14. Jia, Z., Ren, D., Xu, L., Generalized preparation of metal oxalate nano/submicro-rods by facile solvothermal method and their calcined products, Mater. Lett. 76 (2012) 194–197. https://doi.org/https://doi.org/10.1016/j.matlet.2012.02.067.

15. Reddy, M.V.V.S., Lingam, K.V., Rao, T.K.G., Radical studies in oxalate systems: E.S.R. of CO_2^- in irradiated potassium oxalate monohydrate, Mol. Phys. 42 (1981) 1267–1269. https://doi.org/10.1080/00268978100100951.

16. Małecka, B., Drozdz-Cieśla, E., Małecki, A., Mechanism and kinetics of thermal decomposition of zinc oxalate, Thermochim. Acta. 423 (2004) 13–18. https://doi.org/10.1016/j.tca.2004.04.012.

17. Kornienko, V., Influence of cation nature on the thermal decomposition of oxalates, (1957) Ukr. Chem. 23 (1957) 159–167.

18. Hu, C., Mi, J., Shang, S., Shangguan, J., The study of thermal decomposition kinetics of zinc oxide formation from zinc oxalate dihydrate, J. Therm. Anal. Calorim. 115 (2014) 1119–1125. https://doi.org/10.1007/s10973-013-3438-z.

19. Majumdar, R., Sarkar, P., Ray, U., Roy, M. Mukhopadhyay, Secondary catalytic reactions during thermal decomposition of oxalates of zinc, nickel and iron(II), Thermochim. Acta. 335 (1999) 43–53. https://doi.org/10.1016/S0040-6031(99)00128-8.

20. Perulmutter, J.M.D.D., Thermal decomposition of carbonates, carboxylates, oxalates, acetates, formates and hydroxides, Thermochim. Acta. 49 (1981) 207–218.

21. Sabira, K., Muraleedharan, K., Exploration of the thermal decomposition of zinc oxalate by experimental and computational methods, J. Therm. Anal. Calorim. (2020). https://doi.org/10.1007/s10973-019-09169-6.

22. Patterson, A., The Scherrer formula for X-ray paeticle size determination, Phys. Rev. 56 (1939) 978.

23. Ozawa, T., A new method of analyzing thermogravimetric data, Bull. Chem. Soc. Jpn. 38 (1965) 1881–1886. https://doi.org/10.1246/bcsj.38.1881.

24. Joseph, L.A.W., Flynn, H., A quick, direct method for the determination of activation energy from thermogravimetric data, Polym. Lett. 4 (1966) 323–328.

25. Kissinger, H.E., Reaction kinetics in differential thermal analysis, Anal. Chem. 29 (1957) 1702–1706. https://doi.org/10.1021/ac60131a045.

26. Wanjun Tang, C.W., Liu, Y., Zhang, H., New approximate formula for Arrhenius temperature integral, Thermochim. Acta. 408 (2003) 39–43. https://doi.org/10.1007/s10973–009-0323-x.
27. Starink, M.J., The determination of activation energy from linear heating rate experiments: a comparison of the accuracy of isoconversion methods, Thermochim. Acta. 404 (2003) 163–176.

CHAPTER 4

STUDIES ON THE THERMAL BEHAVIOR AND KINETIC PARAMETERS OF THE FORMATION OF BARIUM TITANATE NANOPARTICLES AND ITS BACTERICIDAL EFFECT

N. V. SINDHU and K. MURALEEDHARAN*

Department of Chemistry, University of Calicut, Calicut, India

Corresponding author. E-mail: kmuralika@gmail.com

ABSTRACT

Thermal behavior and kinetic parameters of the formation of barium titanate nanoparticles were determined by differential scanning calorimetry analysis under nonisothermal heating conditions. Different isoconversional methods were used to determine the apparent activation energies of the multistep thermal processes. Kinetic deconvolution procedure was used to perform the overall kinetics of formation of $BaTiO_3$ nanoparticles. In order to predict the physicogeometrical reaction mechanism, empirical kinetic model functions such as phase nucleation and growth-type model, $RO(n)$ was employed. Bactericidal effect of the samples was analyzed on a clinical strain by Kirby–Bauer method using two human pathogenic bacteria *Pseudomonas aeruginosa* (Gram-negative) and *Staphylococcus aureus* (Gram-positive) and found to be effective to some extent. The band structure and density of state (DOS) of stable phase of $BaTiO_3$ were investigated by means of DFT, Vienna Ab-initio Simulation Package (VASP); VASP was used for the DOS and structural optimization calculations. The prepared samples were identified and characterized by means of FTIR, UV–visible spectrometry, X-ray diffraction, scanning electron microscope, and transmission electron microscopy.

4.1   INTRODUCTION

In recent years, a synthesis on "nanomaterials" has become a focal area in materials research, due to their many superior properties than the bulk properties: mechanical strength, thermal stability, catalytic activity, electrical conductivity, magnetic properties, and optical properties [1]. These properties of "nanomaterials" can be different at the nanoscale for the reasons: "nanomaterials" have a relatively larger surface properties when compared to the same mass of material produced in a larger form. Nanoparticles can make materials more chemically reactive and affect their strength, and also quantum effects can begin to dominate the behavior of matter at the "nanometer length scale" and exploitation of novel properties (physical, chemical, biological) at that length scale. These particles have special physical and chemical properties as compared to their bulk materials due to their large reactive and surface area. They are broadly used in many fields, such as chemistry, electronics, photochemical, and biomedicine [2]. Nanoparticles are promising multifunctional molecules and are used for a wide range of imaging and therapeutic functions [3]. Nanoparticles deal with a wide-ranging area of fascinate including electronics, medicine, food industry, environmental, cosmetics, applications [4]. Generally, a physical approach is required to modify or to pick up the pharmacodynamics and pharmacokinetic property of drugs. Scattering of preformed polymerization of monomers and ionic gelation or coacervation of hydrophilic polymer, polymers are a variety of techniques used for the synthesis of nanoparticles [5]. Nanoparticle plays a role in opportunities and possibilities for the growth of sensing tools. In the last decade, sensing of selective biomolecules using functionalized gold nanoparticles has become a major research power. Gold nanoparticles based biosensors are likely to alter the very basics of sensing and detecting biomolecules [6]. Nanoparticles are used for different purposes, from medical treatment, used in a mixture of branches of industry fabrication such as oxide fuel batteries for energy storage and solar, large range materials on a daily basis life such as cosmetics and textiles [7].

The barium titanate nanoparticles (BTNPs) have been extensively applied in various fields such as multilayer ceramic capacitors, integral capacitors in printed circuit boards, dynamic random access memories, resistors with a positive temperature coefficient of resistivity, temperature-humidity-gas sensors, electro-optic devices, piezoelectric transducers, actuators, and PTC thermistors [8]. Ferroelectric properties and a high dielectric constant make $BaTiO_3$ useful in an array of applications such as gate dielectrics, waveguide modulators, IR detectors, and holographic memory [9]. Recently, an increasing number of studies have been focused on the exploitation of BTNPs in the biomedical field, owing to the high biocompatibility of BTNPs, and their peculiar nonlinear optical properties that have encouraged their use as nanocarriers for drug delivery and as label-free imaging probes [10]. The exploration of BTNP potential in biomedical/therapeutic applications, such as cancer therapy through hyperthermia, and drug/gene delivery, has recently started in nanomedicine based on the encouraging results observed in the biocompatibility assessments [11]. In another study, Städler et al. [12] investigated cellular responses to five nonlinear active nanomaterials, including BTNPs, in order to discover the best candidate for biomedical imaging.

Many synthetic methods have been utilized to produce fine $BaTiO_3$ particles including using a solid-state reaction [13], coprecipitation (e.g., citrates) [14], oxalates [15], hydrothermal synthesis [16], a solvothermal method [17], alkoxide hydrolysis [18], a catecholate process [19], and metal-organic processing [20], low-temperature combustion synthesis [21]. The preparation of BTNPs has been done by many authors [21–34]. Anuradha et al. [22] reported the combustion synthesis of cubic nanostructured barium titanate, various samples of $BaTiO_3$ were prepared by the solution combustion of three different barium precursors BaO_2, $Ba(NO_3)_2$, and $Ba(CH_3COO)_2$ and fuels such as carbohydrazide, glycine, or citric acid in the presence of titanyl nitrate. Zhong and Gallagher also prepared cubic $BaTiO_3$ by igniting the spray-dried mixture of $Ba(NO_3)_2$, $TiO(NO_3)_2$, and alanine [23]. Lee et al. [24] studied the preparation of $BaTiO_3$ nanoparticles by combustion spray pyrolysis, by mixing barium nitrate, $Ba(NO_3)_2$ and titanyl nitrate, $TiO(NO_3)_2$ in distilled water. Shaohou et al. [25] studied the nanosized tetragonal barium titanate powders, by the reaction between $TiCl$, $Ba(NO_3)_2$, citric acid, and $NH(NO_4)_3$. Seveyrat et al. [26] reinvestigated the synthesis of $BaTiO_3$ by conventional solid-state reaction and oxalate coprecipitation route for piezoelectric applications.

Wang et al. [27] used two typical wet-chemistry synthesis methods, in the first method, barium acetate, tetrabutyl titanate, isopropyl alcohol, and glacial acetic acid were starting reagents. The average particle size of obtained $BaTiO_3$ nanopowder was 50–80 nm. For the second method, the starting reagents were barium stearate, tetrabutyltitanate, and stearic acid; the obtained nanocrystallites of $BaTiO_3$ were in the range 25–50 nm. Wada et al. [28] reported the preparation of $BaTiO_3$ particle of varying size using a hot uniaxial pressing method and Curie point for a grain size 58 nm was found to be at room temperature. Boulos et al. [29] reported an average particle size of $BaTiO_3$ powders obtained from TiO_2 at 150 °C or 250 °C as 40–70 nm. Li et al. [30] described the oxalic acid precipitation method which is very similar to the sol–gel acetate method. The particle size prepared by this method was 38.2 nm. Recently, a sol-precipitation process was developed, which is quite similar to the sol–gel method. The advantage of this route is that it requires no further thermal treatment of the product, such as calcinations or annealing to enhance the homogeneity of crystals and crystal growth. Prasadarao et al. [31] investigated the formation of a stoichiometric coprecipitated precursor for $BaTiO_3$ from potassium titanyl oxalate and barium chloride.

A simple and cost-effective method for obtaining BTNPs with tetragonal structure and 40–80 nm size was described by Bai et al. [32]. In their work, barium carbonate was used for the first time as a precursor. Gläsel et al. [33] used barium titanium methacrylate as a monomeric metallo-organic precursor of BTNPs, starting from metallic barium, titanium (IV) isopropylate, and methacrylic acid in boiling methanol. Solid-state polymerization and pyrolysis were allowed to occur simultaneously at temperatures over 200 °C. Nucleation and growth of BTNPs were achieved at temperatures between 600 °C and 1400 °C and under inert atmosphere. Depending on temperature and atmosphere, BTNPs could be obtained with sizes ranging from 10 nm (low T and inert atmosphere) to 1 μm. In a work from Ashiri et al. [34], stoichiometric quantities of barium chloride (in CO_2-free water) and titanium chloride (in ethanol) were mixed, kept at pH = 14 with the addition of sodium hydroxide, and sonicated at 50 °C for 45 min in a water bath (53 kHz, 500 W), cubic BTNPs with -10 nm size and spherical shape were obtained, and no carbonate contaminants were found after synthesis at mild temperature. An emerging method for perovskite nanomaterial synthesis is represented by chemical vapor synthesis (CVS). This method enables the obtainment of highly pure, size-controlled, and loosely aggregated nanoparticles in a time-effective and scalable manner. In a work of Mojić-Lanté et al. [35], BTNPs were produced for the first time in the literature using laser-assisted CVS. The set-up for BTNP synthesis comprised a CO_2 laser flash evaporator, a hot-wall tubular reactor, a thermophoretic particle collector, and a gas supply system. BTNP solid precursors were titanium di(i-propoxide)bis(2,2,6,6-tetramethyl-3,5-heptanedionate)   and   barium bis(2,2,6,6-tetramethyl-3,5-heptanedionate) anhydrous. By varying the reactor length and temperature, partially crystalline, spherical particles with a size of a few nanometers could be achieved.

The decomposition kinetics and the formation of BaTiO$_3$ have been extensively studied by many researchers [36–43]. Otta et al. [36] studied the kinetics and mechanism of the thermal decomposition of barium titanyl oxalate (BTO) and they were suggested that BaTiO$_3$ was formed in the temperature range of 500–800 °C. Balek et al. [37] were studied the decomposition reactions of BTO and behavior of the reaction product—BaTiO$_3$ during heating in oxygen, carbon dioxide, argon, and helium during heating up to 1500 °C by thermal analysis, TG, DTA, and EGA. Gallagher et al [38] suggested that the first step in the thermal decomposition of BTO, the 4 moles of water are lost at 20–250 °C, after losing water of crystallization, oxygen is adsorbed to form active BaCO$_3$ and TiO$_2$ which react to form BaTiO$_3$ at temperatures of 500–700 °C and CO$_2$ is released. Kim et al. [39] studied the tetragonality of BaTiO$_3$ derived from BTO various treatments were carried out by considering the thermal decomposition mechanism of BTO in air. The thermal behavior of BaTiO$_3$ prepared at temperatures of 600–900 °C was studied by Swillam et al. [40]. They have shown that when lower temperatures are used for the calcination of BaTiO$_3$ compacts, samples with extremely fine pores are formed, whereas higher temperatures cause the agglomerates to shrink too much smaller particles, eliminating these pores. Ragulya et al. [41] studied the process of synthesis of barium titanate powder from BTO precursor under nonisothermal conditions and the effect of the heating rate on the specific surface area of the powder and established the advantages of nonisothermal conditions, especially the possibility of flexible control over the barium titanate grain size. Gopalakrishnamurthy et al. [42] also investigated the decomposition kinetics of the formation of barium titanate in the temperature range 873–1023 K. Jung et al. [43] examined the formation mechanism of BaTiO$_3$ from the thermal decomposition of BTO.

This work focuses on the kinetic and thermodynamic studies on the formation of BTNPs by means of differential scanning calorimetry (DSC) analysis under nonisothermal heating conditions. The experiments were performed in the temperature range 873 to 1173 K at heating rate of 5, 7, 10, and 15 K min^{-1} in a nitrogen gas atmosphere. Ozawa and Coats and Redfern methods were used to determine the apparent activation energies of the multistep thermal decomposition processes. Kinetic deconvolution analysis was used to perform the overall kinetics of the formation of BaTiO$_3$ nanoparticles. In order to predict the physicogeometrical reaction mechanism, empirical kinetic model functions such as phase nucleation and growth-type model, RO(n) were employed. The antibacterial activity of BTNPs was evaluated by a well diffusion method using two pathogenic bacteria *Staphylococcus aureus* (G+) and *Pseudomonas aeruginosa* (G−). The band structure and density of state (DOS) of stable phase of $BaTiO_3$ were investigated by means of DFT. Vienna Ab-initio Simulation Package (VASP) was used for the DOS calculations. The prepared samples were identified and characterized by means of FTIR, XRD, scanning electron microscope (SEM), and transmission electron microscopy (TEM).

4.2 EXPERIMENTAL

4.2.1 MATERIALS

AnalaR grade barium nitrate $(Ba(NO_3)_2)$ (Merck, India; assay $\geq$ 99.9%), potassium titanyl oxalate $(K_2TiO(C_2O_4)_2)$ (BHO Laboratory England; assay $\geq$ 99.9%), Luria Bertani (LB) broth and Agar plates, Glass speader, *Pseudomonas aeroginosa* and *S. aureus* culture were used in the present investigation.

4.2.1.1 PREPARATION OF BATIO3 NANOPARTICLES

BTNPs were prepared *via* a chemical precipitation method. First BTO was synthesized by the precipitation reaction of mixing of equimolar aqueous solution of barium nitrate and potassium titanyl oxalate with constant stirring. The white BTO precipitate obtained was filtered, washed with deionized water, and dried in an air oven at 50 °C for 24 h. Finally, BTNPs were obtained by the thermal decomposition of BTO in a vacuum oven at 900 °C for 1 h, followed by powdering using an agate mortar. The obtained BTNPs were sieved through the mesh and fixed the particle size in the range of 45–53 μm.

4.2.2 METHODS

4.2.2.1 FTIR SPECTRAL ANALYSIS

The FTIR spectrum of the samples in KBr pellet was recorded using a JASCO FTIR-4100 instrument. The sample was first compressed with KBr into pellet and analyzed as KBr disk from 400 to 4000 cm^{-1}.

4.2.2.2 XRD ANALYSIS

The X-ray diffraction (XRD) measurements of the samples were taken on a RIGAKU MINI FLEX-600 X-ray diffraction spectrophotometer using CuKα (1.5418 Å) radiation.

4.2.2.3 THERMAL ANALYSIS

The thermogravimetric (TG) analysis of the samples was made on a T.A. thermal analyzer, model: TGA Q50 v20.2 Build 27 at a different heating rate. The operational characteristics of the TG system are as follows: atmosphere: flowing air, at a flow rate of 60 mL min^{-1}; sample mass: 5.6 mg; and sample pan: silica. Duplicate runs were made under similar conditions. The DSC measurements of the samples were taken on a Mettler ToledoDSC822e. A heating rate of 5, 7, 10, 15 K min^{-1} from 873 to 1173 K and a sample of mass 5 .1 mg were used in all experiments. The atmosphere was flowing nitrogen with a flow rate of 50 mL min^{-1} and the sample container was made of aluminium. Duplicate runs were made under similar conditions and found that the data overlap with each other, indicating satisfactory reproducibility.

4.2.2.4 SEM ANALYSIS

The SEM analyses of all the samples studied were performed with SEM–EDS combination using JEOL Model JSM-6390LV, JEOL Model JED-2300 used to study specimen topology and morphology.

4.2.2.5 TEM ANALYSIS

The TEM analysis of the particles was achieved by using a JEOL 2100 used to study the particle size in nanometer range.

4.2.2.6 ANTIBACTERIAL ACTIVITY TEST

In order to investigate the antibacterial activity, BaTiO$_3$ nanoparticles were tested against the strains: *P. aeruginosa* (Gram-negative) and *S. aureus* (Gram-positive) LB medium was prepared in distilled water and sterilized. Sterile

LB agar was poured into Petri dishes and allowed to set in a laminar airflow cabinet. The plates were then stored in the refrigerator for further use. A 24 h exponentially grown Pseudomonas culture was further inoculated into LB broth and shake in 37 °C incubator shaker till it reached 0.5 Macfarland's turbidity. The culture was then spread on LB agar plates already prepared, with sterile glass spreader. On drying of the spreaded layer, small wells were dug in the plates according to the number of samples to be screened particles with different concentrations (0.5, 1, 1.5, 1.5, 2 mg mL^{-1}) was pipette into the respective wells and incubated for 16 h at 37 °C. The obtained zone of inhibition of bacterial growth was then analyzed by measuring it and recorded in mm.

4.2.2.7 QUANTUM MECHANICAL CALCULATIONS

Quantum mechanical calculations of DOSs have been performed in the general framework of DFT [44] using projector augmented wave (PAW) method [45]. In this DFT computations, used generalized gradient approximation (GGA) exchange and correlation energy functional as suggested by Perdew and Wang. Kinetic energy cutoff of the electronic wavefunctions was expanded in plane waves up to 225 eV. Integrals over the Brillouin zone were summed on a Monkhorst–Pack mesh [46] of 10×10×10. The 5s^2, 5p^6, 6s^2 levels of Ba, the 3s^2, 3p^6, 3d^2, 4s^2 of Ti, the 2s^2, 2p^4 of O were treated as valence states. The core of atoms was represented by the pseudopotentials of the respective atoms. VASP [47] was used for the (DOS) and band energy calculations and to obtain response function calculations within the recent advances in density functional theory.

4.3 RESULTS AND DISCUSSION

4.3.1 SAMPLE CHARACTERIZATION

The FTIR spectrum (Figure 4.1a) at 510 K shows a broad absorption in a wide range from 2400 to 3800 cm^{-1} indicated the presence of H$_2$O and OH$^-$ in the BaTiO$_3$ nanoparticles. The absorption bands observed near 1600 cm^{-1} are due to the CO group, the sharp bands observed near 1400 cm^{-1} and 1100 cm^{-1} are due to C–O bonded to Ti ions. The IR spectrum of BTO heated to 1173 K in vacuum for 1 h presented in (Figure 4.1b) shows the absorption band at 800 cm^{-1} is due to metal-oxygen ion stretching vibrations and the band observed near 1200 cm^{-1} are due to metal bonded to Ti ions.

The IR bands at 588.5 and 433.5 cm^{-1} are observed corresponding to the pure tetragonal phase [48].

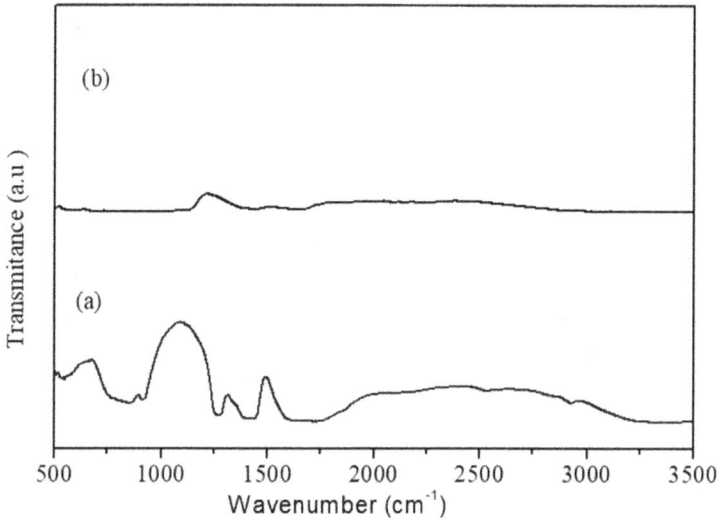

FIGURE 4.1   FTIR spectra of sample calcined at 510 K (a), sample calcined at 1173 K (b) for 1 h.

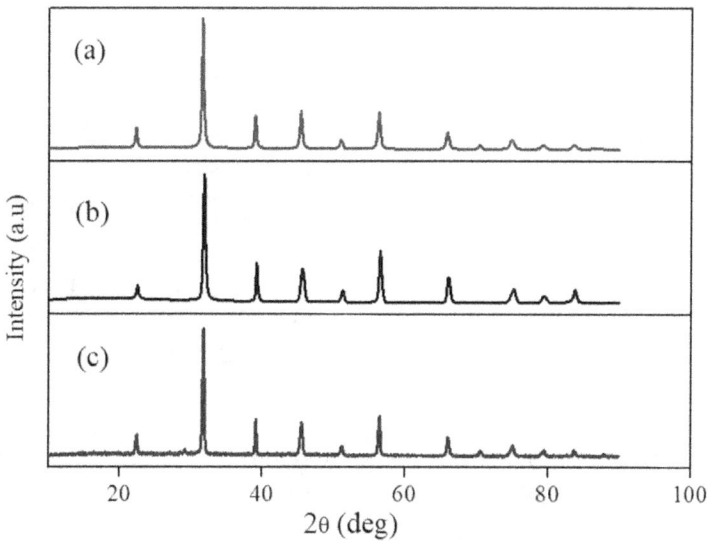

FIGURE 4.2   XRD pattern of sample calcined at 973 K (a), sample calcined at 1023 K (b), sample calcined at1173 K (c) for 1 h.

XRD analysis was employed to affirm the composition of as prepared oxide nanoparticles. Figure 4.2 shows the structural characteristics of barium titanate and their oxide nanoparticles with different temperatures investigated by XRD. The prepared sample clearly indicates the presence of crystalline phase indices, the peak positions with 2θ values of 29.4°, 36.4°, 42.2°, 61.3°, and 73.4° are indexed as (110), (111), (200), (220), and (311) planes, according to the JCPDS file 05-0626. Furthermore, the splitting peak near 45.5° indicates the phase transformed to tetragonal phase and it does not change when the calcinations temperature increases to 1173 K (Figure 4.2b and c). The relative crystalline sizes are determined from the XRD line broadening using the Scherrer equation: $d = 0.9\lambda/\beta\cos\theta$, where d is the crystallite size, λ is the wavelength used in XRD (1.5418A°), θ is the Bragg angle, β is the pure diffraction broadening of a peak at half-height, that is, broadening due to crystallite dimensions. The size of the BTNPs ranges from 22 to 50 nm according to the Scherrer equation [49].

4.3.2 DSC–TG ANALYSIS OF SAMPLE

DSC investigations indicate many endothermic transformations taking place in the prepared sample. The physicogeometrical kinetic behavior and the reaction mechanism under linear nonisothermal conditions were illustrated through kinetic analysis. The sample of 5.01 mg was weighed in an aluminum pan (6 mm in diameter and 2.5 mm in depth); a reference sample was made under similar conditions. This method measures the difference in the amount of heat supplied to the examined sample and reference sample when both are subjected to controlled changes of temperature. The DSC curve of the sample was recorded using a DSC instrument in N_2 flowing atmosphere at a rate of 50 mL min^{-1}, in the range 873–1173 K at a different heating rate (Figure 4.3a). Figure 4.3b shows the result of TG analysis of the sample, which confirms the mass change in the various stages while heating up to 1173 K (Table 4.1). TG curve shows a sequence of five steps corresponding to different mass loss involving dehydration and decarboxylation. For the sample, the first major mass loss of about 17.51% is within the temperature range from room temperature to 450 K which may be attributed to the dehydration of $BaTiO(C_2O_4)_2\cdot4H_2O$ to $BaTiO(C_2O_4)$. The second mass loss of about 3.01% appearing in the temperature range 450–550 K can be attributed to the initial low-temperature decomposition of BTO. The third major mass loss of about 20.81% is found in the temperature range 550–780 K attributed to the complete decomposition of the oxalate groups, resulting in the formation of a carbonate with CO_2 and CO. The fourth mass loss of about 5.20%, observed at 780–873 K, is due to the evolution of entrapped CO_2, the final decomposition of carbonate takes place between 873 and1173 K with a mass loss of about 7.25%, which is due to the formation of $BaTiO_3$ nanoparticles [50]. The α–T curve (Figure 4.3c) shows the multistep thermal decomposition reaction and the formation of BTNPs, considered to have three steps.

TABLE 4.1   Mass Loss Data of the Sample at Different Temperature Range

Decomposition Steps	Temperature Range (K)	(Mass Loss %)
1	310–450	17.51
2	450–550	3.01
3	550–780	20.81
4	780–873	5.20
5	873–1173	7.25

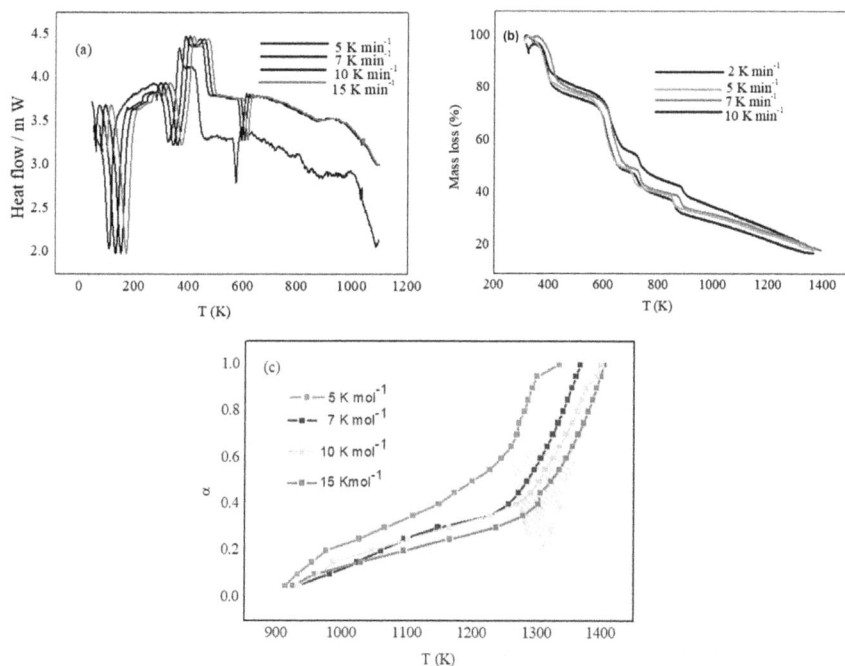

FIGURE 4.3   DSC (a), TG (b) curves for the thermal decomposition of samples and α–T curve (c).

4.3.3   KINETIC ANALYSIS

The DSC curve for the thermal decomposition of BTO recorded under linear nonisothermal conditions at the temperature range 973–1173 K. It (Figure 4.3a) was separated into three partially overlapping peaks with satisfactory fit using Fraser–Suzuki function; it is an effective tool in the investigation of phase transformations and enables very accurate determination of the temperature range and reaction heat in all transformations. It is used for recording the kinetic rate data of the thermal decomposition (Figure 4.4). To evaluate the overall kinetics of the formation of $BaTiO_3$ nanoparticles by thermal decomposition of BTO, we assumed the following derivative kinetic equation for a single step in a physicogeometrical mechanism with constant apparent Arrhenius parameters during the reaction. The deconvoluted DSC curve for the thermal decomposition of the sample at different β was subjected to the formal kinetic analysis by using the fundamental kinetic equation [51]:

$$\frac{d\alpha}{dt} = Ae^{(-E_a/RT)} f(\alpha) \tag{4.1}$$

where t is the time, T the temperature, R the ideal gas constant, A the apparent Arrhenius pre-exponential factor, E_a the apparent activation energy, and $f(\alpha)$ the apparent kinetic model function used to describe the physicogeometrical reaction mechanism as a function of the fractional reaction, α. Using the kinetic data for different β, apparent E_a as a function of the reacted fraction can be determined from the Ozawa [52] and the Coats and Redfern methods [53].

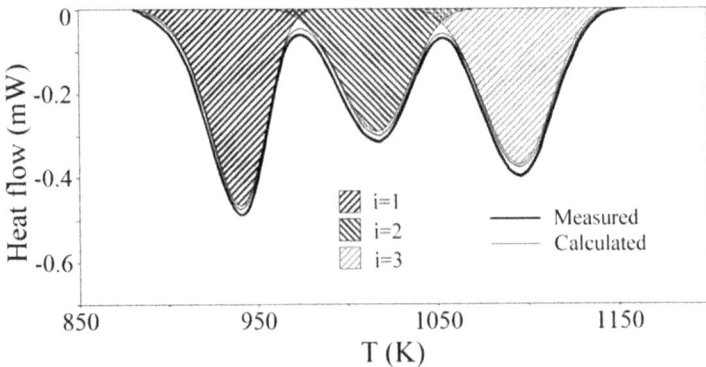

FIGURE 4.4   The kinetic rate data at different β derived from the mathematically separated DSC peaks: first peak ($i = 1$), second peak ($i = 2$), and third peak ($i = 3$).

4.3.4 CALCULATION OF ACTIVATION ENERGY BY INTEGRATED ISOCONVERSIONAL METHODS

4.3.4.1 OZAWA METHOD

The Ozawa method has been widely used for kinetic data analysis [54]. This integral method compared heating rates with temperatures under the same conversion rates. The kinetic parameters of decomposition process were evaluated by using the Ozawa equation

$$\log \beta = \log\left(\frac{AE}{R}\right) - 2.315 - 0.4567\left(\frac{E}{RT}\right) - \log(\alpha) \qquad (4.2)$$

where β is the heating rate (K min^{-1}), A is the pre-exponential factor (min^{-1}), R is the gas constant (8.314 J mol^{-1} K^{-1}), E is the energy of activation, and α is the fraction of decomposition. By plotting graphs between log β versus $1/T$ into a different straight line at different conversion rates (Figure 4.5a), the activation energy E and frequency factor log A could thus be obtained from the slope and the intercept, respectively.

4.3.4.2 COATS AND REDFERN METHOD

Coats and Redfern method is an integral method [55, 56]; the kinetic parameters of decomposition process can be evaluated by using the relation as [53]

$$\log_{10}\left[\frac{-\log(1-\alpha)}{T^2}\right] = \left[\log_{10}\left\{\left(\frac{AR}{\beta E}\right)\left(1 - \frac{2RT}{E}\right)\right\} - \left(\frac{E}{2.303RT}\right)\right] \qquad (4.3)$$

To determine the value of activation energy, a plot of $\left[\dfrac{-\log(1-\alpha)}{T^2}\right]$

versus $1/T$ (Figure 4.5a), and a straight-line graph is obtained. The value of activation energy can be calculated from the slope of a linear plot. The value of the frequency factor is also obtained from Equation (4.3). The average calculated activation energy, frequency factor were depicted in Table 4.2. The other kinetic analysis parameters such as enthalpy of activation (ΔH^*), entropy of activation (ΔS^*), and free energy change of decomposition (ΔG^*) were evaluated [57] using equations

$$\Delta H^* (\text{kJ}^{-1}\text{mol}^{-1}) = E + \Delta nRT \qquad (4.4)$$

where Δn = number of moles of product—number of moles of reactant in the reaction

$$\Delta S^* (\text{kJ}^{-1}\text{mol}^{-1}) = 2.303R \left[\log \left(\frac{Ah}{KT} \right) \right] \qquad (4.5)$$

$$\Delta G^* (\text{kJ}^{-1}\text{mol}^{-1}) = \Delta H^* - T\Delta S^* \qquad (4.6)$$

where A is (Arrhenius constant) determined from the intercept, K is Boltzmann, and h is Plank's constant.

TABLE 4.2   Initial Values of Kinetic Parameters Used for the Kinetic Deconvolution Analysis of the Thermal Decomposition of Sample Under Linear Nonisothermal Conditions

Methods	i	C_i	E_{ai} (kJ mol^{-1})	A_i (s^{-1})	$f(\alpha)$ RO(n)
Ozawa	1	0.45	58.79	0.47×10^8	$n = 1$
	2	0.40	39.14	0.88×10^4	$n = 1.01$
	3	0.15	52.29	0.24×10^4	$n = 0.99$
Coats and	1	0.45	57.20	0.75×10^8	$n = 1$
Redfern	2	0.40	38.80	0.42×10^4	$n = 1.01$
	3	0.15	51.80	0.12×10^4	$n = 0.99$

From Figure 4.5a, the best linear fitted plot for Ozawa and Coats and Redfern methods, the slopes of these straight lines have been used to calculate activation energy (E_a), and their intercepts were used to calculate frequency factor ($\log A$). The variation of the apparent E_a values as a function of α was almost similar for the above two methods (Figure 4.5b). Figure 4.6 shows the comparison of activation energy needed for each resolved stage of the thermal decomposition was calculated by Ozawa and Coats and Redfern methods in the temperature range 873–1173 K by column graph. From this graph, it is shown that during each step the estimation of E_a values by these methods are very close to each other. The other thermodynamic parameters such as enthalpy of activation (ΔH^*), entropy of activation (ΔS^*), and free energy change of decomposition (ΔG^*) were calculated using Equations (4.5)–(4.7). The calculated thermodynamic parameters were also were given in Table 4.3.

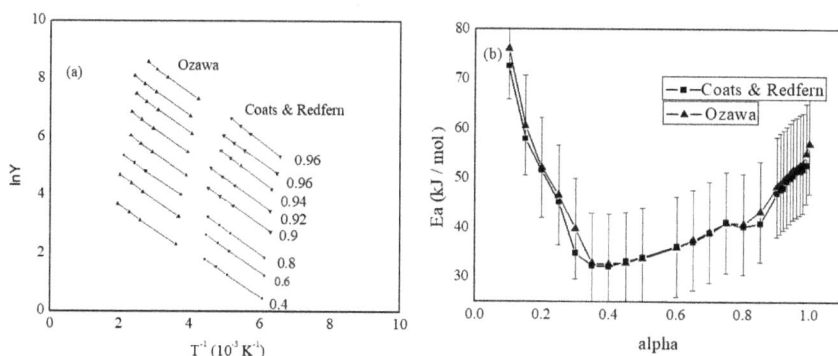

FIGURE 4.5 Typical linear least-squares plot for Ozawa and Coats and Redfern methods of the sample at different conversion values (a), the dependence of activation energy on conversion for two different methods.

TABLE 4.3 Thermodynamic Parameters Calculated for the Thermal Decomposition of Sample Under Linear Nonisothermal Conditions

Methods	i	ΔH	ΔS	ΔG
		(kJ mol^{-1})	(J mol^{-1})	(kJ mol^{-1})
Ozawa	1	52.88	−106.96	147.03
	2	33.68	−190.13	222.71
	3	47.03	−192.61	259.21
Coats and	1	52.88	−114.11	142.93
Redfern	2	33.68	−166.56	200.32
	3	47.03	−177.63	245.45

Since the thermal decomposition is an independent kinetic process, the overall reaction consists of multistep thermal decomposition which is expressed by the following cumulative kinetic equation under linear noniso-thermal conditions [58–59].

$$\frac{d\alpha}{dt} = \sum_{i=1}^{n} c_i A_i \exp\left(\frac{-E_{a,i}}{RT}\right) f_i(\alpha_i) \tag{4.7}$$

with $\sum_{i=1}^{n} c_i = 1$ and $\sum_{i=1}^{n} c_i \alpha_i = \alpha$.

where n and c are the number of component steps and the contribution ratio of each reaction step to the overall process, respectively, and the subscript i denotes each component reaction step. A_i and E_{ai} are the Arrhenius pre-exponential factor and the apparent activation energy, respectively, of the process i. The kinetics of each component process of the overall reaction can be characterized by optimizing all the kinetic parameters in Equation (4.6) using nonlinear least-square analysis. Empirical kinetic model functions such as RO(n) [60] were employed for $f_i(\alpha_i)$ in Equation (4.7) in order to accommodate any possible mechanistic feature of each reaction process

$$f_i(\alpha) = (1-\alpha)^n \tag{4.8}$$

where n is the character for a particular decomposition process, which corresponds to empirically obtain kinetic exponents. The number of component steps for all the samples is obtained through kinetic deconvolution of DSC-peaks. The initial values of the kinetic parameters were determined through a formal kinetic analysis of the kinetic data, first subjected to mathematical deconvolution using a statistical function known as Fraser–Suzuki function. From the ratio of the peak areas for the separated first, second, and third peaks, the contribution c_i of each peak to the overall reaction was determined as $(c_1, c_2, \text{and } c_3) = (0.45 \pm 0.03, 0.40 \pm 0.03, 0.15 \pm 0.03)$. The mathematically separated peaks at different β were used as the kinetic rate data. For all peaks, the kinetic rate data indicated systematic shifts to higher temperatures with increasing β.

The alpha dependent changes in the apparent activation energies (E_a) determined by applying the Ozawa and the Coats and Redfern methods were found for the kinetic rate data of the separated first, second, and third DSC peaks (Figure 4.5b), for the first- and second-DSC peak, a systematic decrease in the E_a value during the course of the reaction was observed, and for the third DSC peak, increase in the Ea value was obtained. The systematic change in the E_a value with α observed for all peaks indicate that the integration isoconversional relationship is one of the best methods to point out the changes in the self-generated conditions as the reaction advanced and depending on β. In these methods Ozawa and the Coats and Redfern, the average values for E_a are 58.79 kJ mol^{-1} and 57.20 kJ mol^{-1} (0.05 $\leq \alpha \leq$ 0.4), respectively, was tentatively used for the first peak. The average value for E_a, for the second peak was 39.14 kJ mol^{-1} and 38.80 kJ mol^{-1} (0.45 $\leq \alpha \leq$ 0.85), and also third peak was 52.29 kJ mol^{-1} and 51.80 kJ mol^{-1} (0.9 $\leq \alpha \leq$ 0.99), respectively. During the first stage, the value of activation energy obtained by Ozawa method is close to that obtained by Coats and Redfern method, and in the second and third stages, the values of E_a obtained by these methods show small variation. The values of activation energy calculated using these methods studied show a similar trend. These E_a values are nearly coincident with one another and also in agreement with that determined for the overall reaction under isothermal and controlled rate conditions. For the thermal decomposition reaction, the empirical kinetic deconvolution of the overlapping reaction steps, all the kinetic parameters in Equation (4.6) should be determined by graphically comparing $(d\alpha/dt)_{exp}$ versus time $(d\alpha/dt)_{cal}$ versus time. The data from thermal analysis curves in the decomposition range 0.05 $\leq \alpha \leq$ 0.4 for first step, 0.45 $\leq \alpha \leq$ 0.85 for the second step, and 0.9 $\leq \alpha \leq$ 0.98 for third steps were used to determine the value of kinetic model functions of this multistep process, by using the linear form of modified integral Coats and Redfern equation

$$\ln\left[\frac{g(\alpha)}{T^2}\right] = \left(\frac{AR}{\beta E}\right) - \frac{E}{RT} \tag{4.9}$$

The integral form of the conversion function $g(\alpha)$ depends on the kinetic model of the occurring reaction. If the correct $g(\alpha)$ is used, the plot of $\ln(g(\alpha)/T^2$ against $1/T^2$ should give a straight line. The straight-line plot with suitable model and regression values is shown in Figure 4.6. The comparison of E_a values calculated by Ozawa and Coats and Redfern methods at different stages is given in Figure 4.7.

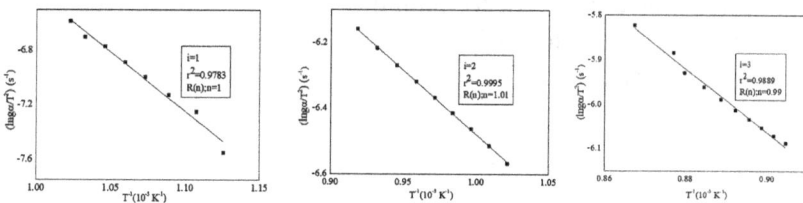

FIGURE 4.6 Typical model fitting least-square plot for the first stage ($i = 1$), second stage ($i = 2$), and third stage ($i = 3$).

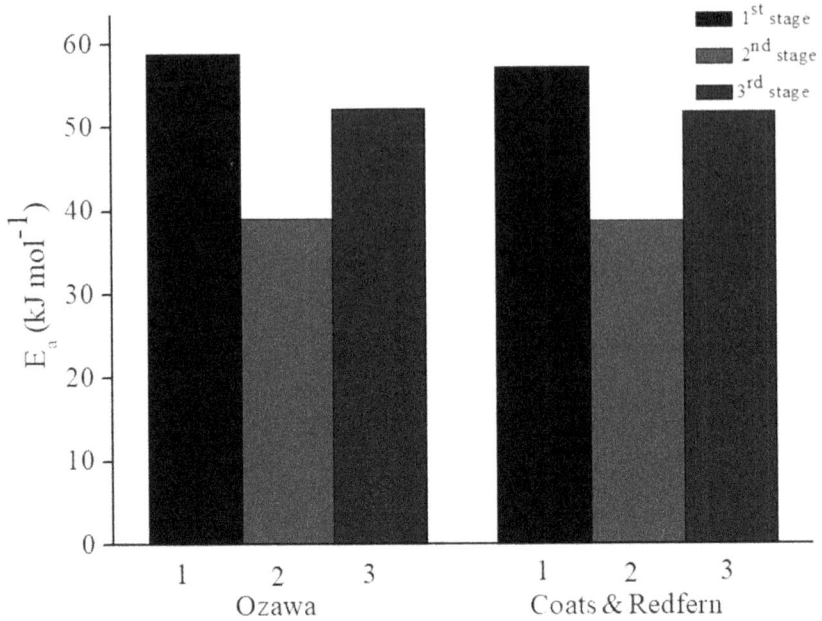

FIGURE 4.7 Comparison of E_a values calculated by Ozawa and Coats and Redfern methods at different stages.

The initial values of kinetic parameters are given in Table 4.2, according to this the activation energy (E_a), frequency factor (A) calculated from Ozawa and the Coats and Redfern methods, and the values are found to be compatible with each other [57]. And the thermodynamic parameters from Table 4.3, the positive value of ΔH^* show that the decomposition processes are endothermic in nature and ΔG^* values are positive, thus the processes are nonspontaneous [61]. The negative values of ΔS^* indicate that the activated complex has a high ordered structure than the reactants and further the high values of A indicate the fast nature of the reaction [57, 62].

After setting all of the initial values of kinetic parameters, a parameter optimization was carried out to minimize the squares of the residues (F) when fitting the calculated curve $(d\alpha/dt)_{cal}$ versus time to the experimental curve $(d\alpha/dt)_{exp}$ versus time [63].

$$F = \sum_{j=1}^{n} \left[\left(\frac{d\alpha}{dt} \right)_{exp,j} - \left(\frac{d\alpha}{dt} \right)_{cal,j} \right]^2 \qquad (4.10)$$

where n is the number of data points.

To determine the initial values for all kinetic parameters, a statistical deconvolution method [64] was applied to the experimental kinetic rate data, and the separated kinetic rate data were analyzed using the formal kinetic analysis method on the basis of each single-step reaction under linear nonisothermal conditions using the mathematical peak deconvolution procedure.

Formation of barium titanate nanoparticles from BTO by oxalate decomposition reaction can be brought through Equations (4.11) and (4.12)

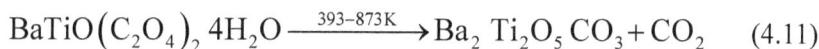

$$\text{BaTiO}\left(\text{C}_2\text{O}_4\right)_2 4\text{H}_2\text{O} \xrightarrow{\text{393-873K}} \text{Ba}_2\,\text{Ti}_2\text{O}_5\,\text{CO}_3 + \text{CO}_2 \quad (4.11)$$

$$\text{Ba}_2\,\text{Ti}_2\,\text{O}_5\,\text{CO}_3 \xrightarrow{\text{873-1173K}} 2\,\text{BaTiO}_3 + \text{CO}_2 \quad (4.12)$$

Barium titanate nanoparticle formation from BTO comes about through the evolution of H_2O, CO, and CO_2 and forming the carbonate intermediate involves the decomposition to form titanate nanoparticles that reaction was resolved into three steps. Figure 4.8 shows the result of the kinetic deconvolution analysis for the thermal decomposition of the sample using Ozawa and the Coats and Redfern methods on the basis of Equation (4.10) after establishing the initial values for the kinetic parameters (Table 4.1), through mathematical deconvolution and the subsequent formal kinetic analysis of each resolved reaction step. Under the linear nonisothermal conditions, the sample was studied by Ozawa method (Figure 4.8a) was resolved into three steps and the similar way by Coats and Redfern method (Figure 4.8b); both go through more complex reaction pathways.

TABLE 4.4   Average Values of Kinetics Parameters Optimized for Each Reaction Step of the Thermal Decomposition of Sample Under Linear Nonisothermal Conditions

Methods	i	C_i	E_{ai}	A_i	$f(\alpha)$
			(kJ mol^{-1})	(s^{-1})	RO(n)
Ozawa	1	0.45 ± 0.01	58.21 ± 0.48	$(0.31 \pm 0.51)\times10^8$	1.0 ± 0.1
	2	0.40 ± 0.10	39.31 ± 0.02	$(0.11 \pm 0.02)\times10^4$	1.01 ± 0.10
	3.	0.15 ± 0.81	52.51 ± 0.20	$(0.22 \pm 0.20)\times10^4$	0.99 ± 0.10
Coats and	1	0.45 ± 0.04	57.16 ± 0.42	$(0.86 \pm 0.50)\times10^8$	1.0 ± 0.1
Redfern	2.	0.40 ± 0.18	38.61 ± 0.02	$(0.39 \pm 0.02)\times10^4$	1.10 ± 0.48
	3.	0.15 ± 0.08	51.42 ± 0.21	$(0.12 \pm 0.20)\times10^4$	0.99 ± 0.20

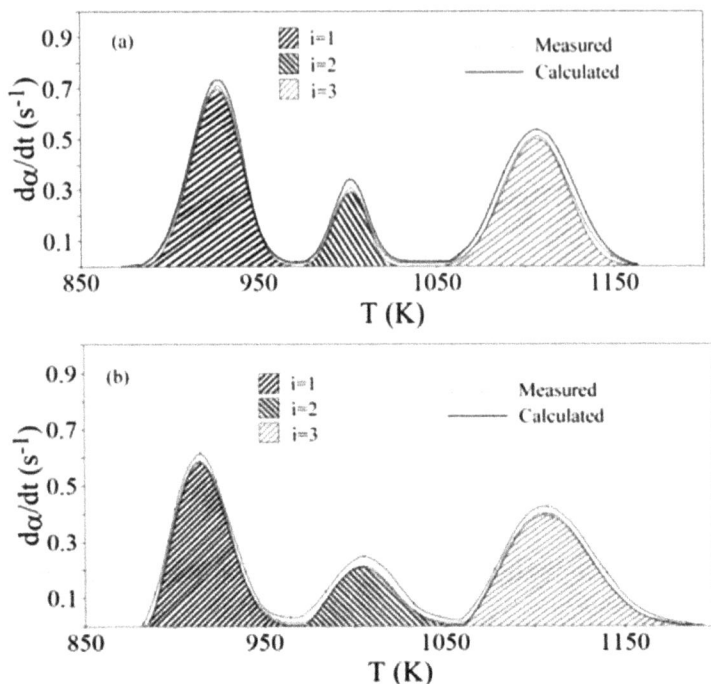

FIGURE 4.8 Results of kinetic deconvolution analysis for the thermal decomposition of the sample using Ozawa method (a), and Coats and Redfern method (b) under the linear nonisothermal condition at $\beta = 5$ K min^{-1}.

The average values of the kinetic parameters optimized for each reaction stage at different β values are summarized in Table 4.4. For each resolved reaction step, the value of Ea calculated for each reaction step is nearly in agreement with the respective corresponding values estimated by the Ozawa and the Coats and Redfern plots for the overall reaction under nonisothermal conditions [65]. The physicogeometrical behavior of each sample was best described empirically by the RO(n) kinetic model functions are also shown in Table 4.4. The required average value of Ea in Ozawa and the Coats and Redfern methods are 58.21 ± 0.48 kJ mol^{-1}, 57.16 ± 0.42 kJ mol^{-1} (for $i = 1$), 39.31 ± 0.02 kJmol^{-1}, 38.61 ± 0.02 kJ mol^{-1}, (for $i = 2$) and 52.51 ± 0.20 kJ mol^{-1}, 51.42 ± 0.21 kJ mol^{-1}, (for $i = 3$), respectively. Using the respective Ea values, experimental master plots of ($d\alpha/dt$) versus temperature for the separated first, second and third DSC peaks for each method were drawn (Figure 4.8) and satisfactorily fitted the RO(n) model were used as the empirical kinetic model function [66]. The values for the pre-exponential factor A for the first and second and third peaks determined through nonlinear regression analysis for the fitting using the RO(n) model for each methods, Ozawa and the Coats and Redfern were $A_1 = (0.31 \pm 0.51) \times 10^8\,\mathrm{s}^{-1}$, $(0.86 \pm 0.50) \times 10^8\,\mathrm{s}^{-1}$ (for $i = 1$), $A_2 = (0.11 \pm 0.20) \times 10^4\,\mathrm{s}^{-1}$, $(0.39 \pm 0.02) \times 10^4\,\mathrm{s}^{-1}$ (for $i = 2$) and $A_3 = (0.22 \pm 0.20) \times 10^4\,\mathrm{s}^{-1}$, $(0.12 \pm 0.20) \times 10^4\,\mathrm{s}^{-1}$ (for $i = 3$), respectively.

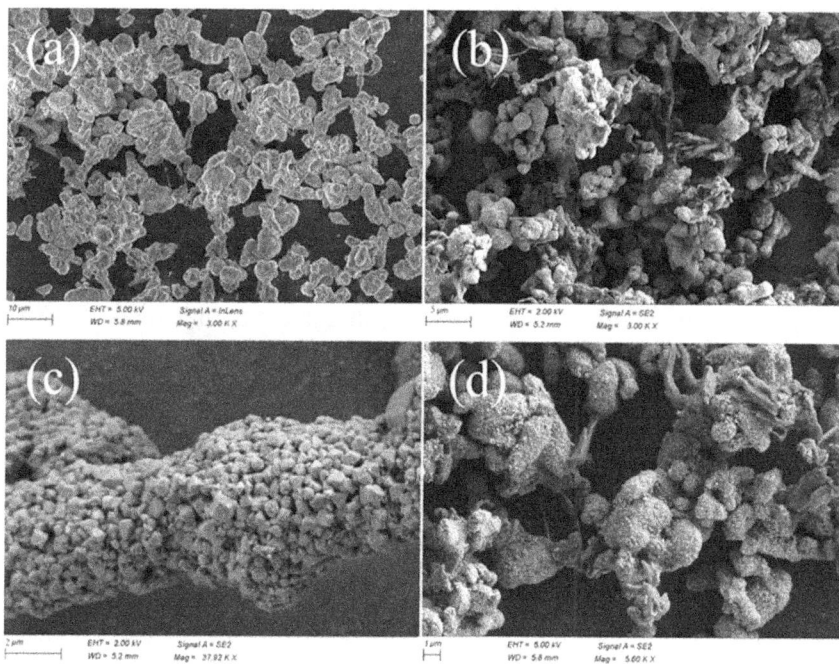

FIGURE 4.9 SEM images showing the surface microstructure of the sample calcined at 1023 K for 1 h at 10 µm (a), 5 µm (b), 2 µm (c), and 1 µm (d).

4.3.5 SEM ANALYSIS

Figure 4.9 shows the SEM images of BaTiO$_3$ synthesized at different temperatures resulted in BaTiO$_3$ particles with a spherical shape; however, some particles exhibit considerably pointed edges, and also most of the particles have smooth surfaces. From the SEM images, it is evident that the particles with different morphologies showed spherical grain agglomerations with smaller grain sizes with a porous surface containing particles with spherical shapes.

4.3.6   TEM ANALYSIS

TEM image (Figure 4.10) of $BaTiO_3$ shows a small depression in the nanospheres, which points out the complexity of the formation of the final product, BTNPs. Formation of cracks and holes in the surface of the product act as the channels for the diffusional removal of gaseous products such as H_2O, CO_2, CO, etc., the synthesized $BaTiO_3$ nanoparticles can be used for a wide range of applications in the electronic as well as optoelectronic field.

4.3.7   ANTIBACTERIAL STUDY

The antibacterial activity of BTNPs on the microorganisms *S. aureus*, *P. aeruginosa*, the inhibition data are given in Table 4.5, and zone of inhibition is shown in Figure 4.11. The BTNPs showed remarkable antibacterial activity against the two studied bacterial strains. The presence of the amount of BTNPs in different ratios increased the antibacterial activity also increased significantly [67]. From the above tests, it can be assumed that the BTNPs are effective in killing a range of bacterial growth. However, higher concentrations of nano $BaTiO_3$ are significant in bactericidal effect. One of the possible reasons for this could be direct interaction between BTNPs and the external membrane surface of the bacteria and also these nanoparticles form stable complexes with vital enzymes inside cells which hamper cellular functioning resulting in their death.

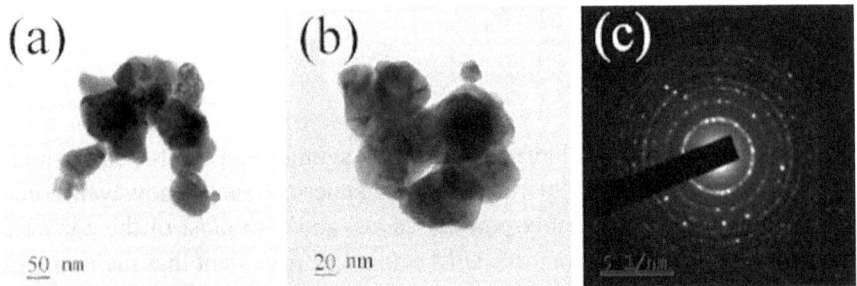

FIGURE 4.10   The TEM images of the sample calcined at 1023 K for 1 h at 50 nm (a), 20 nm (b), and SEAD pattern (c).

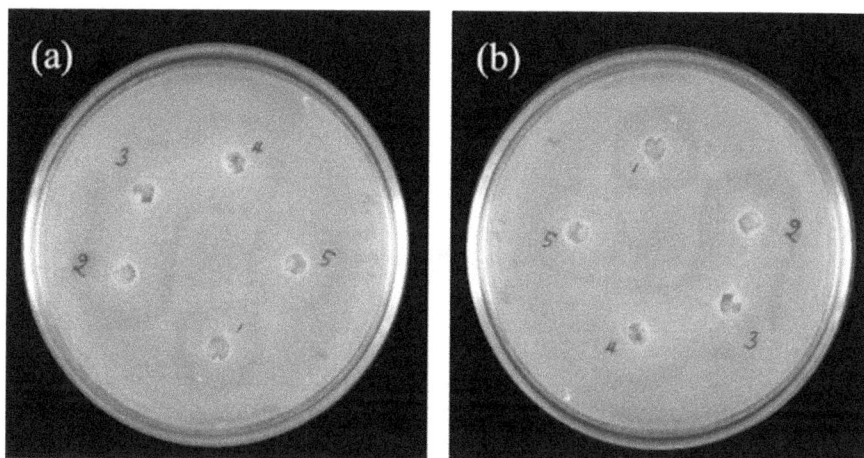

FIGURE 4.11   The zones of inhibition against *P. aeruginosa* (a), *S. aureus* (b), when treated with the prepared BTNPs at different concentrations.

TABLE 4.5   Comparison of Activities of BTNPS on Gram (−ve) and Gram (+ve) Bacteria

No.	Concentration of BTNPs (mg mL^{-1})	*Pseudomonas aeruginosa*	*Staphylcous aureus*
	Inhibition Zone of BTNPs (mm)		
	Bacteria		
1	1	22	24
2	1.5	28	26
3	2	29	28
4	2.5	30	30
5	3	32	32

4.3.8 BAND STRUCTURES AND DOSS OF BATIO3

The calculated electronic band structures of BaTiO$_3$ along the direction of high-symmetry points are shown in Figure 4.12. The band structure was separated by four portions in the valence band [68], the narrow band positioned at −17 eV was derived from O2s states, and the one positioned at −10 eV was originated from the Ba5p states. There is a manifold in the valence band near Fermi level which was attributed to the O2p states; moreover, the conduction band near Fermi level has a strong Ti 3d characteristic. These results could be confirmed by the partial DOSs, the calculated total DOSs in the valence-band region and spin-resolved DOSs are also shown in Figure 4.13. The band gap observed was 2.24 eV. In the experimental study, the band gap of $BaTiO_3$ has been reported around 3.40 eV [69].

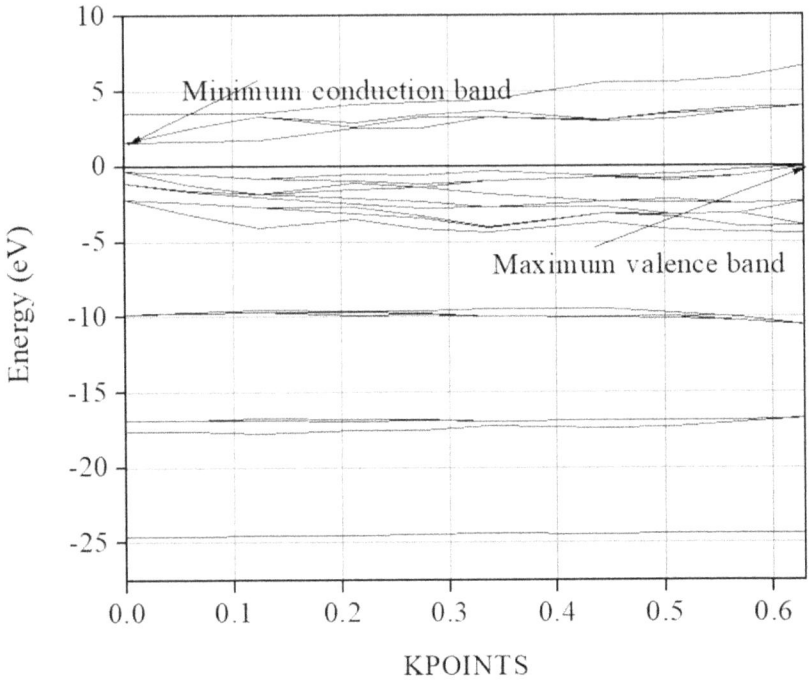

FIGURE 4.12   The calculated energy-band structure for $BaTiO_3$.

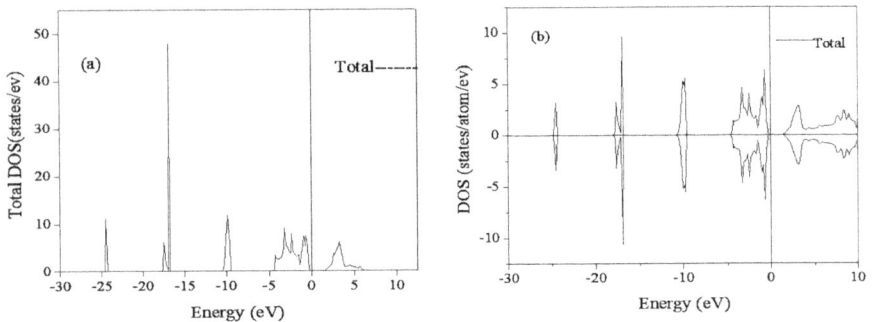

FIGURE 4.13   Total DOS of atoms in $BaTiO_3$ (a) spin-resolved DOS of atoms in $BaTiO_3$ (b).

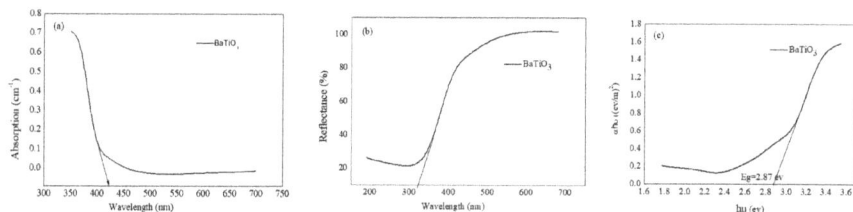

FIGURE 4.14   The absorption spectra (a) reflectance spectra (b) and Tauc plot (c) of BaTiO$_3$ within the range of wavelength.

For the determination of band gap the UV–vis diffuse reflectance spectrum of BaTiO$_3$ was measured experimentally (Figure 4.14). It absorbs the UV light below 400 nm (Figure 4.14a) and Figure 4.14b shows the UV–vis diffuse reflectance spectrum of BaTiO$_3$. An estimate of the optical band gap, E_g, can be determined by the Tauc-equation $(\alpha h\upsilon)^n = B(h\upsilon-E_g)$, where $h\upsilon$ is the photon energy, α is the absorption coefficient, B is a constant relative to the material and n is either 2 for direct transition, or 1/2 for an indirect transition. The $(\alpha h\upsilon)^2$ versus $h\upsilon$ curve (Tauc-plot) is shown in Figure 4.14c, which studied that the band gap obtained as 2.87 eV. The reported value of E_g corresponding to bulk BaTiO$_3$ is 3.1 eV [70].

4.4. CONCLUSION

The kinetic, thermodynamic behavior of the formation of BTNPs from barium titanium oxalate via thermal decomposition under linear noniso-thermal conditions has been studied using DSC. The decomposition showed multistage kinetics through complex reaction pathways. The comparison of the results obtained with different calculation procedures, Ozawa, and the Coats and Redfern methods depend on the empirical kinetic model function known as phase-boundary-controlled model RO(n) of the process. The kinetic deconvolution method is used to yield dependency of the E_a on the extent of conversion and it is established that the final stage decomposition of the sample show three-stage mechanism and can be described by reaction order (RO(n)) method. From the result of kinetic deconvolution of the sample, the different phase changes of the reactant mixtures and the kinetic parameters of the respective thermal decomposi-tion stages are evaluated accurately. The average calculated activation energy, pre-exponential factor, and other thermodynamic parameters were calculated. The antibacterial activity of BTNPs was evaluated by well diffusion method using two pathogenic bacteria *S. aureus* (G+), *P. aeruginosa* (G−). The mixed metal oxides showed remarkable antibacterial activity against the two studied bacterial strains. In DFT study, structure and DOS of stable phase of $BaTiO_3$ were investigated by means of VASP. GGA exchange and correlation functional as suggested by Perdew and Wang (PWGGA) within DFT have been used to found the band structure and total DOS described.

KEYWORDS

- **decomposition kinetics**
- **kinetic deconvolution**
- **Ozawa method**
- **Coats and Redfern method**
- **Kirby–Bauer method**
- **Vienna Ab-initio Simulation Package**

REFERENCES

1. Seifert, A., Sagalowicz, L., Muralt, P., Setter, N., Combustion synthesis of nanostructured barium titanate, J. Mater. Res. 14(1997) 158.

2. Hao, R., Xing, R., Xu, Z., Hou, Y., Gao, S., Sun, S., Synthesis, functionalization, and biomedical applications of multifunctional magnetic nanoparticles, Adv. Mater. 6(2010) 2729–2742

3. Kim, D., Jon, S., Gold nanoparticles in image-guided cancer therapy, Inorgan. Chim. Acta. 393(2012) 154–164.

4. Doane, T.L., Burda, C., The unique role of nanoparticles in nanomedicine: imaging, drug delivery and therapy, Chem. Soc. Rev. 41(2012) 2885–2911.

5. Xie, J., Lee, S. S., Chen, X., Nanoparticle-based theranostic agents, Adv. Drug Del. Rev. 62(2010) 1064–1079

6. Zhao, J., Wallace, M., Melancon, M.P., Cancer theranostics with gold nanoshells, Nanomedicine UK, 9(2014) 2041–2057

7. Yong, K.-T., Ouyang, Q., Zeng, S., Li J., Review on functionalized gold nanoparticles for biosensing applications, Plasmonics. 6(2011) 491–506.

8. Huang, C., Chen, K., Chiu, P., Sze, P., Wang, Y., The novel formation of barium titanate nanodendrites, J. Nanomater. 2014 (2014) 718918–718923.

9. Smith, M.B., Page, K., Siegrist, T., Redmond, P.L., Walter, E.C., Seshadri, R., Brus, L.E., Steigerwald, M.L., Barbara, S., crystal structure and the paraelectric-to-ferroelectric phase transition of nanoscale BaTiO₃, J. Am. Chem. Soc. 130(2008) 6955–6963.

10. Genchi, G. G., Marino, A., Rocca, A., Mattoli, V., Ciofani, G., Barium titanate nanoparticles: promising multitasking vectors in nanomedicine, Nanotechnology 27(2016) 23

11. Ciofani, G., Danti, S.D., Alessandro, D., Moscato, S., Petrini, M., Menciassi, A., Barium titanate nanoparticles: highly cytocompatible dispersions in glycol-chitosan and doxorubicin complexes for cancer therapy, Nanoscale Res. Lett. 5(2010) 1093–1101

12. Städler, D., Harmonic nanocrystals for biolabeling: a survey of optical properties and biocompatibility, ACS Nano 6(2012) 2542–2549

13. Templetona, L.K., Pask, J.A., Formation of BaTiO₃ from BaCO₃ and TiO₂ in air and in CO₂, J. Am. Ceram. Soc. 42(1959) 212–216.

14. Mulder, B.J., Surface and colloid chemistry in advanced ceramics processing, J. Am. Ceram. Soc. Bull. 49(1970) 990–993.

15. Stockenhuber, M., Mayer, H., Lercher, J.A., Preparation of barium titanates from oxalates, J. Am. Ceram. Soc. 76(1993) 1185–1190.

16. Kumazawa, H., Annen, S., Sada, E., Hydrothermal synthesis of barium titanate fine particles from amorphous and crystalline titania, J. Mater. Sci. 30(1995) 4740–4744.

17. Chen, D., Jiao, X., Solvothermal synthesis and characterization of barium titanate powders, J. Am. Ceram. Soc. 83(2000) 2637–2639.

18. Phule, P.P., Raghavan, S., Risbud, S.H., Comparison of Ba(OH)₂, BaO, and Ba as starting materials for the synthesis of barium titanate by the alkoxide method, J. Am. Ceram. Soc. 70(1987) 109.

19. Maison, W., Kleeberg, R., Heimann, R.B., Phanichphant, S., Phase content, tetragonality, and crystallite size of nanoscaled barium titanate synthesized by the catecholate process: effect of calcinations temperature, J. Eur. Ceram. Soc. 23(2003) 127–132.

20. Shaikh, A.S., Vest, G.M., Dielectric properties of ultrafine grained BaTiO₃, J. Am. Ceram. Soc. 69(1986) 682–688.

21. Luo, S., Tang, Z., Yao, W., Zhang, Z., Low-temperature combustion synthesis and characterization of nanosized tetragonal barium titanate powder, J. Microelectron. Eng. 66(2003) 147–152.

22. Anuradha, T.V., Ranganathan, S., Mimani, T., Patil, K.C., Combustion synthesis of nanostructured barium titanate, Scripta Mater. 44(2001) 2237–2241.

23. Zhong, Z., Gallagher, P.K., Combustion synthesis and characterization of BaTiO₃, J. Mater. Res. 10(1)(1995) 945–952.

24. Lee, S., Son, T., Yun, J., Kwon, H., Jun, B., Messing, G.L., Preparation of barium titanate nano particles by combustion spray pyrolysis, Mater. Lett. 58(2004) 2932–2936.

25. Lue, S., Tang, Z., Yao, W., Zang, Z. Low-temperature combustion synthesis and characterization of nanosized tetragonal barium titanate powders, J. Microelectron. Eng. 66 (2014) 147–152.

26. Seveyrat, L.S., Hajjaji, A., Emziane, Y., Guiffrad, B., Re-investigation of synthesis of BaTiO₃ by conventional solid-state reaction and oxalate coprecipitation route for piezoelectric applications, Ceram. Int. 33(2005) 35–40.

27. Wang, L., Liu, L., Xue, D., Kang, H., Liu, C., Wet routes of high purity BaTiO₃ nanopowders, J. Alloys Comp. 440(2007) 78.

28. Wada, S., Suzuki, T., Noma, T., Preparation of barium titanate ceramics from amorphous fine particles of the Ba-Ti-O system and its dielectric properties, J. Mater. Res. 10(1995) 306–311

29. Boulos, M., Fritsch, S.G., Mathieu, F., Durand, B., Lebey, T., Bley, V., Hydrothermal synthesis of nanosized BaTiO₃ powders and. dielectric properties of corresponding ceramics, Solid State Ionics.176(2005) 1309

30. Li, B., Wang, X., Li, L., Synthesis and sintering behavior of BaTiO₃ prepared by different chemical methods, Mater. Chem. Phys. 78(2002) 292–298

31. Prasadarao, A.V., Suresh, M., Komarneni, S., pH dependent coprecipitated oxalate precursors: a thermal study of barium titanate, Mater. Lett. 39(1999) 359–366.

32. Bai, H., Liu, X., Low temperature solvothermal synthesis, optical and electric properties of tetragonal phase BaTiO₃ nanocrystals using BaCO₃ powder, Mater. Lett. 100(2013) 1–3.

33. Gläsel, H.-J., Hartmann, E., Hirsch, D., Böttcher, R., Klimm, C., Michel, D., Semmelhack, H.-C., Hormes, J., Rumpf, H., Preparation of barium titanate ultrafine powders from a monomeric metallo-organic precursor by combined solidstate polymerisation and pyrolysis, J. Mater. Sci. 34(1999) 2319–2323.

34. Ashiri, R., Moghtada, A., Shahrouzianfar, A., Ajami, R., Low temperature synthesis of carbonate free barium titanate nanoscale crystals: toward a generalized strategy of titanate based perovskite nanocrystals synthesis, J. Am. Ceram. Soc. 97(2014) 2027–2031.

35. Lanté, B. M., Djenadic, R., Srdić, V., Hahn, H., Direct preparation of ultrafine BaTiO₃ nanoparticles by chemical vapor synthesis, J. Nanopart. Res. 16 (2014) 2618.

36. Otta, S., Bhattamisra, S.D, Kinetics and mechanism of the thermal decomposition of bariumtitanyl oxalate, J. Therm. Anal. 41(1994) 419–433.

37. Balek, V., Kaisersberger, E., Preparation of BaTiO₃ by thermal decomposition of BTO simultaneously investigated by emanation thermal analysis, TG-DTA and EGA, Thermochim. Acta. 85(1985) 207–210.

38. Gallagher, P.K., Thomson, J. Jr, Thermal analysis of some barium and strontium titanyl oxalates, J. Am. Ceram. Sot. 48(1965) 644–647.

39. Kim, J., Jung, W., Kim, H., Yoon, D., Properties of BaTiO₃ synthesized from barium titanyl oxalate. Ceram. Int. 35(2009) 2337–2342.

40. Swilam, M.N., Gadalla, A.M., Decomposition of barium titanyl oxalate and assessment of barium titanate produced at various temperatures, Trans. J. Brit. Ceram. Sot. 74(1975) 159.

41. Ragulya, A.V., Vasylkiv, O.O., Skorokhod, V.V., Synthesis and sintering of nanocrystalline barium titanate powder under nonisothermal conditions. I. Control of dispersity of barium titanate during its synthesis from barium titanyl oxalate, Powder Metall. Metal Ceram. 36(1997) 170–175.

42. Gopalakrishnamurthy, H.S., Rao, M.S., Kutty, T.R.N., Thermal decomposition of titanyl oxalates—I: Barium Titanyl Oxalate, J. Inorg. Nucl. Chem. 37(1975) 891–898.

43. Jung, W., Min, B., Park, J., Yoon, D., Formation mechanism of barium titanate by thermal decomposition of barium titanyl oxalate, Ceram. Int. 37(2011) 669–672.

44. Hohenberg, P., Kohn, W., Inhomogeneous electron gas, Phys. Rev. 136(1964) 864A.

45. Kresse, G., Joubert, J., From ultrasoft pseudopotentials to the projector augmented-wave method, Phys. Rev. B. 59(1999) 1758.

46. Monkhorst, H.J., Pack, J.D., Special points for Brillouin-zone integrations, Phys. Rev. B. 13(1976) 5188.
47. Kresse, G., Furthmuller, J., Efficient iterative schemes for ab initio total energy calculations using a plane wave basis set, Phys. Rev. B 54(1996) 11169.
48. Nakamoto, K., Infrared spectra of inorganic and co-ordination compounds, 2nd ed. New York, NY: Wiley, (1969) 245.
49. Patterson, A.L., The Scherrer formula for X-Ray particle size determination. Phys. Rev. 56(1939) 978–982.
50. Bera, J., Sarkar, D., Formation of $BaTiO_3$ from barium oxalate and TiO_2, J. Electroceram.11(2003) 131–137.
51. Kotru, P.N., Raina, K.K., Koul, M.L., The kinetics of solid-state decomposition of neodymium tartrate, Indian J. Pure Appl. Phys. 25(1987) 220.
52. Ozawa, T., A new method of analyzing thermogravimetric data, Bull. Chem. Soc. Japan 38(1965) 1881–1886
53. Coats, A.W., Redfern, J.P., Kinetic parameters from thermogravimetric data, Nature. 201(1964) 68–69
54. Cooney, J.D., Wiles, D.M., Thermal degradation of poly (ethylene terephthalate): a kinetic analysis of thermoravimetric data, J. Appl. Polym. Sci. 28(1983) 2887–2902.
55. Laidler, K.J., Chemical Kinetics, 3rd ed., New York, NY: Harper & Row, 1987.
56. Dabhi, R.M., Joshi, M.J., Thermal studies of gel grown zinc tartrate spherulites, Indian J. Phy. 76(2002) 211–213.
57. Dalal, P.V., Saraf, K.B., Shimpi, N.G., Shah, N.R., Pyro and kinetic studies of barium oxalate crystals grown in agar gel, J. Cryst. Process Technol. 2(2012) 156–160.
58. Wada, T., Koga, N., Kinetics and mechanism of the thermal decomposition of sodium per carbonate: role of the surface product layer. J. Phys. Chem. A. 117(2013) 1880–1889.
59. Wada, T., Nakano, M., Koga, N., Multistep kinetic behaviour of the thermal decompositionof granular sodium per carbonate: hindrance effect of the outer surface layer. J. Phys. Chem. A. 119(2015) 9749–9760.
60. Koga, N., Tanaka, H., Accommodation of the actual solid-state process in the kinetic model function. 1. Significance of the non-integral kinetic exponents. J. Therm. Anal. 41(1994) 455–469.
61. Mallakpour, S., Dinari, M., Eco-friendly fast synthesis and thermal degradation of optically active poly-amides under microwave accelerating conditions, Chin. J. Polym. Sci. 28(2010) 685–694.
62. Mallikarjun, K.G., Thermal decomposition kinetics of Ni (II) chelates of substituted chalcones, E-J. Chem. 1(2004) 105–109.
63. Barmak, K., A Commentary on: reaction kinetics in processes of nucleation and growth, Metall. Mater. Trans. A. 41(2010) 271–2775.
64. Ferriol, M., Gentilhomme, A., Cochez, M., Oget, N., Mieloszynski, J.L., Thermal degradation of poly(methyl methacrylate) (PMMA): modelling of DTG and TG curves, Polym. Degrad. Stab. 79(2003) 271–281.
65. Sanchez-Jimenez, P.E., Perejon, A., Criado, J.M., Dianez, M.J., Perez-Maqueda, L.A., Kinetic analysis of complex solid-state reactions. A new deconvolutionprocedure, J. Phys. Chem. B. 115(2011) 1780–1791.
66. Koga, N., Ozawa's kinetic method for analyzing thermoanalytical curves, J. Therm. Anal. Calorim. 113(2013) 1527–1541.

67. Raja, S., Bheeman, D., Rajamani, R., Pattiyappan, S., Sugamaran, S., Bellan, C., Synthesis, characterization and remedial aspect of $BaTiO_3$ nanoparticles against bacteria, J. Nanomed. Nanobiol. 2(2015) 16–20.
68. Feng, H., Liu, F.M., Electronic structure of barium titanate: an abinitio DFT study, Department of Physics, School of Sciences, Beijing University of Aeronautics & Astronautics, Beijing, (2007)
69. Hongwei, G., Theoretical investigation on the structure and electronic properties of barium titanate, J. Mol. Stru. 1003 (2011) 75–81
70. Uchino, K., Sadanaga, E., Hirose, T., Dependence of the crystal structure on particle size in barium titanate, J. Am. Ceram. Soc. 72(1989) 1555–1558.

CHAPTER 5

SYNTHESIS OF LDH FOR PHOTOCATALYTIC REMOVAL OF TOXIC DYES FROM AQUEOUS SOLUTION

RASNA DEVI,* DIPSHIKHA BHARALI, and
RAMESH CHANDRA DEKA*

Department of Chemical Sciences, Tezpur University, Napaam, Assam, 784028, India

*Corresponding authors. E-mail: rasnadevi@gmail.com, ramesh@tezu.ernet.in.

ABSTRACT

Recent advancement of nanocatalysis has lead researchers to develop new techniques for synthesis of nanocatalysts. One such technique is the reverse microemulsion method, which is recently used for the synthesis of layered double hydroxides (LDHs). This chapter deals with the reverse microemulsion technique and coprecipitation technique for synthesizing nano LDH, their structural characterization and uses for photocatalytic removal of cationic and anionic dye pollutants from aqueous solutions. Synthetic methodology for a series of CoZn LDH nanosheets using sodium dodecyl sulfate (SDS), isooctane, and water will be discussed. We will also discuss the effect of water to SDS molar ratio in the synthesis mixture toward the formation of final nanostructure followed by discussion of synthesis of ZnFeLDH via coprecipitation method.

5.1 INTRODUCTION

Environmental pollution caused by dye pollutants is one of the most challenging and serious problems in the modern world. Textile industries use large quantities of water for fabrication processes and thus release large quantities of liquid effluents containing toxic dyes into environment causing serious damage to the ecosystem. The history of natural dyes dates back to 3500 BC [1]. In ancient times natural dyes were extracted from flowers, fruits, vegetables, fishes, and insects for enhancing self appearances as well as enriching the surroundings. Today's dye industries have been evolved as a result of discovery of the first synthetic dye named as "aniline purple" by W. H. Perkins in 1856 [2]. Consequently, more and more dyes are being synthesized in massive sale in the following decades and used by industries like textile, cosmetics, paint, and paper for various purposes. The untreated dyestuff from these industries is released into large water bodies like rivers and lakes every year, which cause a severe threat to aquatic ecosystems as well as human health due to their high toxicity and nonbiodegradability [3–5]. This dissolved dyestuff drastically reduces the amount of dissolved oxygen in water, gas solubility, and transparency of water surface, which thus block light penetration through water surface and damage aquatic ecosystem by hindering photosynthesis [6]. Moreover, industrial waste effluents involve not only one single dye at a time but a complex mixture of dyes and their fragments that make the removal processes extremely hard and tedious. Therefore, taking measure to prevent contamination or treating contaminated water prior to release into large water bodies is quite beneficial. For this purpose, different dye removal techniques such as adsorption, coagulation, photocatalysis, flocculation, and electrochemical methods are used in the recent years [7, 8]. Out of these, the adsorption process has found its place in the broad spectrum because of the availability of excellent adsorbents such as zeolites, activated carbon, layered double hydroxides (LDHs), clays, fly ash, etc. Photocatalytic degradation route again is rapidly growing field through which toxic dye pollutants can be degraded at low cost in presence of sunlight or ultraviolet (UV) light at room temperature [9]. Among various types of photocatalysts, LDH can act as effective visible light photocatalysts due to their tunable structures. LDH are layered anionic clays consisting of positively charged brucite-like layers and negatively charged interlayer anions, which can act as host–guest catalysts for various applications. By tuning their properties during synthesis procedure and after synthesis modifications LDHs can be made excellent adsorbent and photocatalyst. Incorporation of photoactive metal cations in octahedral sites of brucite layers and altering M(II) versus M(III) ratio, LDH structure can be switched into a doped semiconductor for photocatalysis [10, 11]. In addition, presence of surface –OH groups, large surface area, exchangeable gallery anions, easy preparation of bulk quantities of sample at one single synthesis and capacity to absorb both UV, and visible lights makes LDHs advantageous material in the recent years [12]. One step ahead, nano-LDH having larger active surface area, varied shapes, and sizes has become promising photocatalysts and show enhanced activity in this area compared to bulk form [13]. Recently, photocatalysis has been carried out using bulk LDH as well as LDH nanoparticles to enhance the rate of adsorption, rate of degradation, and reduce energy used in the reaction compared to normal adsorption process.

In this chapter, we will cover the concept of LDH structure, concept of photocatalysis, importance of LDH nanoparticles for photoadsorption, and synthetic strategies for LDH by coprecipitation and microemulsion method. We will also describe the structural characterizations and applications of LDHs for removal of cationic dyes rhodamine B (RhB), methylene blue (MB), and crystal violet (CV) from aqueous solutions via adsorption and degradation routes. Here, we have also cited most of the important literatures covering all the aforementioned topics.

5.1.1   STRUCTURE AND PROPERTIES OF DYES

Textile dyeing industry dates back to almost 4000 years ago when the color used was only from natural sources. Natural coloring agents are mainly of inorganic and organic origin. Inorganic colors were mainly extracted from inorganic sources such as clays, earth, metal salts, minerals, and stones, while organic dyes were extracted from animal and plant sources. Some of the dyes used in ancient time are tyrian purple, cochineal (animal dyes); Alzarin (madder), Yellow, and indigo (plant dyes). Some dyes were extracted from organisms like lichens, insects and shellfish. For example, tyrian purple or mauveine, whose structure was first identified as 6, 6'-dibromoindigo (Scheme 5.1) by Paul Friedländer in 1909 required 12,000 *Murex brandaris* snails to produce 1.4 g of pure pigment. This may be the reason why the clothes dyed with tyrian purple were exclusively kept for high royalty people like kings or emperors. Even today although tyrian purple can be synthesized in laboratories, it is as expensive as the natural one.

The era of synthetic dyes began from W. H. Perkins' discovery of "aniline purple" in 1856. Anilline purple or mauveine is a mixture of four related forms of organic compounds differing in the number of methyl groups and their positions (Scheme 5.1). Since then progressive synthesis of dyes over the last 164 years had led to evolution of massive dye industries all over the world.

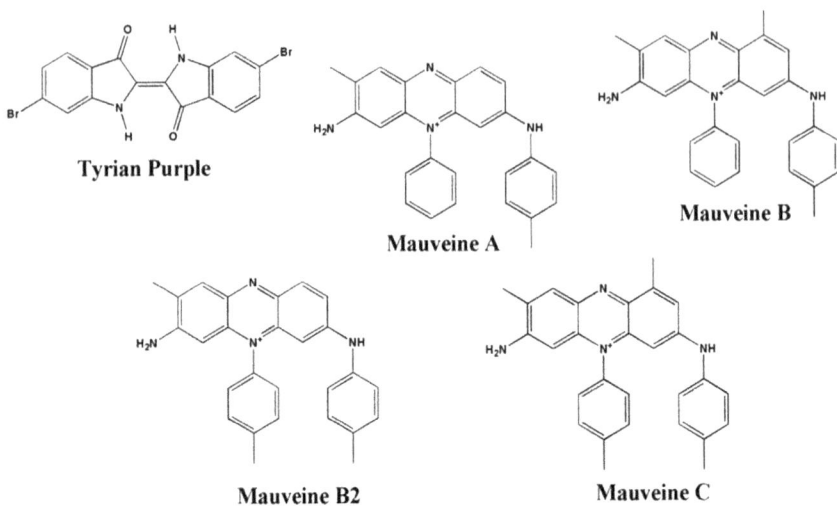

SCHEME 5.1   Structures of first synthetic dyes Aniline purple or mauveine (A, B, B2, and C) and ancient dye Tyrian purple.

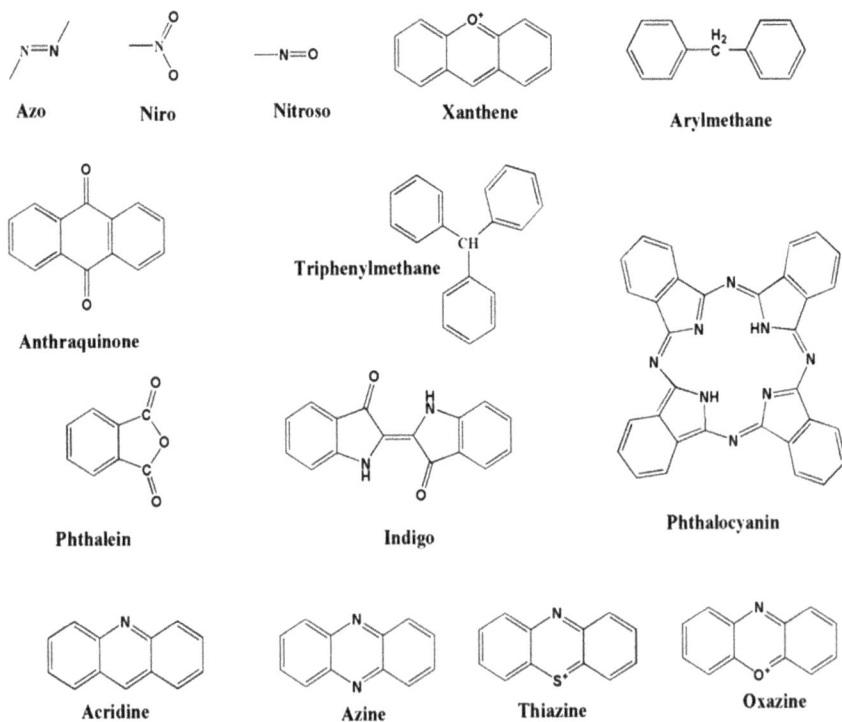

SCHEME 5.2   Different chromophores found in dyes.

Dyes can be defined as colored substances, which bind to the substrate to which it is being applied. Dyes possess colors because they absorb light in the visible region of the electromagnetic spectrum (400–800 nm). The basic structural component responsible for the color is a chromophore group. A chromophore can be defined as substance that has nearly same absorption wavelength as that of the considering dye or as that of the substructure, which is electronically related to the dye [14]. Chromophores possess conjugated system having alternate double and single bonds. Electrons are delocalized in this conjugated system and therefore resonance of electrons occurs. Generally, chromophores contain heteroatoms like N, O, S that have nonbonding electrons. Some of commonly found chromophores in dyes are shown in Scheme 5.2. Apart from the chromophores, most dyes have another structural group, which is called auxochromes or color helpers. Although auxochromes are not directly responsible for furnishing colors of dyes, their presence can enhance the color of a colorant and influence solubility of the dye. For example, benzene itself does not show any color, but when attached to $-NH_2$ it imparts yellowish color. Some commonly known auxochromes are $-NH_2$, $-OH$, $-NHR$, $-NR_2$, $-COOH$, $-HSO_3$, etc.

Based on the chromophore group present in dyes, they are classified according to the name of the chromophore such as azo, nitro, nitroso, xanthenes, acridine, azine, thiazine, oxazine, arylmethane, triarylmethane, anthraquinone, phthalein, indigo, and phthalocyanin dyes. However, from the basis of applications dyes are classified as direct, reactive, acid, basic, disperse, sulfur, and vat dyes.

In this chapter, we are dealing with three different dyes with different chromophores namely, MB, RhB, and CV, which fall under categories thiazine, xanthenes, and triarylmethane dyes, respectively. However, the common chemistry of these three dyes is that they all are cationic in nature and thus fall under the same class of dye, that is basic dyes or cationic dyes. A cationic dye is a dye that ionizes in solution giving positively charged ions and therefore attracts negative charges. Since, we have done all experiments in aqueous solutions and used similar catalysts, these three dyes can be studied together.

5.1.2   LDH AND THEIR STRUCTURES

LDHs are two-dimensional anionic clays having general formula as given below:

$$[M^{II}_{1-x} M^{III}_{x}(OH)_2]^{x+}(A^{n-})_{x/n} \cdot yH_2O$$

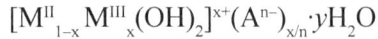

Here, M(II) are divalent cations (M = Mg, Fe, Co, Ni, Cu, Ni, etc.), M(III) are the trivalent cations (M = Al, Cr, Mn, Fe, Cr, etc.), A^{n-} are the charge balancing anions such as CO_3^{2-}, NO_3^{-}, Cl^{-}, SO_4^{-}, and OH^{-} in the interlayer regions and x is the ratio of M(II)/[M(II) + M(III)], which generally lies in the range 0.2–0.33 [15]. LDH structure can be easily understood from brucite structure. In brucite structure, divalent Mg^{2+} cations are octahedrally linked to hydroxyl groups and each –OH group is linked to three Mg^{2+} cation. In this case, the +2 charge of Mg^{2+} divided by six –OH groups (+2/6 +1/3) and −1 charge of OH^{-} divided by three Mg^{2+} cation (−1/3) cancels each other forming neutral, staked, infinite sheets of octahedrons of $Mg(OH)_2$. The sheets are held together by weak residual forces and therefore it is a soft mineral. When some of the divalent Mg^{+2} cations of brucite layers are isomorphously substituted by trivalent cations like Al^{+3}, some extra positive charge generates in the structure that are compensated by charge balancing anions in the hydrated interlayer region to form hydrated layered structures or hydrotalcite structure or LDH [16]. The water molecules and loosly bound anions in the interlayer regions such as CO_3^{2-} can be eleminated by heating or can be exchanged with other anions through ion-exchange method without disturbing the layered structure. The word *hydrotalcite* was derived from the words "talc" that means powder and "hydro" that means water [17]. Hydrotalcite can be easily crushed into powder-like "talc" and it possesses high water content in the interlayers. Discovery of LDH dates back to 1842 when Carl Christian Hochstetter described it from samples from Dypingdal serpentine-magnesite deposit in Snarum, Norway [18, 19]. LDH possesses different properties in three different forms as mentioned below:

1. *As synthesized form*: This form is the one that is obtained right after the synthesis and possesses low surface area due to the presence of water and charge compensating anions in the interlayer galleries. This form having Cl^{-}, Br^{-}, or I^{-} ions in the interlayer act as catalysts for halide ion-exchange reactions between alkyl chloride and Br^{-} or I^{-} ions [20]. Other applications include flame retardant and adsorbent for wastewater remediation in the industries [21].
2. *Activated form*: The *activated form* of hydrotalcites is obtained upon thermal treatment of the *as synthesized* form in temperature ranges from 450 to 500 °C. As a result, the hydrotalcite phase converts to the most widely used mixed oxide phase. The mixed oxide form

possesses the most important properties as catalytic materials, that is high surface area, strong basic properties as well as memory effect. This form act as solid superbase and can be utilized for large number of organic reactions like transesterifications, condensations, epoxidation, isomerization, etc. [22, 23].

3. *Rehydrated form*: When calcined hydrotalcites are rehydrated directly in water or in presence of water vapor, *rehydrated froms* of hydrotalcite occurs and the OH⁻ ions occupy the interlayer galleries [24]. This property of hydrotalcites is called the *memory effect*. When activated LDHs are applied in adsorption purposes in aqueous solutions, simultaneous adsorption and rehydration take place.

Apart from natural existence, LDH can be prepared synthetically. Tuning different structural constituents such as M(II)/M(III) ratio, the interlayer gallery anions, activation temperature, and rehydration temperature, LDH can be synthesized with varied morphologies and surface area [25]. This leads to application in various fields such as catalysis, photochemistry, pharmaceuticals, adsorption, electrochemistry, etc. Besides, LDH can be used as host for proper incorporation of selective guest species such as metal oxides, photoactive metals, and enantioselective reagents inside the structure to carry out variety of reactions.

5.1.3   GENERAL SYNTHETIC STRATEGIES OF 2DLDH NANOSHEETS AND BULK LDH

LDHs comprise an extensive class of materials that can be easily synthesized in the laboratory and industrial scales. Wide varieties of methods are available for synthesis of LDHs and can be engineered with desirable physical and chemical properties for various needed applications [26]. Generally, 2D LDH nanosheets can be synthesized by delamination or "top-down" approach and direct synthesis or "bottom-up" approach.

In the "top-down" approach LDH layers are exfoliated into 2D nanosheets in an appropriate solvent by controlled change in the interlamellar environment. However, exfoliation of LDH into 2D nanosheets is quite challenging due to high intralayer charge density and strong interlayer attraction. Therefore, organic modifiers such as anionic surfactant, lactate, or amino acids are used. Some commonly used solvents include butanol, formamide, toluene, CCl₄, acrylates, and water. It was Adasi-Pagano et al. who reported for the first time total delamination of $Zn-Al-NO_3-LDH$ using sodium dodecyl sulfate (SDS) as intercalating surfactant and butanol as exfoliating solvent [27]. The attempt was successful only when the SDS–LDH sample was dried in vacuum at room temperature and butanol was taken as solvent, while all other attempts with freshly prepared wet LDH, thoroughly dried LDH at 80 °C and other solvents such as water, methanol, propanol, and hexane were failed. Therefore, selection of proper intercalating anion and exfoliating solvent is the key factor in this approach.

In the bottom-up approach, 2D LDH nanosheets are directly synthesized by combining individual components, that is, salts, base, and a solvent. The components in proper ratio are allowed to combine in such an environment in which they later crystallize out in nano form. The first bottom-up synthesis of LDH nanosheets was reported by Hu et al. in a reverse microemulsion system [28]. In the reverse emulsion method, an oil phase containing an organic solvent such as isooctane and a surfactant such as SDS is mixed with aqueous coprecipitation system (mixture of salts and base at pH around 8–10) with the aid of a cosurfactant to form water in oil reverse microemulsion system. The surfactants form micelles surrounding the water droplets by the hydrophilic polar ends and thus act as microreactors for the nutrients to grow [29]. Hence, providing limited space to grow, desired nano LDH can be synthesized. The ratio of water to surfactant and pH of the aqueous systems are the deciding factors of stability of emulsion, dispersion, size of microreactors and final nano LDH formed. Since after the first report of synthesis of LDH by reverse emulsion method [30], it has been utilized frequently to tailor size, shape and morphologies at nanometer scale [31]. Besides, variation of surfactants, aging temperature, cosurfactant, base, and aging method can give hydrotalcite with varied shapes and sizes [32–36]. Followed by the first report on reverse emulsion method, O'Hare et al. successfully synthesized nanocrystals of MgAl [37, 38], ZnTi , NiTi [39, 40], and CoAl [41] LDH by this method using SDS, isooctane, and 1-butanol. Moreover, successful synthesis of nano LDH by varying surfactants (CTAB, Olylamine) cosurfactants (hexanol, isopropanol, Oleic acid), solvents (*n*-hexane, toluene, xylene) or no-surfactant at all is reported in the last few years [42–45]. Although, this method has been widely used for synthesis of inorganic nanocrystals [46–48] only few reports are found for synthesize LDH. This is due to some disadvantage associated with this method: the stability of microemulsion requires large amount of surfactant and cosurfactant, combination of surfactant and cosurfactant is not too obvious and utmost care is needed with proper ratio of all the ingredients. Yet, this technique can be used wisely because microemulsion can be prepared at room temperature, they are thermodynamically stable and hold water droplets having high interfacial surface area [49].

The coprecipitation method is one of the most commonly used techniques for preparation of LDHs. In a typical method, LDHs are prepared by simultaneous addition of two solutions, one containing metal salts of divalent and trivalent cations (M^{2+} and M^{3+}); and another containing base (such as NaOH, Na_2CO_3, NH_4OH, etc) solution. In this method, initially metal hydroxides are formed in presence of base followed by the conversion of metal hydroxides into hydrotalcite-like LDH by further addition of base through mechanism of precipitation. In coprecipitation method, pH is the key factor as it adversely affects the structural as well as chemical properties of the LDH phases. During the addition, the pH of the solution mixture is kept at constant in the range of 8–10 in order to attain high chemical homogeneity in LDH [50]. The pH value of the reaction mixture is maintained at constant by checking the solution and simultaneous addition of a base. The resulting mixture is allowed to undergo aging for a long period of time in order to acquire a material of well-crystallized structure. The obtained solid precipitate is collected by filtration, then washed thoroughly with deionized water and finally dried overnight. Followed by drying, the LDHs are generally activated in the range of 30–450 °C to suit a particular application.

1.4   PHOTOCATALYSIS BY LDH

Photocatalysis involves absorption of photons by a semiconductor metal and excitation of an electron (e^-) from valence band to conduction band by creating a hole (h^+) in the valence band and an extra electron in the conduction band. Existence of this high energy electron–hole pair occurs when the metal is excited with light energy equal or greater than the bandgap energy of the semiconductor metal. As a result, h^+ act as powerful oxidizing agent and oxidizes many organic compounds forming H_2O and CO_2 along with other oxidized products. Besides •OH radical formed as a result of reaction between h^+ and H_2O react with electron-rich organic molecules forming again H_2O and CO_2 as the elementary products. The basic reactions involving photocatalysis are shown in the following equations [51]:

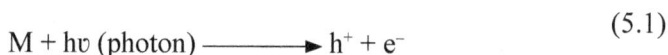

$$M + h\upsilon \text{ (photon)} \longrightarrow h^+ + e^- \tag{5.1}$$

$$h^+ + R \longrightarrow \text{Intermediates} + H_2O + CO_2 \qquad (5.2)$$

$$h^+ + H_2O \longrightarrow \bullet OH + H^+ \qquad (5.3)$$

$$R + \bullet OH \longrightarrow \text{Intermediates} + H_2O + CO_2 \qquad (5.4)$$

$$e^- + O_2 \longrightarrow O_2\bullet^- \qquad (5.5)$$

$$O2\bullet^- + R \longrightarrow HO_2\bullet \qquad (5.6)$$

$$2\,HO_2\bullet \longrightarrow H_2O_2 + O_2 \qquad (5.7)$$

$$H_2O_2 + h\upsilon \longrightarrow 2\bullet OH \qquad (5.8)$$

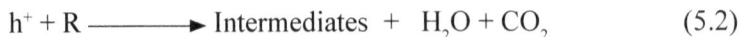

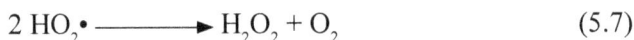

where M = metal catalyst and R = organic compound or reagent.

Here, simultaneous oxidation and reduction reaction is very crucial for stability of electron–hole pair and photocatalysis to happen. For this purpose, the presence of dissolved oxygen in the reaction mixture is necessary to start Equation (5.5), which as a result make the recombination process difficult [52].

One of the most common techniques of removal of dyes from aqueous solutions is the sorption method. Catalysts having high surface area such as zeolites, activated carbon, LDH, and so forth. This method does not require special setup and particular metals like photocatalysts and therefore any materials having high active surface area can be utilized for removal purposes. Enhanced adsorption of RhB under light irradaiation has been reported [53]. Although, this method effectively removes many organic pollutants, the toxic dye molecules still attach to the catalysts and must be dumped into environment safely. Therefore, degradation of the toxic dyes through photocatalytic route has been considered as green technology in these days.

In spite of excellent photoactive capacity of TiO$_2$ nanocatalysts over the years, their application is limited because TiO$_2$ can absorb only UV light (upto 380 nm) due to its high bandgap energy. In LDH, metal ions in the brucite layers can be shuffled with a large number of photoactive transition metals such as Zn, Ti, Fe, Cr, and so forth, and the ratio of these metals can be altered. These metals have moderate to high bandgaps and can absorb both UV and visible lights. As a result, excellent photocatalysts active in visible region as well as UV region can be synthesized. Thus, cation doping in octahedral positions of brucite type layers in LDH displays the properties of a doped semiconductor, which can act as doped semiconductor similar to TiO_2-based catalysts. However, absorption of light by a LDH may not necessarily qualify it as photocatalyst. The absorption process must stabilize the electron–hole pair to turn it to a photocatalyst. The MO_6 octahedron in brucite-like layers plays crucial role in stabilizing the electron–hole pair. High dispersion of MO_6 octahedron enables electron transfer and prevents recombination of electron–hole pair. Through proper dispersion of photoactive metals like Zn, Ti, Fe, Cr, etc. LDH structure can exhibit improved photocatalytic activity than most widely used commercial TiO_2[54, 55]. Another advantage of using LDH over TiO_2 catalysts is because LDH can be synthesized in large scale at one single synthesis.

Recently, many reports have come out showing LDH as potential photocatalysts, which can strengthen this field along with TiO_2 catalysts and open up many possibilities in future. Garcia and his co-workers reported coprecipitation method for the first time to synthesize ZnCrLDH photocatalysts using urea and NaOH as precipitator [56] for oxygen evolution reaction, which showed 1.6 more efficiency than that of WO_3 under similar reaction conditions. Similarly, Wei et al. [57] found that Zn–Ti–LDH synthesized by coprecipitation method showed superior activity for degradarion of MB under visible light compared to TiO_2, ZnO, and P25. One step ahead, nanosized LDH as photocatalsts have shown enhanced activity in this area. It has been found recently that LDH nanosheets or nanocomposites with photoactive metals can be more active than the bulk form. Lan et al. synthesized ZnCr–LDH nanoplatelets over graphene oxide by one step coprecipitation method and found improved photocatalytic activity under visible light toward degradation of RhB compared to pure ZnCr–LDH [58]. In another report, Lu et al. synthesized TiO_2/CuMgAlTi-RLDH by homogeneous distribution of anatase type TiO_2 nanoparticles over selectively reconstructed LDH precursors containing Cu^{2+}, Mg^{2+}, Al^{3+}, and Ti^{4+} cations [59]. This nanocomposite showed superior catalytic activity than single-phase TiO_2 for photodegradation of MB under UV or visible light. Seftel et al. synthesized nanocomposite by embedding MgAl–LDH crystallites into nano TiO_2 anatase phase and applied them for photocatalytic degradation of MB in aqueous solutions, which showed superior catalytic activity in basic environment than commercial TiO_2 nanoparticles Degussa P25 [60]. Zhao et al. synthesized NiTi–LDH nanosheets by reverse microemulsion method, which showed extraordinarily high photocatalytic activity (2148 mmol $g^{-1} h^{-1}$) for oxygen evolution from water . Qiao et al. reported synthesis of ZnCo–LDH nanosheets by microwave irradiation and found that they contain more active sites than ZnCo–LDH nanoparticles and showed higher activity and better stability towards oxygen evolution reactions [61]. Gunjakar et al. synthesized ordered layer-by-layer nanohybrids of LDH and layered metal oxides by assembling oppositely charged 2D nanosheets of Zn–Cr–LDH and TiO_2 [62], which showed higher catalytic activity for visible light-induced O_2 generation reaction than pristine Zn–Cr–LDH material, which is one of the most effective catalyst for this reaction. Zhao et al. reported synthesis of NiO nanosheets stabilized by TiO_2 by calcinations of a monolayer LDH precursor. In another report, particle interaction of LDH nano particles is described by Gursky et al. [63]. Thus, continuous research on photocatalytic applications of bulk and nanostructured LDH has been going on over the years. The goal is to synthesize more benign and effective catalysts, which can be photoactive under both visible light as well as UV light irradiation.

5.1.5   REMOVAL OF DYE TOXICITY FROM AQUEOUS SOLUTION BY LDH

5.1.5.1   PHOTOADSORPTION ROUTE

Removal of dissolved dyestuff from aqueous solutions has been successful in the recent years via photocatalytic degradation as well as adsorption routes. We have explained in Section 5.1.4 that LDHs nanoparticles can be converted to excellent photocatalysts, which can compete with the most widely used TiO_2 photocatalysts and show enhanced activity under visible and UV light irradiation. Keeping this in mind, CoZnLDH nanosheets were synthesized by reverse microemulsion method for photocatalytic removal of two cationic dyes, that is, MB and CV from aqueous solutions (Scheme 5.3) [64]. Initially, a transparent emulsion (w/o) was prepared by mixing SDS, water, and isooctane with the aid of cosurfactant 1-butanol, which follows the addition of $Zn(NO_3)_2 \cdot 6H_2O$, $Co(NO_3)_2 \cdot 6H_2O$ and urea in continuous stirring condition. When the mixture was allowed to age at 100 °C for 24 h in an oil bath under reflux condition, LDH nanosheets were crystallized out

inside the micelles. The product was then filtered, washed with ethanol–
water mixture, and dried in an air oven to get the final LDH nanosheets. The
resulting sample is named as CoZnLDH

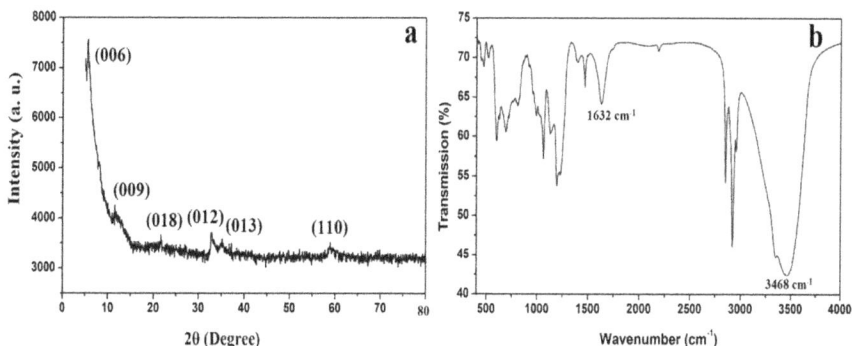

FIGURE 5.1   The powder XRD (a) and FTIR (b) patterns ZnCo–LDH.

We have observed well developed CoZnLDH layered structure from
powder X-ray diffraction (PXRD) pattern as shown in Figure 5.1a. The
peaks at 2θ positions 5.56°, 11.66°, 21.63°, 32.84°, 34.90°, and 59.21° can
be assigned to (006), (009), (018), (012), (013), and (110) plans, respectively,
of CoZn LDH structure [65]. Again, shift of peak position of (006) plan from
2θ value 18° to 5.56° is a direct indication of intercalation of SDS in between
LDH layers [66]. Shifting of 2θ positions to lower value can be understood
from Bragg's diffraction formula $n\lambda/2d = \sin\theta$. As the "d" value increases as
a result intercalation of long chain SDS molecules in the interlayer region,
the $n\lambda/2d$ value decreases thus lowering 2θ value.

The formation of LDH structure was further confirmed from FTIR spectral analysis as depicted in Figure 5.1b. Characteristics bands for LDH at
around 3468, 1632, 2921, 2952, and 1396 cm^{-1} clearly indicate the presence
of hydroxyl groups in the brucite layers, interlayer water molecules, long-
chain SDS molecules, and CO_3^{2-} molecules in the structure, which are typical
for LDH type materials. The bands ranging from 400 to 850 cm^{-1} arise due to
M–O and M–OH vibrations of LDH lattice. The spectral pattern resembles
with previous literature patterns.

To check whether the desired nano morphology was formed or not we
investigated the CoZnLDH through scanning electron microscopy (SEM),
transmission electron microscopy (TEM), high-resolution transmission electron microscopy (HRTEM), and selected area electron diffraction (SAED)

analysis. The results are illustrated in Figure 5.2a–e. The SEM images (Figure 5.2a and b) magnified at ×2000 and ×10,000 could not extract information regarding shape and size of particles in micrometer ranges, which thus indicates that the particle sizes may fall in nanometer ranges. However, nonuniform sheet-like agglomeration of the particles was observed. The TEM micrograph (Figure 5.2c) showed thin sheet-like morphologies, which are nearly transparent to the electron beam. This further confirmed formation of ultrathin nanosheets having quite larger longitudinal dimension than the lateral dimension [67]. Thus, the nanosheets are folded in their self-assembly. The SAED pattern (Figure 5.2e) showed diffraction rings with small and indistinct bright spots showing polycrystalline nature of the catalysts having low crystallinity. The HRTEM image (Figure 5.2d) reveal interplanar distance as 0.20 nm.

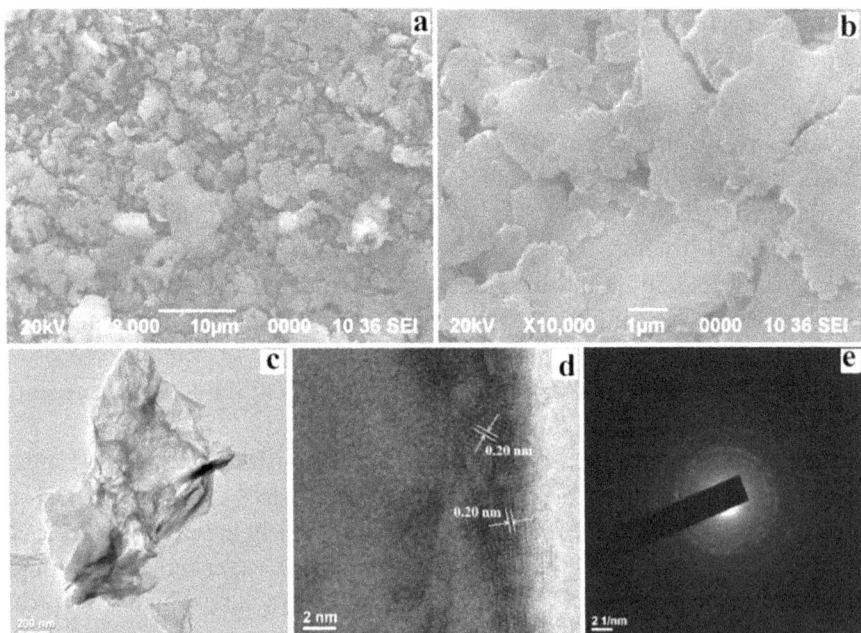

FIGURE 5.2 The SEM (a) and (b), TEM (c), HRTEM (d) and SAED (e) images of ZnCo–LDH nanosheets.

It is customary to know that the surface area of a catalyst is a crucial property for its adsorption capacity. Therefore, we obtained specific surface area, pore volume, pore radius, and pore size distributions of CoZn–LDH nanosheets by Brunauer–Emmett–Teller (BET) method and Barrett–Joyner–Halenda equations using N_2 adsorption–desorption isotherms as shown in Figure 5.3a and b. The LDH shows type IV isotherm indicating mesoporous material as expected for LDH type materials. Moreover, wide H3 type hysteresis loop from 0.4 to 1 relative pressure (p/p_0) indicates mesoporous material having slit-shaped pores and aggregates of plate or sheet-like particles [68, 69]. Sharp and short-range pore size distribution in the range 17–22 can be attributed to the nano-slit formed by the LDH nanosheets. To our expectation, the results showed BET surface area 111.909 m²/g having pore volume 0.656 cc/g and average pore radius 20.254 Å **similar to a typical LDH**-type structure.

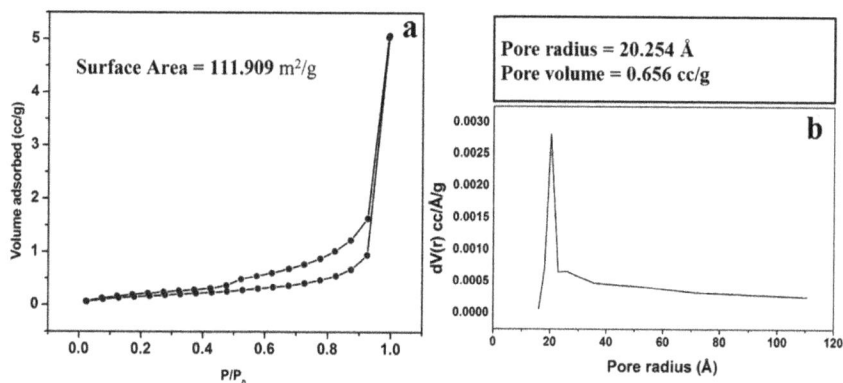

FIGURE 5.3 The N_2 adsorption–desorption isotherms (a) and pore size distribution curve (b) of CoZnLDH nanosheets.

Followed by the aforementioned structural characterizations, catalytic activity of CoZn–LDH nanosheets were evaluated by removal of MB and CV from their aqueous solutions via adsorption route. The study was carried out by adsorption of 0.02 M aqueous solution of MB and CV dyes under controlled condition, that is, without catalyst in dark, without catalyst in light, with catalyst in dark, and with catalyst in light. At first blank reactions were performed under dark as well as visible light irradiations in order to check any loss of concentration of the dye in absence of catalyst. To our expectation, no amount of degradation or any instability was noticed. Thus, light irradiation does not affect the dye in absence of catalyst and therefore catalyst has got crucial role in this experiment. For a typical dark reaction in presence of catalysts, 10 mg CoZnLDH was added to 20 mL MB solution in a dark chamber under continuous stirring condition, and concentrations were recorded at specific time intervals. Concentration recorded prior to addition of catalyst was termed as initial concentration.

In case of light experiments, similar amount of catalysts and dye solutions were exposed to visible light irradiation for required time. For this purpose, Phillips 200 W incandescent tungsten lamp was used as visible light source. At specific time intervals, 3 mL dye solution was taken out and concentrations were recorded through UV–visible spectrophotometer. The dye removal percentage of both dark and light experiments were calculated by using the following equation,

$$\text{Dye removal } (\%) = (C_0 - C_t)/C_0 \times 100\%$$

where C_0 is the initial concentration at time $t = 0$ and C_t is the final concentration at time "t."

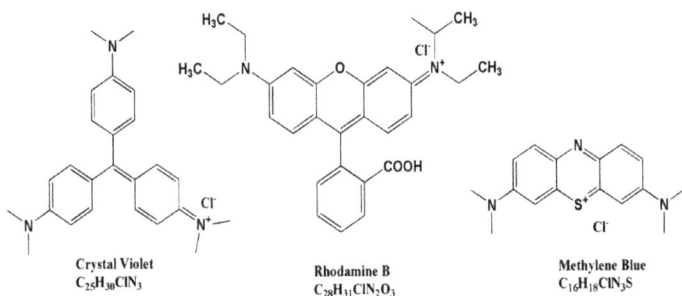

SCHEME 5.3   Structure of crystal violet and Methyl violet dyes.

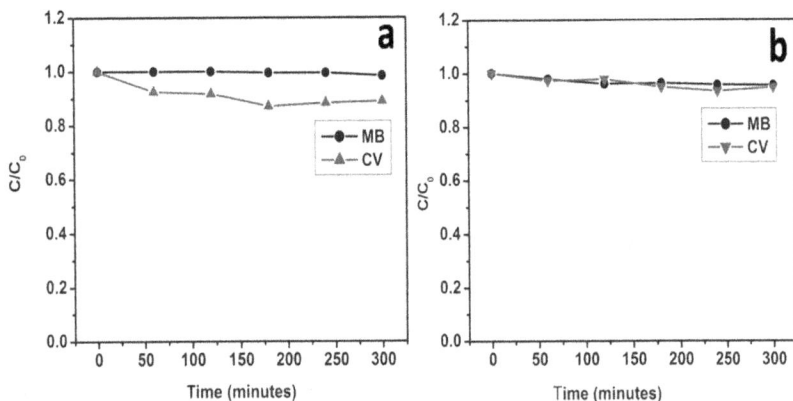

FIGURE 5.4   Adsorption MB and CV under dark (a) and visible light irradiation (b) without catalysts (conditions: $C_0 = 10$ mg/L, $V_{solution} = 20$ mL, catalyst amount = 10 mg, pH = 7).

Figure 5.4a and b shows controlled adsorption experiments without catalysts in dark as well as in visible light irradiation for MB and CV. From the characteristic absorption peaks (λ_{max} = 664 nm for MB and 578 nm for CV), it was observed that both the dyes showed no loss or negligible loss of concentrations up to 300 min. This reveals that light irradiation without catalyst does not affect the stability of dyes. Therefore, an effective catalyst will be required for further experiments. Here, CoZnLDH nanosheets were used for this purpose.

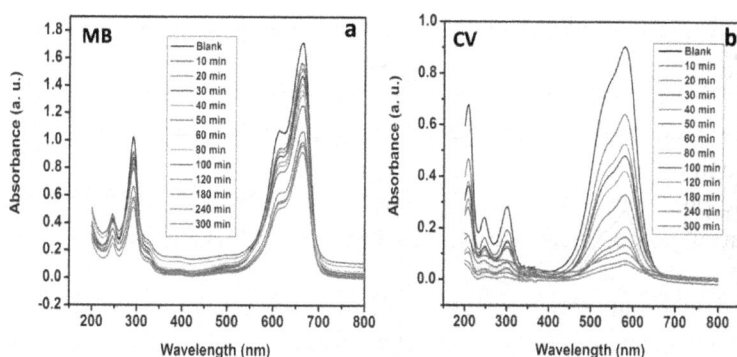

FIGURE 5.5 Adsorption MB and CV over CoZnLDH nanosheets in dark chamber for 300 min (conditions: C_o = 10 mg/L, $V_{solution}$ = 20 mL, catalyst amount = 10 mg, pH = 7).

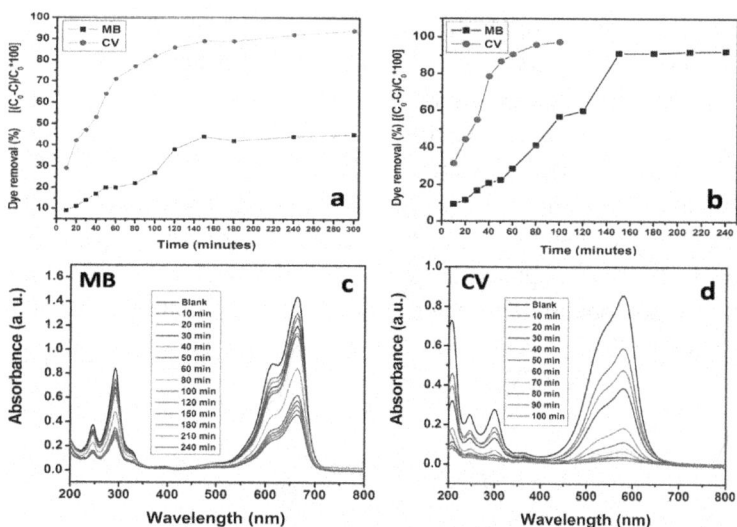

FIGURE 5.6 UV–visible absorption spectra MB and CV over CoZnLDH nanosheets under visible light irradiation (c) and (d) and dye removal (%) graphs in dark (a) and light (b) (conditions: C_o = 10 mg/L, $V_{solution}$ = 20 mL, catalyst amount = 10 mg, pH = 7).

Figure 5.5 depicts UV–visible adsorption spectra of MB (a) and CV (b) over CoZnLDH nanosheets in dark chamber for 300 min. Removal of MB gradually increases to 44% till 150 min, while it becomes nearly constant beyond 150 min (Figures 5.5 and 5.6a). On the other hand, removal of CV goes on increasing gradually showing 92% in 300 min. However, beyond 100 min (81% removal) the increase of dye removal becomes slower similar to MB. It is because, as the time increases the active surface area of CoZnLDH becomes less available for dye molecules to get adsorbed than the initial stage when the whole surface area was available and thus adsorption took place rapidly. Thus, it is clear from dark experiment that although both dyes are cationic in nature, the removal percentage of CV is double in comparison to MB for the same reaction with same reaction conditions. It may be due to difference in the structure of the dyes, which adsorb to LDH surface in different ways.

The photocatalytic adsorption study was carried out under visible light irradiation as depicted in Figure 5.6b–d. Here, MB displays 91% removal in 150 min, which is double compared to 44% removal in dark condition. Beyond 150 min, MB displays steady increase similar to dark reaction and shows 92% removal in 240 min. Thus, light irradiation doubles the adsorption of MB for same reaction conditions as in dark. In case of CV, 91% removal was achieved within 60 min, which is five times faster compared to dark reaction (91% removal in 300 min). The increase was rapid till 60 min, while it slowly increases beyond 60 min and gives 97% dye removal in 100 min. This shows that irradiation of light displays improved dye removal at half the reaction time than dark reactions.

In summary, CoZn–LDH nanosheets synthesized by reverse emulsion method act as excellent adsorbent under dark as well as visible light irradiation for removal of cationic dye CV and MB from aqueous solutions. The result gives scope to utilize the nanosheets for broad spectrum of dye adsorption and dye degradation study. Adventuring in this field and optimizing reaction parameters such as light source, irradiation time, amount of catalysts as well as pH can lead to improved removal of dye toxicity from aqueous solutions in the future.

5.1.5.2   PHOTODEGRADATION ROUTE

A coprecipitation method was employed for the synthesis of ZnFeLDH with Zn/Fe molar ratio of 3. An aqueous solution of M^{2+} salt-containing $Zn(NO_3)_2 \cdot 6H_2O$ and M^{3+} salt-containing $Fe(NO_3)_3 \cdot 9H_2O$ was added dropwise to an another aqueous solution of NaOH and Na_2CO_3 in a two necked round bottom flask under vigorous stirring at room temperature. The pH of the mixture was kept constant at 10 ± 2 by using 1 M NaOH. The resulting solution was then allowed to stir for 6h at room temperature. The obtained precipitate was filtered, washed several times with deionized water until pH of the filtrate was 7, and dried overnight in oven at 80 °C.

The formation of well crystalline hydrotalcite-like LDH can be evidenced from the powder XRD analysis. Figure 5.7a shows the XRD patterns of ZnFeLDH exhibiting characteristic sharp and intense diffraction peaks at lower 2θ angles and broad, less intense peaks at higher side of 2θ angles. The peaks at lower side of 2θ angles (i.e., at $2\theta = 11.6°$, $23.4°$, and $34.7°$) corresponds to the (003), (006), and (009) reflection planes and diffraction peak at higher 2θ angles (i.e. at $2\theta = 39.0°$, $46.1°$, $60.7°$, and $62.1°$) corresponds to (015), (012), (110), and (113) reflection planes, respectively.

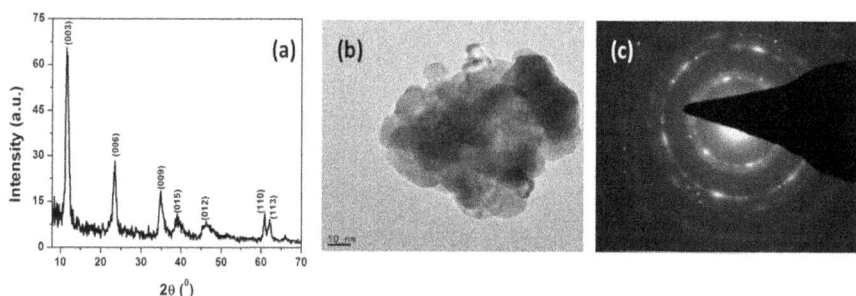

FIGURE 5.7 (a) Powder XRD patterns, (b) TEM image, and (c) SAED patterns of ZnFeLDH.

The morphological study of the prepared LDH sample was performed through TEM analysis. Figure 5.7b and c shows the TEM image and corresponding SAED patterns of ZnFeLDH. The TEM image shows that crystallites with flat plate-like morphology are lying one above of each other and thereby resulting in the irregular particle size. Moreover, the presence of different crystal planes can be evidenced from the formation of the concentric rings that are visualized from the bright dots and thus suggesting the uniform and crystalline nature of the LDH.

The light absorption behavior of the prepared sample was determined through UV–visible diffuse reflectance spectroscopy and the result is presented in Figure 5.8a. The absorption band around 200–300 nm (UV range) corresponds to the ligand to metal charge transfer transitions in the brucite-like LDH. The absorption band in the visible ranges of 500–800 nm is due to the metal to metal charge transfer transitions of oxo bridge bimetallic linkage (M^{II}–O–M^{III}). The optical bandgap energy (E_g) of the sample was determined by using Tauc/Davis–Mott expression as

$$(\alpha h\upsilon)^{\frac{1}{n_t}} = A\left(h\upsilon - E_g\right) \tag{5.1}$$

where $h\upsilon$ is photon energy; α is absorption coefficient and A denotes constant of proportionality. The "n_t" value determines the characteristics of the transition in a semiconductor [70]. If the value of $n_t = 1$, it is a direct transitions and when the value of $n_t = 4$, the transition is an indirect transition. Here, the n_t value is 1 indicating a directly allowed transition. The extrapolation of the curve to the x-axis has done in order to calculate the value of E_g and the value obtained is 2.18 eV (Figure 5.8b).

The N_2 adsorption–desorption study of the LDH sample was performed and shown in Figure 5.9. The material exhibits adsorption–desorption isotherm of type IV with hysteresis loop of H2 type; characteristic of a mesoporous material with regular narrow pore size distribution correlated with the aggregation of particles with plate-like morphology [71–73]. The upper closer point of the hysteresis loop appeared at higher value of relative pressure signifies the porous nature of the material with large and open pores, thereby making easier for the reactant molecules to diffuse through the material. The calculated specific BET surface area of the LDH material is 90.7 m^2/g with pore volume of 1.84 cm^3/g along with average pore diameter of 3.22 nm as calculated from the desorption branch of the isotherm.

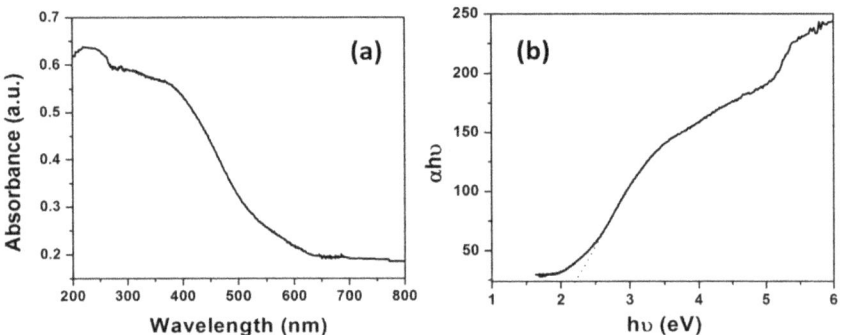

FIGURE 5.8 (a) UV–vis diffuse reflectance spectra and (b) corresponding bandgap energy of ZnFeLDH.

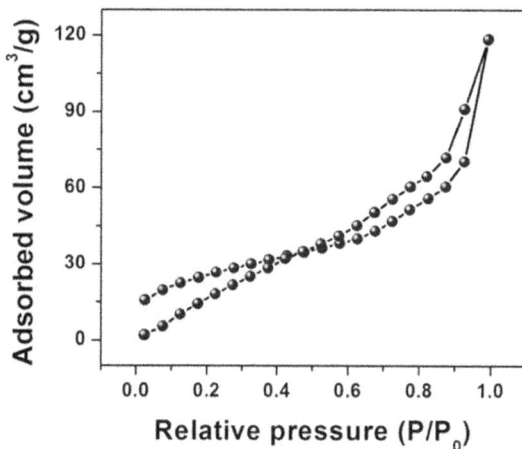

FIGURE 5.9 N_2 adsorption–desorption isotherms of ZnFe LDH.

The photodegradation study was carried out for dye removal using MB and RhB as model dye using both UV and visible light sources. First, the blank experiments were carried out under both UV and visible light irradiations in order to test the stability of dyes using no catalyst. In absence of catalyst, photolysis takes place with zero or negligible amount of degradation of dye and thus indicating the dye stability under different light irradiations. Prior to expose under light irradiations, the dye suspensions were kept in dark for 120 and 240 min in order to achieve an adsorption–desorption equilibrium for MB and RhB, respectively. A state of equilibrium has been attained within first 30 min with percentage dye uptake of only 4.5% and 2.9% for MB and RhB, respectively (Figure 5.10). Thus, an effective photocatalyst is needed for efficient dye degradation under appropriate reaction conditions and light illuminations.

In a typical photoatalytic reaction, catalyst amount of 5 and 10 mg was suspended to 50 mL solution of MB and RhB, respectively, with initial dye concentration of 10 mg/L under both UV and visible light irradiations. The light sources used were 125 W medium pressure mercury lamp (UV) and 250 W tungsten lamp (visible). In order to reduce the heat eliminated from the light source, a distance of 16 cm was maintained between the light source and dye suspension. Prior to expose under light irradiation, all the experiments were performed in dark for 30 min in order to attain adsorption–desorption equilibrium. The blank experiment was also performed without using catalyst for both MB and RhB under similar conditions. At a particular time intervals, samples were withdrawn and centrifuged to separate the catalyst and monitored on a UV–visible spectrophotometer at maximum absorbance (λ_{max} = 663 and 554 nm for MB and RhB, respectively). The % degradation of dye pollutants was determined by using the following equation

$$D = \frac{(C_O - C_t)}{C_O} \times 100\%$$ (5.2)

where C_o is initial concentration and C_t is final concentration of dye pollutants. For accuracy, all the experiments were done in duplicate. The reaction kinetics was analyzed by using Langmuir–Hinshelwood first-order kinetics as expressed follows:

(5.3)

$$\ln(C_O / C) = k_{app}t$$

where C_0 and C are the concentration of dye pollutants at time 0 and t, respectively; and is the apparent rate constant in min^{-1}.

At first dye suspensions were subjected under UV light irradiation for photodegradation process and results are depicted in Figure 5.10. The catalyst exhibits excellent activity with degradation of 99.9% and 99.8% in 120 and 240 min for MB and RhB, respectively, under UV light irradiation. Figure 5.11 shows the corresponding UV–vis spectral changes for photocatalytic degradation of MB and RhB under UV light irradiation. The characteristic absorption peaks (λ_{max} = 663 nm for MB and 554 nm for RhB) of the dyes goes on decreasing and diminished completely after a particular time indicating complete degradation of dyes under UV light irradiations.

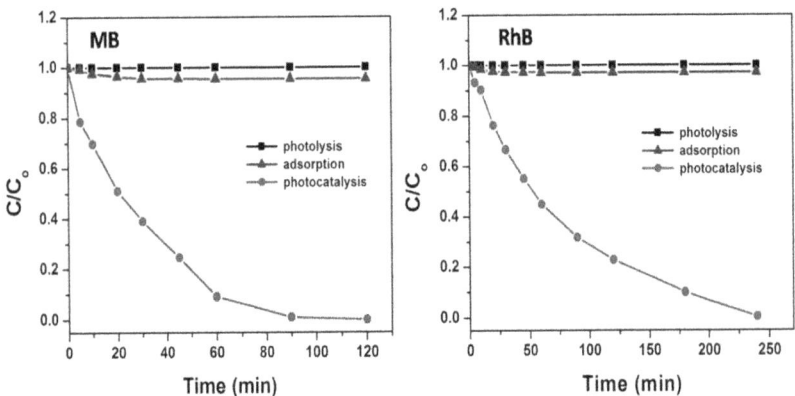

FIGURE 5.10 Photolysis, adsorption, and photocatalysis of MB and RhB over ZnFe LDH under UV light irradiation (conditions: C_o = 10 mg/L, $V_{solution}$ = 50 mL, catalyst amount = 5 mg for MB and 10 mg for RhB, pH = 7).

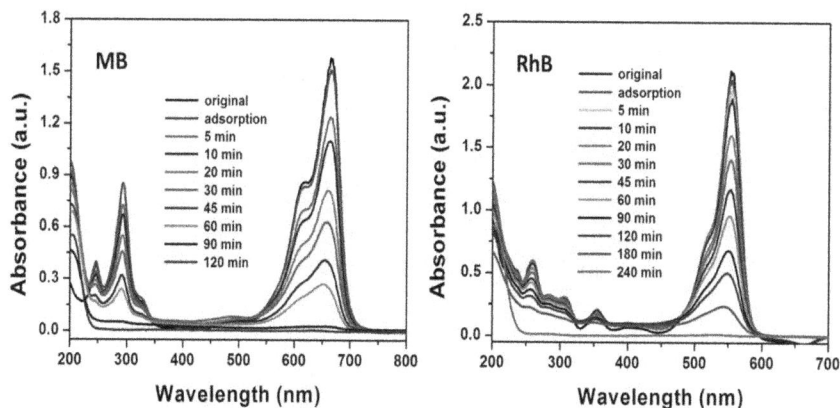

FIGURE 5.11 UV–visible spectra for phocatalytic degradation of MB and RhB over ZnFeLDH under UV light irradiation (C_o = 10 mg/L, $V_{solution}$ = 50 mL, catalyst amount = 5 mg for MB and 10 mg for RhB, pH = 7).

The photocatalytic study was also carried out under visible light irradiation but the process is slow for both cases with percentage degradation of only 73.6% and 57.7% in 120 and 240 min for MB and RhB, respectively (Figure 5.12). The corresponding UV–vis spectral changes for photocatalyticdegradation of both dyes under visible light irradiation are displayed in Figure 5.13.

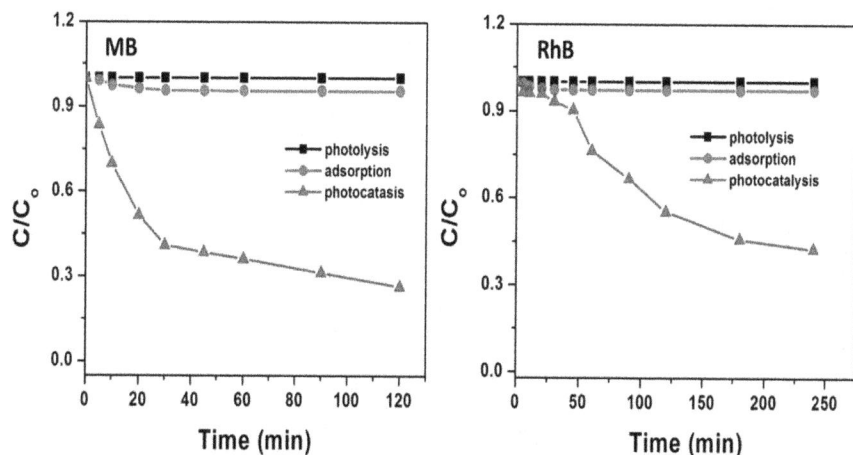

FIGURE 5.12 Photolysis, adsorption, and photocatalysis of MB and RhB over ZnFeLDH under visible light irradiation (conditions: C_o = 10 mg/L, $V_{solution}$ = 50 mL, catalyst amount = 5 mg for MB and 10 mg for RhB, pH = 7).

The solution pH is an important factor influencing the photodegradation efficiency [74]. Figure 5.14 shows the effect of solution pH on % degradation of MB and RhB under UV light irradiation. For both cases, the photocatalyst shows efficient activity at pH of 7. At pH<7, the highly positively charged catalyst surface thus prevents the dye adsorption, resulting the lower in degradation percentage. While at pH>7, the higher adsorption of cationic dyes on the more negatively charged catalyst surface inhibits the light penetration and consequently, lower in percentage of degradation.

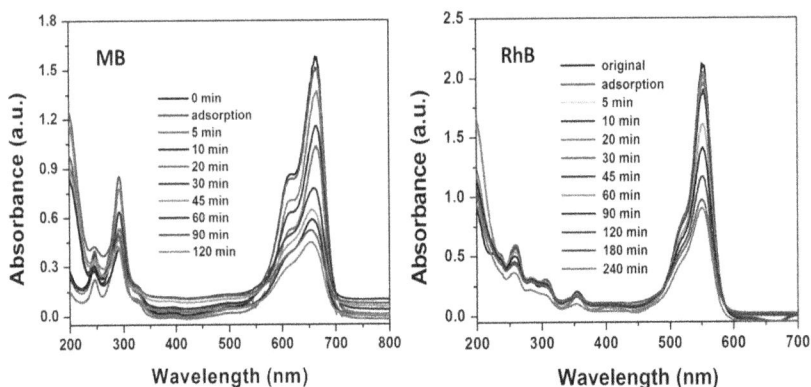

FIGURE 5.13 UV–visible spectra for photocatalytic degradation of MB and RhB over ZnFeLDH under visible light irradiation (C_o = 10 mg/L, $V_{solution}$ = 50 mL, catalyst amount = 5 mg for MB and 10 mg for RhB, pH = 7).

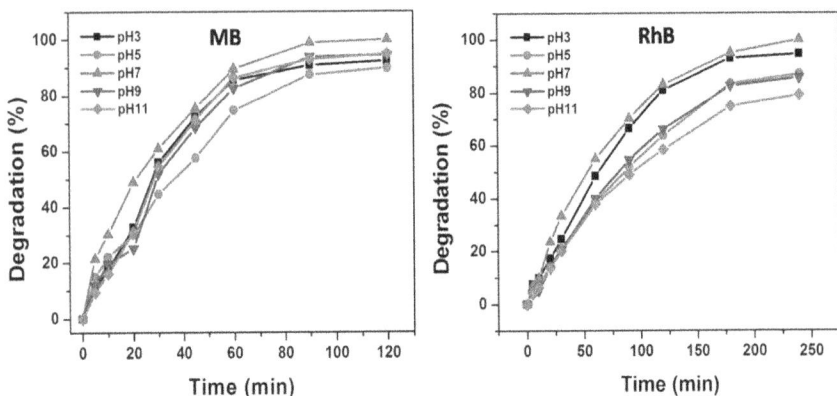

FIGURE 5.14 Effect of pH on photocatalytic degradation of MB and RhB with irradiation time under UV light irradiation (conditions: C_o = 10 mg/L, $V_{solution}$ = 50 mL, catalyst amount = 5 mg for MB and 10 mg for RhB, pH = 7).

The kinetics study was also performed for both dyes and the corresponding R^2 values are obtained as 0.992 and 0.996 with values 0.033 and 0.0162 for MB and RhB, respectively indicating that the degradation kinetics follows Langmuir–Hinshelwood first-order kinetics. The electron and hole $(e^-–h^+)$ pair mechanism is a key factor enhancing the photocatalytic performance. The hydroxyl groups present in the LDH surface play a significant role in the photodegradation process by capturing the h^+so that the hole and electron recombination do not take place and thus leading to the enhanced catalytic activity [75, 76]. The presence of interlayer carbonate anions in ZnFe LDH is another factor enhancing the photocatalytic activity. Due to its synergetic effect, the carbonate intercalated LDHs exhibit stronger photoabsorption properties and significantly enhances the activity of the photocatalyst towards the dye degradation. Moreover, the optimal bandgap energy (2.8 Å) of the catalyst significantly increases the electron–hole pair generation and hence adds up to the enhanced photocatalytic activity. The proposed mechanism for photocatalytic degradation of MB and RhB overZnFe LDH can be expressed as follows:

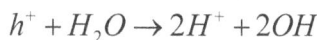

$$Zn^{II} Fe\ LDH + hv \rightarrow h^+ + e^-$$

$$h^+ + H_2O \rightarrow 2H^+ + 2OH$$

$$e^- + O_2 \rightarrow O_2^-$$

$$O_2^- + H^+ \rightarrow HO_2$$

$$HO_2 + HO_2 \rightarrow H_2O_{(2)} + O_2$$

$$H_2O_2 \rightarrow 2OH$$

$$Dye + OH \rightarrow Degraded\ Products$$

The degradation pathway of MB takes place via the attack of $^{\cdot}OH$ radicals on C–S$^+$=C functional group of the dye molecule [77, 78]. This result in the opening of aromatic ring structure leads to the formation of various smallest aliphatic compounds and finally mineralization of the harmful organic dye pollutants into the harmless CO_2, H_2O, SO_4^{2-}, NH_4^+ and NO_3^- ions [79]. Similarly, degradation of RhB takes place through the aromatic ring opening resulting in the formation of smallest nontoxic compounds. These smallest compounds then further break down to mineralized products such as CO_2 and H_2O [80].

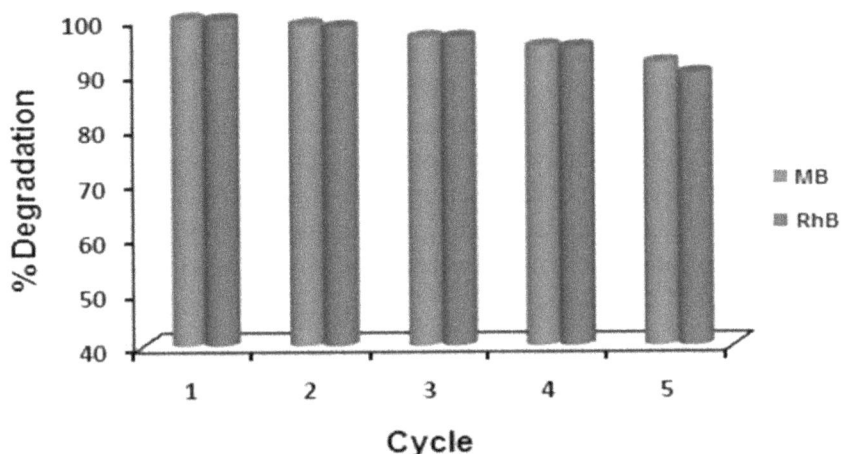

FIGURE 5.15   Recyclability test for photocatalytic degradation of MB and RhB over ZnFe LDH. (Conditions: C_0 = 10 mg/L, $V_{solution}$ = 50 mL, catalyst amount = 5 mg for MB and 10 mg for RhB, pH = 7).

The stability of the photocatalyst was also determined from the use of the regenerated catalyst for new and fresh reaction cycles and the efficiency of the photocatalyst is retained upto 90% even after five cycles for both cases (Figure 5.15).

In summary, ZnFeLDH synthesized through a simple coprecipitation method has been employed as photocatalyst for degradation of MB and RhB in aqueous solution using both UV and visible light irradiations. The catalyst displays excellent activity with 100% degradation of the dyes under UV light. The photocatalytic process followed first-order kinetics for both cases. The catalyst could be reused upto five cycles without any significant loss in the activity.

1.6   CONCLUSIONS

Photocatalytic removal of toxic organic pollutants from the environment via degradation as well as adsorption routes are promising technology in this field due to its capacity to drastically increase the adsorption and degradation in presence of light compared to normal conditions. Vast number of dissolved dyestuff can be removed and degraded from water thorough these routes. Although large number of publications are found on TiO_2 photocatalysts, research is focused on searching catalysts having lower bandgap than TiO_2, which is photoactive under both UV and visible light irradiation and thus can stabilize the electron–hole pair. Besides, searching of excellent adsorbents having high active surface area, better reusability, stability at adverse conditions is another field of continuous research. Herein, visible light active CoZn–LDH nanosheets were successfully synthesized by reverse emulsion system of SDS, water, and isooctane with the aid of cosurfactant 1-butanol and then characterized by physicochemical methods. Spectroscopic results exhibited well-developed LDH layered structure and SDS intercalation in between the layers. The nanosheets showed removal of MB and CV from aqueous solutions in dark as well as under visible light irradiation via adsorption route. We have seen that MB displays 91% removal in 150 min and CV displays 97% removal in 100 min. Again, irradiation of light displays improved dye removal at half the reaction time than dark reactions. Photocatalytic degradation route has its importance because this route not only removes the toxic dyes from aqueous solutions but degrade them to other simple molecules like CO_2 and H_2O, which otherwise may create secondary pollution after dumping the catalysts into environment. In this chapter, ZnFe LDH synthesized through a simple coprecipitation method has been employed as photocatalyst for degradation of MB and RhB in aqueous solution using both UV and visible light irradiations. The catalyst displays excellent activity with 100% degradation of the dyes under UV light. The photocatalytic process followed first-order kinetics for both cases. The catalyst could be reused upto five cycles without any significant loss in the activity

ACKNOWLEDGMENTS

This work was supported by a women scientist—a project under the Department of Science and Technology, India.

KEYWORDS

- **layered double hydroxide (LDH)**
- **microemulsion**
- **coprecipitation**
- **dye degradation**
- **photocatalysis**
- **photoadsorption**

REFERENCES

1. Kant, R., *Natur. Sci.* 2012, **4**, 22–26.
2. Johnston, W., *Biotech. Histochem.* 2009, **83**, 83–87
3. Lachheb, H., Puzenat, E., Houas, A., Ksibi, M., Elaloui, E., Guillard, C., Herrmann, J. M., *Appl. Catal. B: Environ.* 2002, **39**, 75–90.
4. Brown, M. A., De Vito, S. C., *Crit. Rev. Environ. Sci. Technol.* 1993, **23**, 249–324.
5. Luan, J. F., Hao, X. P., Zheng, S. R., Luan, G. Y., Wu, X. S. *J. Mater. Sci.* 2006, **41**, 8001–8012.
6. Sukumar, M., Sivasamy, A., Swaminathan, G., *Appl. Biochem. Biotechnol.* 2007, **136**, 53–62.
7. Xia, S. J., Liu, F. X., Ni, Z. M., Xue, J. L., Qian, P. P., *J. Colloid. Interf. Sci.* 2013, **407**, 195–200.
8. Kadirova, Z. C., Katsumata, K., Isobe, T., Matsushita, N., Nakajima, A., Okada, K., *Appl. Surf. Sci.* 2013, **284**, 72–79.
9. Deka, P., Deka, R. C., Bharali, P., *New J. Chem.* 2016, **40**, 348–357.
10. Parida, K., Mohapatra, L., *Dalton Trans.*, 2012, **41**, 1173–1178.
11. Parida, K., Mohapatra, L., Baliarsingh, N., *J. Phys. Chem. C*, 2012, **116**, 22417–22424.
12. Liua, J., Zhang, G., *Phys Chem.Chem. Phys.*, 2014, **16**, 8178–8192
13. Ge, X., Gu, C. D., Wang, X. L., Tu, J. P., *J. Mater. Chem. A* 2014, **2**, 17066–17076.
14. Fabian, J., Mehlhorn, A., Dietz, F., Tyutyulkov, N., *Monatshefte fur Chemie-Chemical*
15. Zhang, F. Z., Xiang, X.., Li, F., Duan, X., *Catal. Surv. Asia.* 2008, **12**, 253–265.
16. Reichle, W.T., *Solid State Ionics* 1986, **22**, 135–141.
17. Cavani, F., Trifiro, F., Vaccari, A., *Catal. Today* 1991, **11**, 173–301.
18. Hochstetter, C., *J. für Praktische Chemie* 1842, **27**, 375–378.
19. Mills, S.J., Christy, A.G., Schmit, R.T., *Mineralogical Magazine* 2016, 80, 1023–1029.
20. Suzum, E., Okamato, M., Ono, Y., *J. Mol. Catal.*1990, **61**, 283–294.
21. Shi, L., Li, D., Wang, J., Li, S., Evans, D.G., Duan, X., *Clays and Clay Minerals* 2005, **53**, 294–300.

22. Devi, R., Begum, P., Bharali, P., Deka, R.C., *ACS Omega* 2018, **3**, 7086–7093.

23. Bharali, D., Devi, R., Bharali, P., Deka, R.C., *New J. Chem.*2015, **39**, 172–178.

24. Corma, A., Hamid, S.B.A., Iborra, S., Velty, A., *J. Catal.* 2005, **234**, 340–347.

25. Evans, D.G., Slade, R.C.T., Structural aspects of layered double hydroxides, in Layered Double Hydroxides, X. Duan and Evans, D.G. Eds., Springer, Berlin, 2006, volume **119**, pp 1–87.

26. Galvão, T. L. P., Neves, C.S., Caetano, A. P. F., Maia, F., Mata, D., Malheiro, E., Ferreira, M.J., Bastos, A.C., Salak, A.N., Gomes, J. R. B., Tedim, J., Ferreira, M. G. S., *J. Colloid Interface Sci.* 2016, **468**, 86–94.

27. Adachi-Pagano, M., Forano, C., Besse, J.P., *Chem. Commun.* 2000, 91–92

28. Hu, G., and O'Hare, D., *J. Am. Chem. Soc.* 2005, **127**, 17808–17813.

29. Xu, J., Deng, H., Song, J., Zhao, J., Zhang, L., Hou, W., *J. Colloid Interf. Sci.* 2017, **505**, 816–823.

30. Hu, G., O'Hare, D., *J. Am. Chem. Soc.* 2005, **127**, 17808–17813.

31. Wang, C.J., Wu, Y.A., Jacobs, R.M.J., Warner, J.H., Williams, G.R., O'Hare, D., *Chem. Mater.* 2011, **23**, 171–180.

32. Zhao, Y., Wang, C.J., Gao, W., Li, B., Wang, Q., Zheng, L., Wei, M., Evans, D.G., Duana, X., O'Hare, D., *J. Mater. Chem. B* 2013, **1**, 5988–5994.

33. Hu, G., Wang, N., O'Hare, D., Davis, J., *Chem. Commun.* 2006, 287–289.

34. Wang, C. J., O'Hare, D., *J. Mater. Chem.* 2012, **22**, 21125–21130

35. Wu, H., Jiao, Q., Zhao, Y., Huang, S., Li, X., Liu, H., Zhou, M., *Mater. Characteriz.* 2010, **61**, 227–232.

36. Xu, J., Zhang, L., Li, D., Zhao, J., Hou, W., *Colloid. Polym. Sci.* 2013, 291, 2515–2521.

37. Hu, G., Wang, N., O'Hare, D., Davis, J., J. Mater. Chem. 2007, **17**, 2257–2266.

38. Hu, G., Wang, N., O'Hare, D., Davis, J., *Chem. Commun.* 2006, 287–289.

39. Zhao, Y., Li, B., Wang, Q., Gao, W., Wang, C.J., Wei, M., Evans, D.G., Duana, X., O'Hare, D., *Chem. Sci.* 2014, **5**, 951–958.

40. Zhao, Y., Jia, X., Chen, G., Shang, L., Waterhouse, G. I. N., Wu, L.Z., Tung, C.H., O'Hare, D., Zhang, T., *J. Am. Chem. Soc.* 2016, **138**, 6517–6524.

41. Wang, C. J., Wu, Y.A., Jacobs, R.M.J., Warner, J.H., Williams, G.R., O'Hare, D., *Chem. Mater.* 2011, **23**, 171–180.

42. Wu, H., Jiao, Q., Zhao, Y., Huang, S., Li, X., Liu, H., Zhou, M., *Mater. Characteriz.* 2010, **61**, 227–232.

43. Wang, C.J., O'Hare, D., *J. Mater. Chem.* 2012, **22**, 21125–21130.

44. Sim, H., Jo, C., Yu, T., Lim, E., Yoon, S., Lee, J.H., Yoo, J., Lee, J., Lim, B., *Chem. Eur. J.* 2014, **20**, 14880–14884.

45. Xu, J., Zhang, L., Hou, W., *Colloid Polym. Sci.* 2013, **291**, 2515–2521

46. Liu, D., Yates, M.Z., *Langmuir* 2006, **22**, 5567–5569.

47. Wang, H., Schaefer, K., Moeller, M., *J. Phys. Chem. C* 2008, **11**, 3175–3178.

48. Kind, C., Popescu, R., Muller, E., Gerthsen, D., Feldmann, C., *Nanoscale* 2010, **2**, 2223–2229.

49. Sarciaux, J.M., Acar, L., Sado, P.A., *Pharmaceutics* 1995, **120**, 127–136.

50. Forano, C., Hibino, T., Leroux, F., Taviot-Gueho, C., Layered double hydroxides, in *Handbook of clay science,* Bergaya, F. B. K. G. Theng and Lagaly, G. Eds., Elsevier, Oxford, 2006,volume 1, pp 1021–1095.

51. Kisch, H., *Angew. Chem. Int. Ed.* 2013, **52**, 812–847.

52. Hoffmann, M.R., Martin, S.T., Choi, W., Bahnemann, D.W., *Chem. Rev.* 1995, **95**, 69–96.

53. Wu, Y., Du, X., Kou, Y., Wang, Y., Teng, F., *Ceramic Int.*, 2019, **45**, 24594–24600.

54. Lee, Y., Choi, J.H., Jeon, H.J., Choi, K.M., Lee, J.W., Kang, J.Q., *Energ. Environ. Sci.* 2011, **4**, 914–920.

55. Fan, G., Li, F., Evans, D.G., Duan, X., *Chem. Soc. Rev.* 2014, **43**, 7040–7066.

56. Gomes Silva, C., Bouizi, Y., Fornes, V., Garcia, H., *J. Am. Chem. Soc.* 2009, **131**, 13833–13839.

57. Shao, M., Han, J., Wei, M., Evans, G.D., Duan, X., *Chem. Eng. J.* 2011, 168, 519–524.

58. Lan, M., Fan, G., Yang, L., Li, F., *Ind. Eng. Chem. Res.* 2014, **53**, 12943–12952.

59. Lu, R. Xu, Chang, J., Zhu, Y., Xu, S., Zhang, F., *Appl. Catal. B: Environ.* 2012, **111–112,** 389–396.

60. Seftel, E.M., Popovici, E., Beyers, E., Mertens, M., Zhu, H.Y., Vansant, E.F., Cool, P., *J. Nanosci. Nanotechnol.* 2010, **10**, 8227–8233.

61. Qiao, C., Zhang, Y., Zhu, Y., Cao, C., Bao, X., Xu, J., *J. Mater. Chem.* 2015, **3**, 6878–6883

62. Gunjakar, J.L., Kim, T.W., Kim, H.N., Kim, I.Y., Hwang, S.J., *J. Am. Chem. Soc.* 2011, **133**, 14998–15007.

63. Gursky, J.A., Blough, S.D., Luna, C., Gomez, C., Luevano, A.N., Gardner, E.A. *J. Am. Chem. Soc.* 2006, **128**, 8376–8377.

64. Zhao, Y., Jia, X., Chen, G., Shang, L., Waterhouse, G.I.N., Wu, L., Tung, C., O'Hare, D., Zhang, T., *J. Am. Chem. Soc.*, 2016, **138**, 6517–6524.

65. Lv, F., Meng, Z., Li, P., Zhang, Y., Lv, G., Zhang, Q., Zhang, Z., *Bull. Mater. Sci.* 2015, 38, 1079–1085.

66. Zou, X., Goswami, A., Asefa, T., *J. Am. Chem. Soc.* 2013, **135**, 17242–17245.

67. Qiao, C., Zhang, Y., Zhu, Y., Cao, C., Bao, X., Xu, J., *J. Mater. Chem. A*, 2015, **3**, 6878–6883.

68. Gamil, S., El Rouby, W.M.A., Antuch, M., Zedana, I.T., *RSC Adv.* 2019, **9**, 13503–13514.

69. Puscasu, C.M., Seftel, E.M., Mertens, M., Cool, P., Carja, G., *J. Inorg. Organomet. Polym.* 2015, **25**, 259–266.

70. Jothivenkatachalam, K., Prabhu, S., Nithyaand, A., Jeganathan, K., *RSC Adv.* 2014, **4**, 21221–21229.

71. Zaghouane-Boudiaf, H., Boutahala, M., Arab, L., *Chemical Eng. J.* 2012, **187**, 142–149.

72. Parida, K.M., Mohapatra, L., *Chemical Eng. J.* 2012, **179**, 131–139.

73. Zhang, M., Yao, Q., Lu, C., Li, Z., Wang, W., *ACS App. Mat. Interf.* 2014, **6**, 20225–20233.

74. Tayade, R.J., Natarajan, T.S., Bajaj, H.C., *Ind. Eng. Chem. Research.* 2009, **48**, 10262–10267.

75. Parida, K., Mohapatra, L., Baliarsingh, N., *J. Phys. Chem. C* 2012, **116**, 22417–22424.

76. Gaya, U.I., Abdullah, A.H., *J. Photochem. Photobiol. C* 2008, **9**, 1–12.

77. Houas, A., Lachheb, H., Ksibi, M., Elaloui, E., Guillard, C., Herrmann, J.M., *Appl. Catal. B: Environ.* 2001, **31**, 145–157.

78. Xia, S., Zhang, L., Pan, G., Qian, P., Ni, Z., *Phys. Chem. Chem. Phy.* 2015, **17**, 5345–5351.

79. Darabdhara, G., Boruah, P.K., Borthakur, P., Hussain, N., Das, M.R., Ahamad, T., Alshehri, S.M., Malgras, V., Wu, K.C.W., Yamauchi, Y., *Nanoscale* 2016, **8**, 8276–8287.

80. Baldev, E., MubarakAli, D., Ilavarasi, A., Pandiara, D., Ishack, K.A.S.S., Thajuddin, N., *Colloids Surf. B* 2013, **105**, 207–214.

CHAPTER 6

THE CHALLENGE OF REALIZING NANOTHIN PEROVSKITE SINGLE-CRYSTALLINE WAFERS: COMPUTATIONAL AND EXPERIMENTAL ASPECTS

M. PRATHEEK,[1] T. A. SHAHUL HAMEED,[2] and P. PREDEEP[1*]

[1]*Laboratory for Molecular Electronics and Photonics (LAMP), Department of Physics, National Institute of Technology, Calicut, Kerala, India*

[2]*Department of Electronics and Communication, TKM College of Engineering, Kollam, Kerala, India*

Corresponding author. E-mail: predeep@nitc.ac.in

ABSTRACT

Organic–inorganic hybrid perovskites had emerged as the most promising photovoltaic materials with comparable or more efficiencies with Silicone in a short span of few years from its discovery. During the less than a decade of its invention in the application of solar cells these materials had recorded a certified efficiency of 21% that promises improving it to even 30%. This has all potential to change the rules of the game. However, thin film forming of perovskites is still a tricky business and solution processing is adding to the vows of instability that is the main hurdle in making perovskites a practical solar energy harvester. In the case of silicon, single-crystalline wafers have much greater efficiency and stability, and the same is expected out of perovskite single crystals also. However, as the diffusion length of carriers in perovskite single crystals will not exceed more than 30–50 μm, only single-crystalline wafers of thickness will have the potential to be used as solar energy harvester. This is a real challenge and despite a lot of reported attempts during these years, growing large area thin single-crystalline wafers of perovskites are still evading success. Computational studies on some important aspects of the single-crystalline perovskites are also attempted and made use by experimentalists, though only few can be seen in the literature. In this chapter, these aspects have been described and discussed. Along with these major facets of single-crystalline perovskite evolution, from growth, properties, and characterization are discussed, followed by the challenges and efforts to address the issue of developing effective technique to grow single-crystalline wafers of perovskite having thickness <50 **μm**. Besides the techniques for fabrication solar cells with such thin wafers of organometallic perovskites will be explained.

6.1   INTRODUCTION

Earth receives kW h of sunlight every year [1]. This is more than hundred times the global energy need. So is the charm and logic of the notion of harvesting solar energy as the most promising solution for global energy crisis. Currently, the solar PV market is dominated by crystalline silicon solar panels with efficiencies near 20 percentages. The new generation photovoltaic technology include organic solar cells, quantum dot solar cells, dye sensitized solar cells (DSSCs), and the most recent perovskite solar cells (PSCs), which can be fabricated through low-temperature solution processing techniques such as, blade coating, screen printing, spin coating and spray pyrolysis, physical and chemical vapor deposition techniques, etc. Although, low-temperature processing reduces the energetic cost and energy payback time, the power conversion efficiencies (PCEs) and stability significantly lags behind the conventional silicon solar cells. An efficient solar cell allows reduction in all costs associated with installation, while requiring much lower area and lower number of panels to be installed. Thus power conversion efficiency is the primary driver of solar cell cost. Organic–inorganic lead halide PSCs are already registered a PCEs approaching that of inorganic silicon solar cells. In a few years of its first reporting by Miyasaka et al., PCE of PSCs had rapidly increased [2] from 3.8% to 22.1%. Therefore PSCs have generated both research and industrial attention in a large scale. Low cost, flexible, lightweight, and high throughput PSCs can be fabricated through solution process without high-temperature processing [3]. The properties of lead halide perovskites such as high carrier mobility, high extinction coefficient, and small bandgap had gained great attention in recent years.

It all started with Miyasaka and co-workers using methylammonium lead iodide (MAPbI$_3$) as sensitizer [4] in DSSC with liquid electrolyte in 2009 for the first time to obtain a PCE of 3.81%. The introduction of *spiro*-OMeTAD instead of liquid electrolyte improved the value PCE to 9.7% and enhanced the stability of perovskite PV devices. Intense research followed to understand the mechanism involved. The advantage is that perovskite allows both solution and physical methods for processing the film. In the planar structure of the PSC, this deposition can be in a single step [5] by spin coating a solution containing stoichiometric amount of methylamine iodide (MAI) and lead iodide (PbI$_2$) in polar solvents like gamma-butyrolactone (GBL) or DMF. With the optimization of spin coating conditions and precursor concentration, perovskite can be deposited in mesoporous layer and to get a uniform layer. A significant step in the development of solution-based fabrication is the application of sequential deposition process [6]solution-processable organic-inorganic hybrid perovskites-such as CH3NH3PbX3 (X = Cl, Br, I. This process consists of spin coating of PbI$_2$ on a substrate coated with TiO$_2$ layers and dipping of this yellow colored substrate in to MAI solution. When this MAI coated substrate is dipped into the lead iodide solution, within a few seconds it gets converted to dark brown film of MAPbI$_3$.

Apart from these types of widely used solution processing techniques, physical vapor deposition also has been used to fabricate PSCs. Snaith et al. showed that efficient solar cells can be fabricated by dual-source thermal evaporation of lead chloride (PbCl$_2$) and MAI. The cell is fabricated in such a way that the evaporated film is sandwiched in between an electron transport layer and a hole transport layer. Fan and co-workers deposited thin films of MAPbI$_3$ by single-source thermal evaporation of ground single crystal [7] an alternative route to fabricating high-quality CH3NH3PbI3 thin films is proposed. Single-source physical vapour deposition (SSPVD and obtained an efficiency of 10.90%. They have used organic materials like PCBM and PEDOT: PSS as electron and hole transport layers, respectively. Another interesting method, which utilizes both vapor phase transformation and solution deposition [8], has been reported by Chen et al. In vapor-assisted solution process, PbI$_2$ was first spin coated on to compact TiO2 layer. Subsequently, vapor of MAI was exposed to the films in nitrogen atmosphere at 1500 C. The slow crystallization process resulted in films of very low surface roughness and exhibited an efficiency of 12.1%. In the case of dual-source thermal evaporation, high precision in rate control is needed to get the perovskite uniform film. And when it was tried to deposit perovskite films using single-source thermal evaporation, it has been found very difficult, as during the evaporation perovskite dissociates and only lead iodide gets deposited on the substrate. To avoid this dissociation the distance between source and substrate is reduced, then perovskite layer is formed but with lot of pin holes.

The commonly used deposition techniques, briefed above, like solution processing and thermal evaporation are problematic in that, they directly affect the performance of the coated perovskite layer. Spin coating is preferred over other techniques because it is facile and of low cost. But the major issues of these methods are (1) poor morphology, (2) presence of pin holes, (3) requirement of scaffolding layer, (4) requirement of postannealing, (5) poor reproducibility, and (6) not suitable for large-area devices. A solution to these problems is to develop single-crystalline wafers of perovskite absorbers directly to the transparent anodes coated with transport layers. In the case of silicon solar cells, single-crystalline wafer solar cell show high efficiency and operational stability; however, in silicon technology, single-crystalline wafer production is rather easy. Wafers of mm size can easily be sliced out of large single-crystalline ingots by sawing. This is feasible just because the diffusion length of charge carriers in silicon is in the range of millimeters. However, for perovskites the diffusion length is only few tens of micrometers and therefore needs wafers of thickness around this thickness, which is not possible by slicing out as in the case of silicon wafers.

The quest for improving diffusion length motivated the researchers to search for strategies to synthesize single-crystal perovskite materials having phase purity and macroscopic dimensions [9]. When single-crystal perovskite material used in solar cells, the efficiency of solar cells improved because carrier mobility and diffusion length increase due [10] two things: (1) the reduced interfacial area associated with large grains suppress charge trapping and eliminate hysteresis and (2) larger grains have lower bulk defects and higher mobility [5]. Crystalline silicon is grown into ingots and then cut into wafers of thicknesses in millimeter sizes. Since the diffusion length of carriers in crystalline silicon is in that ranges this is possible. However, the diffusion lengths in perovskite crystals are much lesser in the range of only a few tens of micrometers. This makes cutting single crystals of perovskite a nearly impossible task and further making operation able interface with electrodes and charge transport layers is also difficult. This necessitates the growing of thin wafers of thicknesses in the range of its diffusion length, that is, in the range of about 10–25 μm.

This poses a great challenge as growing such thin wafers in large scale is difficult. Still investigations are in a global scale and do available some reports, which describe the efforts and methods for this task. Though scarce, efforts to carry out computational studies on some of the aspects of the perovskites crystal growth are available. In this chapter, such efforts and various methods to produce single-crystalline wafers of organo lead halide perovskite for solar cells applications are reviewed and discussed.

6.2   PEROVSKITE SINGLE CRYSTALS

As already mentioned PCE of a solar cell is greatly affected by the parameters, "charge carrier life time, mobility, and diffusion length." The quest for improving diffusion length motivated the researchers to search for strategies to synthesize single-crystal perovskite materials having phase purity and macroscopic dimensions [9]. When single-crystal perovskite material is used in solar cells the efficiency of PSCs improved because [10] carrier mobility and diffusion length increased due two things: (1) the reduced interfacial area associated with large grains suppresses charge trapping and eliminates hysteresis, and (2) larger grains have lower bulk defects and higher mobility [5]. Single-crystal perovskite materials show low trap state, high mobility, large diffusion length, and wide light absorption capabilities compared to its polycrystalline counterparts. This property enhancement is due to [11] the more order and long-range structure. However, synthesis of single crystal is a time-consuming process.

The following are relatively fast methods to grow single-crystal perovskite: (1) temperature lowering method, (2) antisolvent vapor-assisted crystallization (AVAC), and (3) inverse temperature crystallization (ITC). Temperature lowering method is suitable for the growth of single-crystal perovskite since the solubility of perovskite in the corresponding acid halide solvents is moderate at room temperature, but changes considerably with temperature. The decrease in temperature induces saturation of the solute, leading to the formation of centimeter-sized methylammonium halide perovskite crystal from saturated aqueous perovskite precursor solution containing an inorganic metal salt (PbI_2), and organic halide salt (MAI). ITC is a much faster method to grow single crystals of perovskites like $MAPbI_3$, and this is based on retrograde solubility of this perovskite in specific solvents. The two important solvents for methylammonium lead iodide perovskite ($MAPbI_3$) are dimethylformamide (DMF) and GBL. $MAPbI_3$ shows retrograde solubility in GBL but not in DMF. ITC exploits this rather unusual property of the hybrid perovskites for the crystallization of $MAPbI_3$ in GBL. It is seen that [7an alternative route to fabricating high-quality CH3NH3PbI3 thin films is proposed. Single-source physical vapour deposition (SSPVD, 12, 13] perovskite shows retrograde solubility in GBL above 60 °C and when its concentrated solution was kept undisturbed at a temperature above 60 °C it leads to crystallization. Photograph of a single crystal grown by ITC method in the author's laboratory is shown in Figure 6.1. In the AVAC method, a suitable antisolvent [9] is diffused into the solution containing crystal precursors leading to the growth of sizable high-quality single crystals. Here, GBL is used for preparing the precursor solution and dichloromethane (DCM) is used as the antisolvent for the growth of methylammonium lead [9] iodide single crystals.

FIGURE 6.1   Photograph of single-crystal MAPbI$_3$ synthesized by ITC method.

6.3   COMPUTATIONAL STUDIES ON THICKNESS DEPENDENCE OF PEROVSKITE SINGLE CRYSTALS

The widely studied PSCs are based on methylammonium lead iodide (MAPbI$_3$) and the bandgap of which is too large for single-junction solar cells. The bandgap can be reduced by composition engineering by partially replacing MA$^+$ and Pb^{2+} by FA$^+$ and Sn^{2+}, but that leads to new challenges. The easy oxidation of Sn^{2+} to Sn^{4+} causes additional severe chemical instability of

Sn containing perovskite. Instability of excess amount of FA$^+$ ions can cause the deformation of material structure.

Below bandgap absorption has been identified in MAPbI$_3$ in previous studies, which has been attributed to the indirect bandgap absorption transition with a bandgap of 60 meV smaller than the direct bandgap. However, the below bandgap absorption coefficient corresponding to the transition is several orders magnitude smaller than that of above gap transition. Therefore, the thickness of the perovskite films thus needs to be over several micrometers or more, which is however beyond the carrier diffusion length in existing polycrystalline perovskite films. The reported carrier diffusion length in perovskite single crystals is much longer as well as above tens of micrometers due to the absence of grain boundaries and significantly reduced defect density [14].

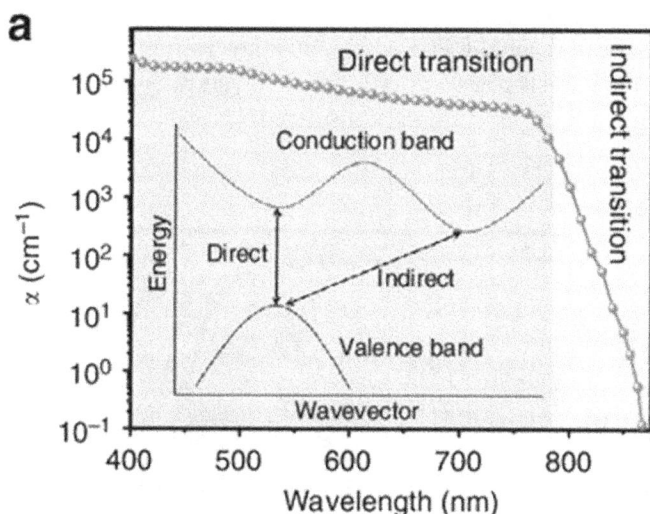

FIGURE 6.2   Schematic illustration of the direct, below bandgap transition and absorption coefficient of polycrystalline MAPbI$_3$ thin film.

Source: Reprinted with permission from Ref. [15]. http://creativecommons.org/licenses/by/4.0/.

Chen et al. simulated the thickness-dependent absorption of MAPbI$_3$ and device efficiency to find out the feasibility of increasing device efficiency by utilizing below bandgap transition in MAPbI$_3$ perovskite [15]. The cell structure used is ITO/PTAA/MAPbI$_3$/PCBM/C$_{60}$/BCP/Cu. A previous study reports that, the below bandgap transition is three-four orders magnitude less than that of direct transition [16]. When the thickness of $MAPbI_3$ films in devices increases from 500 nm to 200 μm, the absorption spectrum expands significantly to near-infrared region (Figures 6.2 and 6.3). Leading to the red-shift of the absorption edge to 850 nm and short circuit current (J_{sc}) increases from 23 to 27.1 mA cm⁻² due to broadened absorption spectrum (Figure 6.4) [15]. Here, J_{sc} was calculated by the integration of

$$J_{SC} = q\int I(E)EQE\left(1-e^{-2\alpha(E)d}\right)dE$$

where q is the charge, I is the incident spectral photon flux density, α is the absorption coefficient, and d is the thickness. A full reflection of light is assumed from electron transport layer and metal back electrode. Due to the large absorption coefficient of perovskite in UV–Visible region and large thickness of the perovskite the UV–Visible light is completely absorbed before approaching the PCBM/C₆₀/BCP layer. PCBM, C₆₀ and BCP have negligible light absorption in near infrared region and the only portion of the light that can reach the PCBM/C₆₀/BCP layer is almost at the band edge. Therefore the parasitic absorption of these layers was not considered in the simulation, but light reflection is considered.

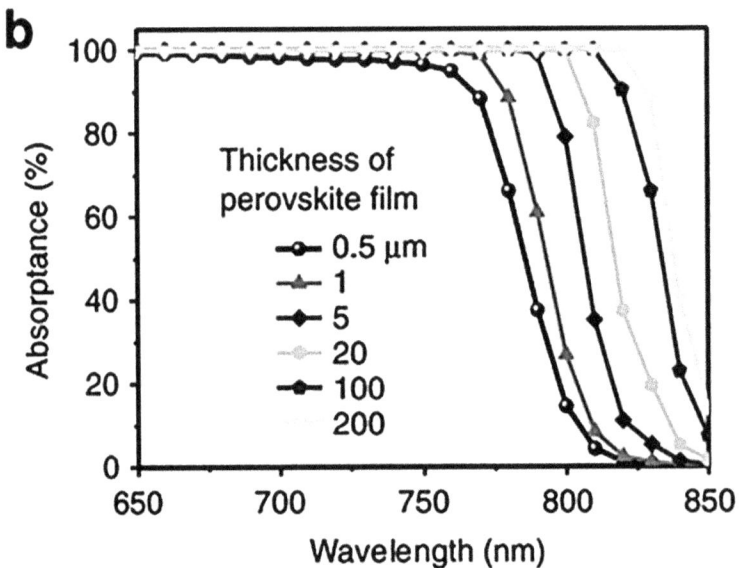

FIGURE 6.3 The calculated absorption of $MAPbI_3$ with different thickness.

Source: Reprinted with permission from Ref. [15]. http://creativecommons.org/licenses/by/4.0/.

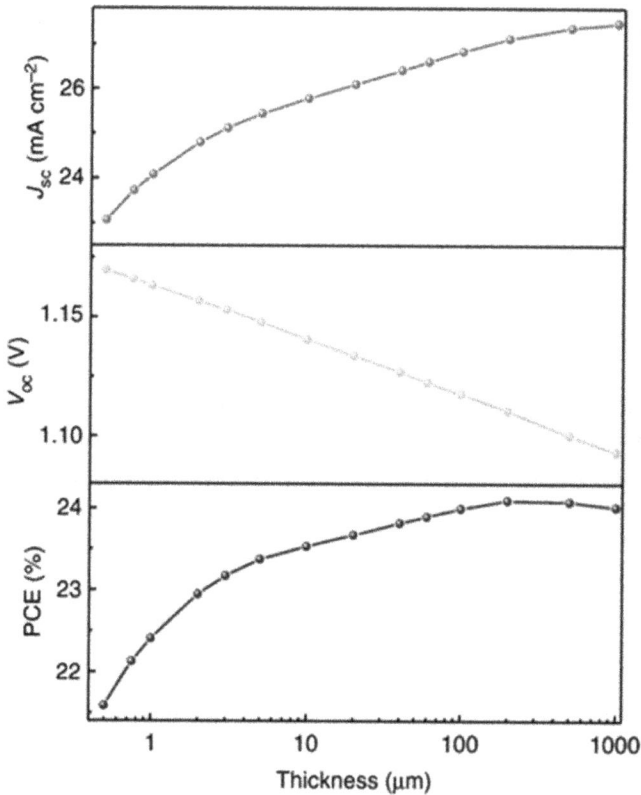

FIGURE 6.4   The calculated ideal dependence of J_{sc}, V_{oc}, and PCE of the single-crystal solar cells on the thickness of the single crystals.

Source: Reprinted with permission from Ref. [15]. http://creativecommons.org/licenses/by/4.0/.

The VOC was calculated from the quasi-Fermi level splitting based on the nonequilibrium carrier concentration, resulting a relation of V_{OC} with J_{SC}, τ_{eff}, and crystal thickness (d):

$$V_{OC} = \frac{kT}{q} \ln \left(\frac{J_{SC} N_D \tau_{eff}}{q n_i^2 d} \right)$$

where T and k are the absolute temperature and Boltzmann constant, respectively, N_D is the concentration of the donor atom, τ_{eff} is effective recombination lifetime, n_i is the intrinsic carrier concentration, and d is the film thickness.

The τ_{eff} was obtained by measured effective lifetime and extracted lifetime at different MAPbI$_3$ film thickness The dependence of τ_{eff} on crystal thickness was derived based on

$$\frac{1}{\tau_{eff}} = \frac{1}{\tau_s} + \frac{1}{\tau_b}$$

$$\tau_b = \frac{1}{(k_b N_D {P}/{Ed}\tau_{eff})}$$

where τ_s and τ_b are surface recombination lifetime and bulk recombination lifetime, respectively, k_b is the bulk recombination rate constant, P is the irradiance of the incident light, and E is the average energy of incident photons. The dependence of τ_{eff} on crystal thickness is derived to be

$$t_{eff} = \frac{-dE + \sqrt{d^2 E^2 + 4k_b N_A PE\tau_s^2 d}}{2k_b N_A I \tau_s}$$

If the τ_s is much higher than the τ_b that is the case for perovskite single crystals, τ_{eff} is found to be proportional to the square root of the crystal thickness, which is consistent with the experimental results V_{OC} of 1.163 V at thickness of 1 µm was obtained from the calculated result of Yin et al. Subsequently, the V_{OC} at other crystal thickness was calculated with reference to the 1-µm-thick device.

The FF was calculated by the empirical equation from single-crystal silicon solar cells:

$$FF = \frac{q{V_{oc}}/{kT} - \ln(q{V_{oc}}/{kT} + 0.72)}{q{V_{oc}}/{kT} + 1}$$

The calculated FF overall remains almost unchanged with value of about 0.89, which is much larger than common values observed in high-efficiency PSCs. The FF was fixed at 0.8 when calculating the PCE. Combining the J_{sc}, V_{OC}, and FF, the dependence of PCE on film thickness was calculated.

6.4   PEROVSKITE SINGLE-CRYSTAL WAFERS: SYNTHESIS AND APPLICATIONS

Even though bulk single crystals of perovskites show exciting solar harvesting properties, fabrication of a solar cell directly from bulk single-crystal perovskite is not feasible. As already mentioned, the two key factors in photoelectronic device design are carrier diffusion length (L_D) and light absorption length (L_A), and the former limits the maximum thickness that carriers can be effectively collected without significant recombination loss and latter determines the minimum thickness of the active material required to harvest sufficient intensity of light. For perovskite, L_A and L_D are determined to be 1 and 175 μm, respectively. This indicates that the layer thickness should be preferred to be <175 μm. Therefore millimeter thick single crystal will not make major improvements in solar cell performance. And also the large thickness will make the fabrication process more complex. The theoretical studies of Chen et al show that when the thickness of the film is increased from 500 nm to 200 μm the absorption spectrum expands significantly to near-infrared region leading to the red-shift of the absorption edge to 850 nm and which in turn improves the short circuit current and PCE. In this context, the objective of this chapter is to introduce the readers to the possibilities and promises in growing a few micrometer thick single-crystalline perovskite wafers. It is heartening to see that different research groups are actively engaged in finding feasible methods to grow such wafers. Though no report is so far available that describes the growth large area single-crystalline wafers of thickness in the range of the diffusion lengths of perovskites, there do available some reports that successful grown such wafers, though only with minute sizes. In the following sections, various approaches and techniques employed to obtain thin wafers of single-crystal organo-lead halide perovskites are critically described and discussed.

6.4.1   TOP-TO-BOTTOM APPROACH

In this method, a large single crystal of perovskite is grown by ITC method and then it is sliced into wafers using advanced cutting tools. Yang et al. synthesized large single crystals of perovskite and then it is sliced into wafers using a diamond saw cutting machine [17] designed for silicon wafer production from LONGI silicon (largest silicon wafer provider). It is found that the wafers show remarkable features like: (1) its trap-state density is a million times smaller than that in the microcrystalline perovskite thin films (MPTF); (2) carrier mobility is 410 times higher than its most popular organic counterpart P3HT; (3) shows extended optical absorption up to 910 nm, while MPTF has absorption only up to 790 nm; (4) the wafer is stable at high temperature up to 270 °C, while MPTF decomposes at 150 °C; (5) when exposed to high humidity, the wafer shows no change for overnight while MPTF decomposes in 5 h; (6) photocurrent response is 250 times higher than its MPTF counterpart. Although, having the above mentioned remarkable features, the thicknesses of the wafers obtained by this method are 190, 380, and 570 μm only. For solar cell applications these thickness are too high and the solar cell fabricated using the wafer produced by diamond saw cutting, showed no promising results. However, it showed good performance in photodetectors than by its polycrystalline counterpart. The other drawbacks inherited with cutting process include high material loss, sludge waste, the need for post polishing processes, and efforts to reduce the surface defects.

6.4.2   SPACE CONFINED SEED INDUCED CRYSTALLIZATION METHOD

In this method, space confined crystal growth is incorporated with seed induced crystal growth to grow large single-crystalline wafers. The confined space is constructed by two glass slides separated by 170-μm-thick glass strip spacers, which defines the thickness of the SC MAPbI$_3$ wafer. In order to get the seed, the assembled glass setup is immersed in precursor solution of MAPbI$_3$ in GBL solution. The solution containing the glass setup is then placed on the hot plate and maintained at 100 °C for 240 h. Small 1–2 mm seeds can be obtained within the confined space. The glass setup containing seed crystal is then transferred in to fresh solution and the growth process is continued. Yanxia et al. used this method [18] and successfully grown a single-crystal wafer of 10 mm size with ~170 μm thickness. Since the thickness of the wafer is too large, it was found only applicable for photodetectors. The as-fabricated photodetector showed a light on/off ratio of 43×10^3, a response time of 770 μs and a linear dynamic range of 119 dB. Moreover, the photocurrent of the device maintained above 80% of its original value after 30 days of storage in the air. These aspects show that grown single crystal is of high quality. A schematic representation of growth process is shown in Figure 6.5.

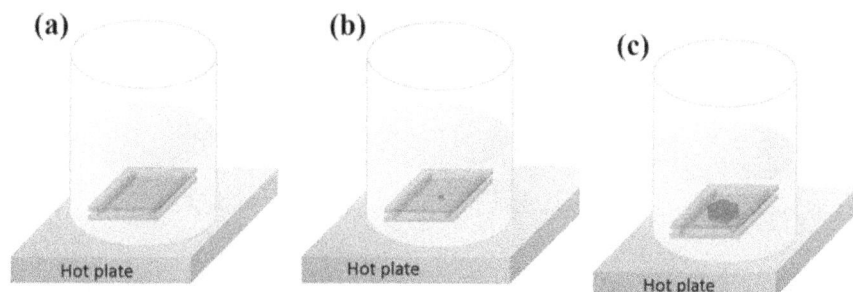

FIGURE 6.5   Schematic representation of single-crystal wafer growth by space confined seed induced crystallization method [18] (reprinted with permission of the Royal Society of Chemistry).

6.4.3   INDUCED PERIPHERAL CRYSTALLIZATION METHOD

In this method, the growth of thin wafer is achieved in a confined space, constructed by two glass plates. And the supersaturation and crystallization are achieved by temperature lowering method. Liu et al. used this method to fabricate a thin membrane of phenyl methylamine lead iodide perovskite of area exceeding 2500 mm^2 with a thickness of 0.6 µm [19]. For the growth of single-crystalline membrane along sideways dimensions while limiting its growth in thickness, the growth rate control is very critical. The crystal growth is directly controlled by two things; first one is the diffusion of solute to the crystallite surface, which is controlled by the solution temperature, second one is the solute deposition to the crystallite, which is difficult to control. In order to maintain the crystal growth process under diffusion-controlled region Liu et al. used relatively low temperature for the crystal growth since, diffusion rate changes [20] exponentially with temperature. On the other hand, the key challenge is to grow the membrane just over the solubility curve or slightly oversaturated. When it is well supersaturated [21] small crystallites forms immediately with lot of defects. When it goes to unsaturated region, the crystal formed may dissolve back to the solution. In the light above mentioned ideas, Liu group used a solution of 2.12 M concentration and the solution is preheated to a temperature of 80 °C to avoid local temperature change and associated crystallization. First, the growth solution is dropped onto a glass slide, which is preheated to 80 °C and the second glass slide is placed above that. Then the temperature is reduced to 30 °C at a rate of 1 °C h^{-1}. When the temperature is lowered below 75 °C, the solution became supersaturated, which may lead to crystal formation. The different stages of crystal growth by this method are shown in Figure 6.6. Using the as-synthesized perovskite membrane, the authors were able to fabricate flexible photosensors of a very high external quantum efficiency of 26,530% , responsivity of 98.17 A W^{-1} and directivity as much as 1.62×10^{15} cm Hz W^{-1}(jones).

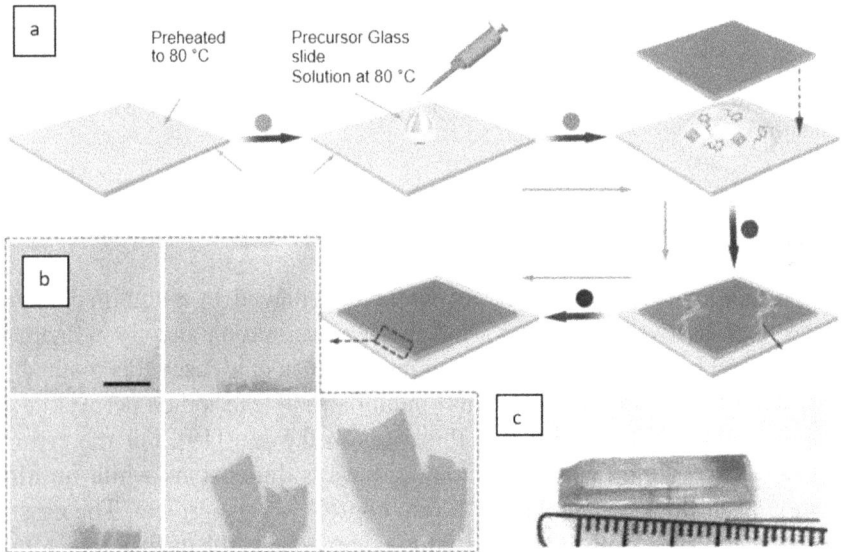

FIGURE 6.6 (a) Schematic representation of induced peripheral crystallization method, (b) photos of single-crystal membrane at different stages, (c) photo of single crystal $_2$PbI$_2$ (reproduced from Ref. [19] Copyright © 2018, Springer Nature).

6.4.4 CAVITATION-TRIGGERED ASYMMETRICAL CRYSTALLIZATION STRATEGY (CTAC)

Bakr's group designed a CTAC strategy [22] by incorporating ultrasound nucleation in antisolvent vapor assisted crystallization method to grow thin single crystals of thickness ranging from one up to several tens of micrometers with lateral dimensions in millimeters. Ultrasound can be used to promote nucleation [23] under small supersaturation levels without using seed crystals. In CTAC method, an experimental setup similar to the setup that is discussed earlier in the AVAC method is used for crystal growth. A perovskite equimolar solution of MABr and PbBr$_2$ in N, N-DMF was loaded to crystallizing dish in which substrates are placed. And the crystallizing dish is covered with aluminum foil having a small hole 0.5 mm diameter, through which the antisolvent is allowed to diffuse into the perovskite solution. Then this crystallizing dish is placed in a larger dish containing DCM as antisolvent. To grow the single-crystalline films, the whole setup was transferred to an ultrasonic bath and when the growth solution reached a low supersaturation a short ultrasonic pulse ($\leq$1 s) was triggered. Due to the large difference in solubility of MAI and PbI$_2$ [22] and possibly the intrinsic anisotropic growth of tetragonal crystal the growth of lead iodide single-crystal films through this method was more challenging. Bakr's group was able to fabricate a solar cell with a PCE of 6.53% using the wafer grown as above with device architecture of FTO/TiO$_2$/perovskite single crystal/Au, and a PCE of 5.49% using the inverted architecture (ITO/perovskite/Au). A variety of solar cells with a varying thickness from 1 to 60 μm and the maximum efficiency is fabricated with 1- and 4-μm-thick wafers. There is large decrease in efficiency as the thickness is increased above 10 μm and when the thickness is 60 μm the efficiency is 0.65% [22]. It can be concluded that for construction of high-efficiency photovoltaics with single-crystalline perovskite, the thickness of the single-crystal wafer should be below 10 μm. Here, the solar cells are fabricated with very basic architecture and without hole transport layers and buffer layers. Further improvements are possible in this method by the addition of suitable layers and surface passivation techniques.

6.4.5 SPACE-CONFINED CRYSTAL GROWTH

In space confined crystal growth method, thin wafers of perovskite are grown by ITC or temperature lowering method in a confined space. Yang et al. successfully grown 150 μm thick wafers of methylammonium lead iodide single crystals using a dynamic flow micro-reactor system. A confined space is constructed using two glass plates, aligned parallel and separated by two spacers. The single-crystalline wafer thickness is defined by the spacer and the wafer shape is delineated by the slit channel. The dynamic flow of the precursor solution through the channel is achieved by a peristaltic pump to warrant the crystal growth (Figure 6.7). Here GBL is used as the solvent and for precursor solution preparation. In this method, also the thickness of the obtained wafers is above 100 μm, and obviously not suitable for solar cell fabrication.

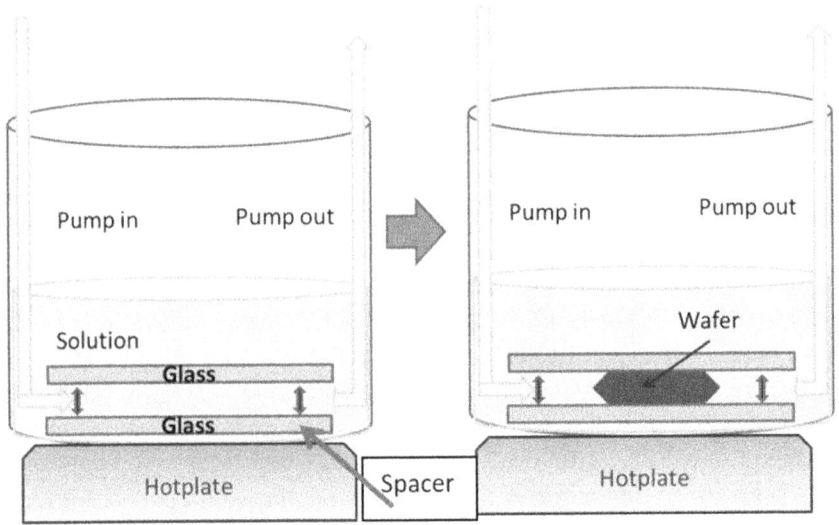

FIGURE 6.7   Space-confined crystal growth.

6.4.6   SPACE LIMITED INVERSE TEMPERATURE CRYSTAL GROWTH (SLITC)

Kuang et al. developed a method called SLITC [24, 25] to grow large wafers of MAPbBr$_3$ on FTO glass. A confined space is constructed by two FTO-glass plates and a thin u-shaped PTFE spacer. The thicknesses of the U-shaped PTFE thin sheets were defined to 0.1, 0.2, 0.4, and 0.8 mm, which determine the thickness of resulting perovskite crystal films. A confined space is constructed by sandwiching a U-shape thin PTFE in between a FTO glass and a PTFE board. The MAPbBr$_3$ precursor solution was inserted into the limited space and a few seconds vacuum pumping was conducted to eliminate the bubbles. The sandwiched setup is heated at the bottom of the glass sheets. For the continuous and dense growth of the thin crystal, a decreased temperature gradient was applied on the module with the aid of aluminum block to conduct heat from a hot plate to the bottom of FTO glass. Around 120 cm^2 bromide perovskite thin single crystals with a thickness range of 0.1–0.8 mm were grown by this setup, which is used for the fabrication of narrow band photodetectors. Although the wafer was of large area, the thickness is not useful for solar cell applications.

In order to achieve single-crystal wafer with lower thickness Kuang's group modified the above mentioned SLITC method and obtained a single crystal of well-regulated thickness of 16 μm with a size of 6 × 8 mm [25]. A precursor solution of 1.1 molar MAPbI$_3$ solution is prepared in DMF, then it is injected into the gap created by the PTFE spacer in between the FTO plates as shown in Figures 6.8 and 6.9. Substrates like TiO$_2$ coated FTO glass, ITO glass or other substrates were also be used. The thickness of the PTFE spacer was 20, 35, and 50 μm,. The typical growing time was mentioned as 2 days.

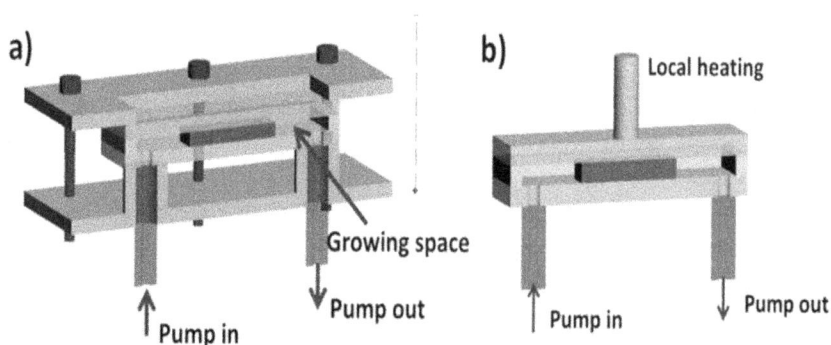

FIGURE 6.8 (a) Schematic diagram of growth module for Space Limited inverse temperature crystal growth (SLITC) and (b) local heating with aluminum block for the laminar MAPbBr$_3$ single-crystal wafer [25] (reprinted with permission of Royal Society of Chemistry).

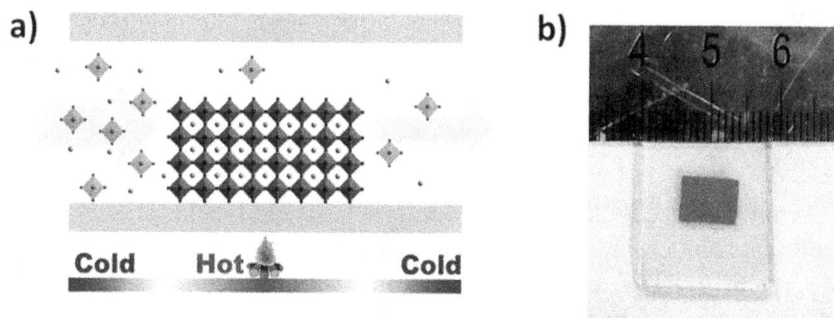

FIGURE 6.9 (a) Schematic representation of growth process and (b) as grown single-crystal wafer of MAPbBr$_3$ [25] (reprinted with permission of Royal Society of Chemistry).

The as grown single crystal by this method was used for the fabrication of solar cell and obtained an intriguing efficiency of 7.11% without any other treatment like passivation. The solar cell structure used here is FTO/TiO$_2$/MAPbBr$_3$/HTM/Au and with this structure the cell held 93% PCE of initial value after storing time of 1000 h. This manifests the excellent stability and potential of single-crystal perovskite for future application.

6.4.7 DIFFUSION FACILITATED SPACE CONFINED INVERSE TEMPERATURE CRYSTAL GROWTH METHOD

In this method, the confined space is assembled by two PTAA coated glass substrates. 0.2 wt.% solution of PTAA in toluene is spin coated over two glass substrates at 4000 rpm for 30 s and annealed at 100 °C for 10 min. One of the PTAA coated glass plate was placed on a hotplate, which is preheated to 110 °C. The perovskite precursor solution was then dropped on to the preheated glass substrate using a syringe. Immediately the other plate was placed on the dropped precursor solution in such a way that perovskite solution is sandwiched in between the PTAA layers. The author was also able to grow 6 mm^2 sized thin single crystals within 48 h using this method.

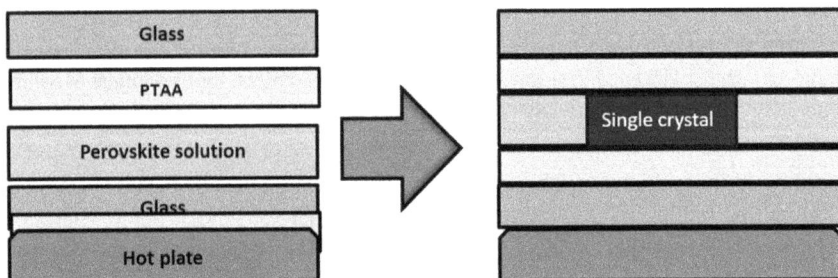

FIGURE 6.10 Schematic representation of thin single crystal growth using diffusion facilitated space confined crystal growth method.

A confined space below 40 μm is constructed by sandwiching the precursor perovskite solution in between two PTAA coated glass substrates as shown in Figure 6.10. PTAA effectively act as an electron blocking layer in PSCs and also, the hydrophobic nature of PTAA reduce the frictional force on the ions by the wetting surface of glass plate. When wetting surfaces like glass or ITO is used, the large surface tension enforces a friction force that slow down the ion diffusion [15]. The perovskite precursor filled growth setup is then placed on a hot plate maintained at 110 °C for 48 h. The growth of single crystals is based on inverse temperature crystallization (ITC) growth method. Since MAPbI$_3$ shows retrograde solubility (decrease in solubility of a solute in a solvent on increase in temperature) in GBL the high temperature of 110 °C induces super saturation leading to seed formation and crystal growth [11, 12, 14]. They used precursor solutions of different concentrations and a 3 mm sized single crystal of 20 μm thickness was obtained by using 1.6 M solution of MAPbI$_3$ (Figure 6.11).

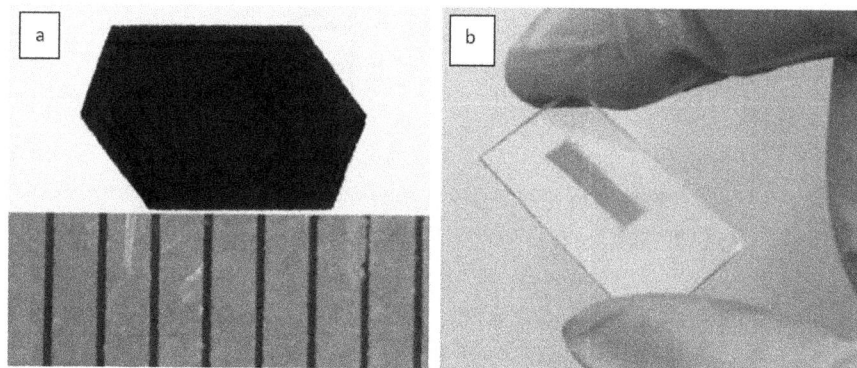

FIGURE 6.11   Thin single crystals grown by diffusion facilitated space confined crystal growth method: (a) methylammonium iodide; (b) methylammonium bromide.

Source: Reprinted with permission from Ref. [15]. http://creativecommons.org/licenses/by/4.0/.

Chen et al. [25] was able to grow thin single crystals of MAPbI$_3$ with thickness of tens of micrometers directly on hole-transport layer (PTAA) coated substrates by diffusion facilitated confined crystal growth method and fabricated a solar cells to obtain a PCE of 17.8% [15] using a cell structure, ITO/PTAA/perovskite single-crystal/PCBM/C60/BCP/Cu. With a different cell structure, ITO/PTAA/perovskite single-crystal/C60/BCP/Cu higher efficiency of 21% [26] with this single crystal of area 6 mm^2 was obtained. The thin single crystal used for the solar cell fabrication had shown a 30 nm broader absorption than the MPTFs. This is attributed to the below bandgap transitions [15, 27] in perovskite single crystals. The performance of the single-crystal PSC is greatly affected by the thickness of the perovskite wafer. Chen's group fabricated solar cells with different wafer thickness and the corresponding solar cell characteristics, which are given in Figure 6.9. When the thickness is increased from 20 to 40 *μm the value of J$_{sc}$* (short circuit current) is reduced from 21 to 15 mA and there is significant loss in efficiency also. Therefor it could be inferred that optimum thickness of the perovskite wafer lies in between 1 and 20 *µm, since the light absorption length is* **1** *µm* (Figure 6.11 and 6.12).

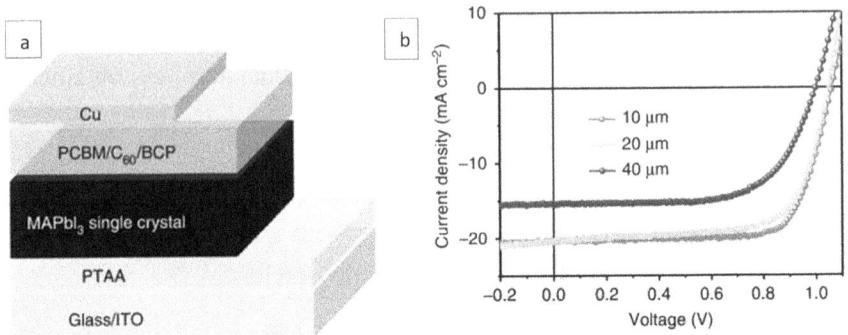

FIGURE 6.12   (a) Structure of single-crystal perovskite solar cell and (b) its characteristics. *Source*: Reprinted with permission from Ref. [15]. http://creativecommons.org/licenses/by/4.0/.

The thin wafers produced using diffusion facilitated confined crystal growth method requires post passivation treatment to reduce the surface traps. These traps are mainly arising from the loss of MAI when single crystal is removed from the hot precursor solution. In order to reduce the large density of surface traps, surface passivation using MAI or PCBM can be used. Chen et al. [15] spin coated a thin layer of MAI on the perovskite wafer and then, the efficiency of the best solar cell increased from 16.1 to 17.8%. The other drawback of this method is that the obtained thin wafers are of very small area- in millimeters only. If organo-lead halide perovskite wafers could be synthesize in large areas with about 20 µm thickness it will have disruptive effect in solar cell technology.

6.5   CONCLUSION

One major success secret of the single-crystalline silicon solar cell is the availability of well matured technology for growing ingots of single crystals and cutting out wafers for direct use in solar cells. However, as the charge carrier diffusion length in perovskite crystals is much smaller, cutting out wafers from bulk crystals won't be feasible. Once developed, perovskite single-crystalline wafer growth technology will be disruptive enough to revolutionize solar energy harvesting. While the technology for growing bulk single crystals of perovskites is well matured, efforts to grow thin single-crystalline perovskites in few tens of micrometer thickness (well within the diffusion length of charge carriers) are yet to succeed in large scales. Methods like SLITC and diffusion facilitated space confined crystal growth are so far the best methods to obtain single-crystal wafer of thickness below 20 μm. But the problem with these methods is the extremely small area of the resulting wafers. This gives ample scope for future investigations for large area single-crystalline wafers of thicknesses around 20 **μm**.

ACKNOWLEDGMENT

The author (P. Predeep) acknowledges the support from ANERT in carrying out this work.

KEYWORDS

- **hybrid perovskites**
- **nanothin single-crystalline wafers**
- **space confined seed induced crystallization**
- **thickness-dependent absorption**
- **below bandgap absorption**

REFERENCES

1. Sum, T.C., Mathews, N., Advancements in perovskite solar cells: photophysics behind the photovoltaics, Energy Environ. Sci. 7 (2014) 2518–2534. doi:10.1039/C4EE00673A.
2. Akihiro Kojima, T.M., Kenjiro Teshima, Yasuo Shirai, Organometal halide perovskites as visible- light sensitizers for photovoltaic cells, J Am Chem Soc. 131 (2009) 6050–6051. https://doi.org/10.1021/ja809598r.
3. Yang, S., Fu, W., Zhang, Z., Chen, H., Li, C.-Z., Recent advances in perovskite solar cells: efficiency, stability and lead-free perovskite, J. Mater. Chem. A. 5 (2017) 11462–11482. doi:10.1039/C7TA00366H.
4. Ono, L.K., Leyden, M.R., Wang, S., Qi, Y., Organometal halide perovskite thin films and solar cells by vapor deposition, J. Mater. Chem. A. 4 (2016) 6693–6713. doi:10.1039/C5TA08963H.
5. Nie, W., Tsai, H., Asadpour, R., Neukirch, A.J., Gupta, G., Crochet, J.J., Chhowalla, M., Tretiak, S., Alam, M.A., Wang, H., High-efficiency solution-processed perovskite solar

cells with millimeter-scale grains, Science. 347 (2015) 522–525. doi:10.1126/science. aaa0472.

6. Burschka, J., Pellet, N., Moon, S.J., Humphry-Baker, R., Gao, P., Nazeeruddin, M.K., Grätzel, M., Sequential deposition as a route to high-performance perovskite-sensitized solar cells, Nature. 499 (2013) 316–319. doi:10.1038/nature12340.

7. Fan, P., Gu, D., Liang, G.-X., Luo, J.-T., Chen, J.-L., Zheng, Z.-H., Zhang, D.-P., High-performance perovskite $CH_3NH_3PbI_3$ thin films for solar cells prepared by single-source physical vapour deposition, Sci. Rep. 6 (2016) 29910. doi:10.1038/srep29910.

8. Chen, Q., Zhou, H., Hong, Z., Luo, S., Duan, H.S., Wang, H.H., Liu, Y., Li, G., Yang, Y., Planar heterojunction perovskite solar cells via vapor-assisted solution process, J. Am. Chem. Soc. 136 (2014) 622–625. doi:10.1021/ja411509g.

9. Shi, D., Adinolfi, V., Comin, R., Yuan, M., Alarousu, E., Buin, A., Chen, Y., Hoogland, S., Rothenberger, A., Katsiev, K., Losovyj, Y., Zhang, X., Dowben, P.A., Mohammed, O.F., Sargent, E.H., Bakr, O.M., Low trap-state density and long carrier diffusion in organolead trihalide perovskite single crystals, Science (80-). 347 (2015) 519–522. doi:10.1126/science.aaa2725.

10. Benli, A.D., Perovskite Solar Cells (Review Article) Department of Metallurgical and Materials Engineering, METU, 06800 Ankara, Turkey ABSTRACT, 2 (n.d.) 40–43.

11. Maculan, G., Sheikh, A.D., Abdelhady, A.L., Saidaminov, M.I., Haque, M.A., Murali, B., Alarousu, E., Mohammed, O.F., Wu, T., Bakr, O.M., $CH_3NH_3PbCl_3$ single crystals: inverse temperature crystallization and visible-blind UV-photodetector, J. Phys. Chem. Lett. 6 (2015) 3781–3786. doi:10.1021/acs.jpclett.5b01666.

12. Saidaminov, M.I., Abdelhady, A.L., Maculan, G., Bakr, O.M., Retrograde solubility of formamidinium and methylammonium lead halide perovskites enabling rapid single crystal growth, Chem. Commun. 51 (2015) 17658–17661. doi:10.1039/C5CC06916E.

13. Liu, Y., Yang, Z., Cui, D., Ren, X., Sun, J., Liu, X., Zhang, J., Wei, Q., Fan, H., Yu, F., Zhang, X., Zhao, C., Liu, S.F., Two-inch-sized perovskite $CH_3NH_3PbX_3$ (X = Cl, Br, I) crystals: growth and characterization, Adv. Mater. 27 (2015) 5176–5183. doi:10.1002/ adma.201502597.

14. Saidaminov, M.I., Abdelhady, A.L., Murali, B., Alarousu, E., Burlakov, V.M., Peng, W., Dursun, I., Wang, L., He, Y., Maculan, G., Goriely, A., Wu, T., Mohammed, O.F., Bakr, O.M., High-quality bulk hybrid perovskite single crystals within minutes by inverse temperature crystallization, Nat. Commun. 6 (2015) 7586. doi:10.1038/ncomms8586.

15. Chen, Z., Dong, Q., Liu, Y., Bao, C., Xiao, X., Bai, Y., Deng, Y., Huang, J., Fang, Y., Lin, Y., Tang, S., Wang, Q., Thin single crystal perovskite solar cells to harvest below-bandgap light absorption, Nat. Commun. 8 (2017) 1–7. doi:10.1038/s41467-017-02039-5.

16. De Wolf, S., Holovsky, J., Moon, S.J., Löper, P., Niesen, B., Ledinsky, M., Haug, F.J., Yum, J.H., Ballif, C., Organometallic halide perovskites: Sharp optical absorption edge and its relation to photovoltaic performance, J. Phys. Chem. Lett. 5 (2014) 1035–1039. doi:10.1021/jz500279b.

17. Liu, Y., Ren, X., Zhang, J., Yang, Z., Yang, D., Yu, F., Sun, J., Zhao, C., Yao, Z., Wang, B., Wei, Q., Xiao, F., Fan, H., Deng, H., Deng, L., Liu, S.F., 120 mm Single-crystalline perovskite and wafers: towards viable application, Sci. China Chem. 1367 (2017) 1367–1376. doi:10.1007/s11426-017-9081-3.

18. Gao, J., Liang, Q., Li, G., Ji, T., Liu, Y., Fan, M., Hao, Y., Liu, S.F., Wu, Y., Cui, Y., Single-crystalline lead halide perovskite wafers for high performance photodetectors, J. Mater. Chem. C. 7 (2019) 8357–8363. doi:10.1039/c9tc01309a.

19. Liu, Y., Zhang, Y., Yang, Z., Ye, H., Feng, J., Xu, Z., Zhang, X., Munir, R., Liu, J., Zuo, P., Li, Q., Hu, M., Meng, L., Wang, K., Smilgies, D.M., Zhao, G., Xu, H., Yang, Z., Amassian, A., Li, J., Zhao, K., Liu, S.F., Multi-inch single-crystalline perovskite membrane for high-detectivity flexible photosensors, Nat. Commun. 9 (2018) 1–11. doi:10.1038/s41467-018-07440-2.

20. Haber, F., Anhang: Zur Theorie der Reaktionsgeschwindigkeit in heterogenen Systemen, Zeitschrift Für Elektrotechnik Und Elektrochemie. 10 (1904) 156–157. doi:10.1002/bbpc.19040100904.

21. Goleman, D., Boyatzis, R., Mckee, A. J. Chem. Inf. Model. 53 (2019) 1689–1699. doi:10.1017/CBO9781107415324.004.

22. Peng, W., Wang, L., Murali, B., Ho, K.T., Bera, A., Cho, N., Kang, C.F., Burlakov, V.M., Pan, J., Sinatra, L., Ma, C., Xu, W., Shi, D., Alarousu, E., Goriely, A., He, J.H., Mohammed, O.F., Wu, T., Bakr, O.M., Solution-grown monocrystalline hybrid perovskite films for hole-transporter-free solar cells, Adv. Mater. 28 (2016) 3383–3390. doi:10.1002/adma.201506292.

23. Thompson, L.H., Doraiswamy, L.K. Sonochemistry: science and engineering, Ind. Eng. Chem. Res. 38 (1999) 1215–1249. doi:10.1021/ie9804172.

24. Rao, H., Li, W., Chen, B., Kuang, D., Su, C., In Situ Growth of 120 cm 2 CH 3 NH 3 PbBr 3 perovskite crystal film on FTO glass for narrowband-photodetectors, 1602639 (2017). doi:10.1002/adma.201602639.

25. Rao, H.S., Chen, B.X., Wang, X.D., Bin Kuang, D., Su, C.Y., A micron-scale laminar MAPbBr3 single crystal for an efficient and stable perovskite solar cell, Chem. Commun. 53 (2017) 5163–5166. doi:10.1039/c7cc02447a.

26. Chen, Z., Turedi, B., Alsalloum, A.Y., Yang, C., Zheng, X., Gereige, I., Alsaggaf, A., Mohammed, O.F., Bakr, O.M., Single-crystal MAPbI 3 perovskite solar cells exceeding 21% power conversion efficiency, (2019) 1258–1259. doi:10.1021/acsenergylett.9b00847.

27. Mekkat, P., Predeep, P., Hybrid perovskite single crystal with extended absorption edge and environmental stability: towards a simple and easy synthesis procedure, Mater. Chem. Phys. 239 (2020). doi:10.1016/j.matchemphys.2019.122084.

CHAPTER 7

CREEP STRESSES AND STRAINS IN CERAMIC DISCS EXHIBITING TRANSVERSELY ISOTROPIC NANO- AND MACRO-STRUCTURAL SYMMETRY SUBJECTED TO CENTRIFUGAL FORCES

SHIVDEV SHAHI,[1*] S. B. SINGH,[2] A. K. HAGHI[3] and S. SARANYA[4]

[1]UIS, Chandigarh University, Gharuan, India

[2]Department of Mathematics, Punjabi University, Patiala, India

[3]Canadian Research and Development Centre of Science and Cultures, Montreal, Canada

[4]Department of Mathematics, Mannar Tirumalai Naicker College, Madurai, India

[*]Corresponding author. shivdevshahi93@gmail.com

ABSTRACT

Predetermination of creep stresses and strains in annular ceramic discs experiencing high centrifugal forces is of much significance in the theory of structural components. In this chapter, transition theory has been incorporated to obtain these stresses and strains in ceramic discs, which exhibit transversely isotropic macro-structural symmetry and having a bore at the center on which it rotates. Yield criterions from the classical theory have not been assumed for the analysis. The creep transition stresses are obtained by transition theory given by B.R. Seth. The analytic solution is applied on two types of ceramics, that is lead zirconate titanate (PZT-5H) and barium titanate, molded as annular discs. The material constants have been taken from available literature. The results of analytic solution are plotted graphically. It is observed that the centrifugal forces increase the magnitude of radial and circumferential stresses at the internal surface of discs. Strains are maximum at internal surface and diminish toward the outer surface. The rise in strains proportional to increasing angular speed infers to the fact that the disc will tend to fracture at the bore adjoining the inclusion when subjected to higher centrifugal forces.

7.1   INTRODUCTION

Ceramic discs are characterized by chemical inertness, high melting point, low electrical, and thermal conductivity. Traditional ceramics included insulating materials, glass, abrasives, and enamels. In the present times, as typical examples, thermal and electrical insulations for various structural components are manufactured using ceramics as a base. Ceramic discs are presently being extensively used in the manufacture of capacitors, filtration discs, automobile breaking systems, gas turbines, etc. A material is said to exhibit transverse isotropy when the physical properties for that material are alike in a single preferential direction. This particular property provides an edge in the manufacture of thin structural components under state of plane stress. Creep strains of such materials are accompanied by changes in their structure, which are irreversible. Creep analysis of such dynamic structural components is necessary for predetermining their deformation, fracture points, and for stable design. In this chapter, we will examine creep deformation of annular ceramic discs experiencing high centrifugal forces and exhibiting transversely isotropic macro-structural symmetry. Numerical values of elastic stiffness constants for two ceramic materials have been used to calculate the trends of creep stresses and strains. The centrifugal forces are added to the analysis by altering the equation of equilibrium accordingly. Wang and Cao [18] determined the elastic constants of lead zirconate titanate [PZT-5H] using ultrasonic wave propagation. Elastic constants for barium titanate have been given by Royer and Dieulesaint [9].

The analysis for structures made of monolithic materials, composites, functionally graded materials exhibiting the respective isotropy, and anisotropy have been given in most standard textbooks [1–5, 8, 19, 20]. Transition is a naturally occurring phenomenon, which is nonlinear in character. Creep analysis assumed certain ad hoc assumptions like incompressibility and yield criterion, which is insufficient to analyze the transition phase. Seth's transition theory [11] bypasses these assumptions. The generalized principal strain measure was first defined by Seth [12]:

$$e_{ii=} \int_{0}^{A e_{ii}} \left[1+2^A e_{ii}\right]^{\frac{n}{2}-1} d^A e_{ii} = \frac{1}{n}\left[1-\left(1+2^A e_{ii}\right)^{\frac{n}{2}}\right],\cdots(i=1,2,3) \quad (7.1)$$

The theory has been used to solve various problems of transitional stress and strain determination in structures modeled in the form of discs exhibiting transverse isotropy and orthotropy [6, 7, 16, 17].

7.2 GOVERNING EQUATIONS

Ceramic annular disc having a perforation of thickness a and external radius b is modeled. This ceramic disc is experiencing high centrifugal forces in the radial direction. There is no variation in the density and thickness of the ceramic disc. The axial stresses will not be considered in this case due to plane stress conditions. The displacement components are given as

$$u = r(1-\beta); v = 0; w = dz \quad (7.2)$$

where β is a function, depending on $r = \sqrt{x^2 + y^2}$ only.

The generalized strain component given by Seth [12] is represented as

$$e_{rr} = -\left[\left(\beta + r\beta'\right)^n - 1\right], e_{\theta\theta} = -\left[\beta^n - 1\right], e_{zz} = -\left[\left(1-d\right)^n - 1\right] \quad (7.3)$$

$$e_{r\theta} = e_{\theta z} = e_{zr} = 0$$

Considering $n = 1$ as the measure and $\beta = d\beta/dr$.

For materials exhibiting transverse isotropy, following relations of stress-strain have been defined [17]:

$$T_{rr} = C_{11}e_{rr} + C_{13}e_{zz} + \left(C_{11} - 2C_{66}\right)e_{\theta\theta}$$

$$T_{\theta\theta} = C_{11}e_{\theta\theta} + C_{13}e_{zz} + \left(C_{11} - 2C_{66}\right)e_{rr}$$

$$T_{zz} = C_{13}e_{rr} + C_{33}e_{zz} + C_{13} - e_{\theta\theta} = 0$$

$$T_{zr} = T_{\theta z} = T_{r\theta} = 0 \tag{7.4}$$

The strain components are obtained in terms of radial and circumferential stresses from Equations (7.4) and (7.3):

$$e_{rr} = \frac{\partial u}{\partial r} - \frac{1}{2}\left(\frac{\partial u}{\partial r}\right)^2 = \frac{1}{2}\left[1 - (r\beta' + \beta)^2\right] = T_{rr} - \left(\frac{C_{11}C_{33} - C_{13}^2 - 2C_{66}C_{33}}{C_{11}C_{33} - C_{13}^2}\right)T_{\theta\theta}$$

$$e_{\theta\theta} = \frac{u}{r} - \frac{u^2}{2r^2} = \frac{1}{2}\left[1 - \beta^2\right] = \frac{1}{E}\left[T_{\theta\theta} - \left(\frac{C_{11}C_{33} - C_{13}^2 - 2C_{66}C_{33}}{C_{11}C_{33} - C_{13}^2}\right)T_{rr}\right]$$

$$e_{zz} = \frac{\partial w}{\partial z} - \frac{1}{2}\left(\frac{\partial w}{\partial z}\right)^2 = \frac{1}{2}\left[1 - (1-d)^2\right] = \frac{1}{E}\left(\frac{C_{11}C_{33} - C_{13}^2 - 2C_{66}C_{33}}{C_{11}C_{33} - C_{13}^2}\right)\left[T_{rr} - T_{\theta\theta}\right]$$

$$e_{r\theta} = e_{\theta z} = e_{zr} = 0 \tag{7.5}$$

where T_{rr} and $T_{\theta\theta}$ are the principal stresses, ρ is the density and $E = 4C_{66}\left(\frac{C_{11}C_{33} - C_{13}^2 - C_{66}C_{33}}{C_{11}C_{33} - C_{13}^2}\right)$.

Substituting Equations (7.4) into Equations (7.5), we get

$$T_{rr} = \frac{2C_{66}}{n}\left[1 - \beta^n\right] + \frac{A}{n}\left[2 - \beta^n\left[1 + [1 + P]^n\right]\right]$$

$$T_{\theta\theta} = \frac{2C_{66}}{n}\left[1 - \beta^n[1 + P]^n\right] + \frac{A}{n}\left[2 - \beta^n\left[1 + [1 + P]^n\right]\right]$$

$$T_{zz} = T_{r\theta} = 0; T_{\theta z} = 0 \tag{7.6}$$

for $A = C_{11} - \left(C_{13}^2 / C_{33} \right)$.

The required equation of equilibrium for ceramic disc is given as

$$\frac{d}{d_r}\left(rT_{rr} \right) - T_{\theta\theta} + \rho\omega^2 r^2 = 0 \qquad (7.7)$$

The boundary conditions at internal and external radius are given as

At internal radius $r = a$; displacement $u = 0$.
At external radius $r = b$; radial stress $T_{rr} = 0$. $\qquad (7.8)$

From Equations (7.6) and (7.7), we get the following nonlinear differential equation:

$$\left[\frac{\rho\omega^2 r^2}{A} + \beta^n \left[\frac{2C_{66}}{nA} \right] \left[1 + nP - [1+P]^n - P\left[1 + [1+P]^n \right] \right] \right] = \beta^{n+1}[1+P]^{n-1}\frac{dp}{d\beta}. \quad (7.9)$$

where $r\beta' = \beta P$ (P is a function of β and β is a function of r only).
The transition points of β in Equation (7.9) are $P \to -1$ and $P \to \pm\infty$.

7.3 SOLUTION FOR CREEP STRESSES THROUGH DIFFERENCE OF PRINCIPAL STRESSES

The transition function C is obtained through the difference of radial and circumferential stresses [6, 7, 11–17]. The creep stresses are obtained from the transition point $P \to -1$.

$$\mathbb{C} \equiv T_{rr} - T_{\theta\theta} = \frac{2C_{66}\beta^n}{n}\left[1 + [P+1]^n \right] \qquad (7.10)$$

Taking the logarithmic differentiation and substitute the value of $dP/d\beta$ from Equation (7.10) in Equation (7.10), we get

$$\frac{d}{dr}\left(\log \mathbb{C} \right) = \frac{1}{r}\left(-2n - \frac{n\rho\omega^2 r^2}{A\beta^n} + C_T\left(n-1 \right) \right) \qquad (7.11)$$

for $C_T = 2C_{66}/A$,

The asymptotic value of Equation (7.11) at $P \rightarrow -1$ gives

$$\mathbb{C}(\text{principal stress differencer}) = A_1 r^\eta \exp(\xi) \tag{7.12}$$

where $\xi = -\dfrac{n\rho\omega^2 r^{n+2}}{AD^n (n+2)}, \eta = 2n + C_T(n+1)$.

From Equations (7.12) and (7.8), we get

$$T_{rr} = A_2 - A_1 \int r^{\eta-1} \exp(\xi) dr - \frac{\rho\omega^2 r^2}{2} \tag{7.13}$$

where A_1 and A_2 are constants, which depend on the boundary conditions. From boundary condition Equations (7.8) and (7.13), we get

$$A_2 = A_1 \int_{r=b} r^{\eta-1} \exp(\xi) dr - \frac{\rho\omega^2 r^2}{2} \tag{7.14}$$

Substituting Equation (7.14) in Equation (7.13), the radial stresses are given as

$$T_{rr} = A_1 \int_r^b r^{\eta-1} \exp(\xi) dr + \frac{\rho\omega^2 (b^2 - r^2)}{2} \tag{7.15}$$

From Equations (7.12) and (7.15), the circumferential stresses are given as

$$T_{\theta\theta} = A_1 \left[\int_r^b r^{\eta-1} \exp(\xi) dr - r^\eta \exp(\xi) \right] + \frac{\rho\omega^2 (b^2 - r^2)}{2} \tag{7.16}$$

Taking asymptotic value $P \rightarrow -1$ for Equation (7.10)

$$\beta = \left[\left(\frac{(T_{rr} - T_{\theta\theta})n}{2C_{66}} \right) \right]^{\frac{1}{n}} \Rightarrow \beta = \left[\left(\frac{\left[A_1 r^\eta \exp(\xi) \right] n}{2C_{66}} \right) \right]^{\frac{1}{n}}. \tag{7.17}$$

From Equations (7.17) and (7.2), the displacement is given as

$$u = r - r\left[\left(\frac{\left[A_1 r^\eta \exp(\xi)\right]n}{2C_{66}}\right)\right]^{\frac{1}{n}}$$
(7.18)

where $A_1 = \dfrac{2C_{66}}{na^{\eta-1}\exp(\xi_1)}$; where $\xi_1 = \dfrac{n\rho\omega^2 a^{n+2}}{AD^\eta(n+2)}$

The principal stresses and displacement are obtained substituting A_1 in Equations (7.15), (7.16), and (7.18), respectively;

$$T_{rr} = \left(\frac{2C_{66}}{na^\eta \exp(\xi_1)}\right)\int_r^b r^{\eta-1}\exp(\xi)dr + \frac{\rho\omega^2\left(b^2 - r^2\right)}{2}$$
(7.19)

$$T_{\theta\theta} = \left(\frac{2C_{66}}{na^\eta \exp(\xi_1)}\right)\left[\int_r^b r^{\eta-1}\exp(\xi)dr - r^\eta \exp(\xi)\right] + \frac{\rho\omega^2\left(b^2 - r^2\right)}{2}.$$
(7.20)

$$u = r - r\left[\frac{r^\eta \exp(\xi)}{a^\eta \exp(\xi_1)}\right]^{\frac{1}{n}}$$
(7.21)

Conversion of principal creep stresses and displacement into nondimensional components for generalizations $R = r/b$, $R_0 = a/b$, $\Omega^2 = \rho\omega^2 b^2/C_{66}$, $\sigma_r = T_{rr}/C_{66}$, $\sigma_\theta = T_{\theta\theta}/C_{66}$ and $u_1 = u/b$. Equations (7.20), (7.21), and (7.22) are given as

$$\sigma_{rr} = \left(\frac{2}{nR_0^\eta \exp(\xi_3)}\right)\int_R^1 R^{\eta-1}\exp(\xi)dR + \frac{\Omega\left(1 - R^2\right)}{2}.$$
(7.22)

$$\sigma_{\theta\theta} = \left(\frac{2}{nR_0^\eta \exp(\xi_3)}\right)\int_R^1 R^{\eta-1}\exp(\xi_2)dR - R^\eta \exp(\xi_2) + \frac{\Omega^2\left(1 - R^2\right)}{2}$$
(7.23)

$$u_1 = R - R \left[\frac{R^\eta \exp(\xi_2)}{R_0^{\ \eta} \exp(\xi_3)} \right]^{1/n} \tag{7.24}$$

where $\xi_2 = \dfrac{n\Omega^2 R^{n+2}}{2(n+2)} \left(\dfrac{b}{D} \right)^n C_T$ and $\xi_3 = -\dfrac{n\Omega^2 R_0^2}{2(n+2)} \left(\dfrac{a}{D} \right)^n C_T$.

7.4   SOLUTION FOR CREEP STRAINS

The following relation is defined for stresses and strains in a disc:

$$e_{ij} = \left(\frac{A - 2C_{66}}{C_R} \right) \delta_{ij} T - \frac{2(A - C_{66})}{C_R} T_{ij} \tag{7.25}$$

where, $C_R = 4C_{66}(C_{66} - A)$ and $A = C_{66} - \dfrac{C_{13}^2}{C_{13}}$.

From generalized strains, we have

$$e_{\theta\theta} = -\beta^{n-1}\beta \tag{7.26}$$

The Swainger measure defines $n = 1$, substituting it in Equation (7.26) we get the relation

$$\varepsilon_{\theta\theta} = -\beta \tag{7.27}$$

We already have the transition value :

$$\beta = \left[T_{rr} - T_{\theta\theta} \right]^{1/n} \left[n/2C_{66} \right]^{1/n} \tag{7.28}$$

From Equations (7.26), (7.27), and (7.28) we obtain the radial and circumferential strains as

$$\dot{\varepsilon}_{rr} = \left[\frac{\left[1 - \dfrac{C_{13}^2}{C_{11}C_{33}} - 2\dfrac{C_{66}}{C_{11}}\right]\sigma_{rr} - \left[1 - \dfrac{C_{13}^2}{C_{11}C_{33}}\right]\sigma_{\theta\theta}}{4\left[\dfrac{C_{66}}{C_{11}} - 1 + \dfrac{C_{13}^2}{C_{11}C_{33}}\right]} \right] \left[\frac{n}{2}\left(\sigma_{rr} - \sigma_{\theta\theta}\right)\right]^{\frac{1}{n}-1}$$

$$\dot{\varepsilon}_{\theta\theta} = \left[\frac{\left[1 - \dfrac{C_{13}^2}{C_{11}C_{33}} - 2\dfrac{C_{66}}{C_{11}}\right]\sigma_{\theta\theta} - \left[1 - \dfrac{C_{13}^2}{C_{11}C_{33}}\right]\sigma_{rr}}{4\left[\dfrac{C_{66}}{C_{11}} - 1 + \dfrac{C_{13}^2}{C_{11}C_{33}}\right]} \right] \left[\frac{n}{2}\left(\sigma_{rr} - \sigma_{\theta\theta}\right)\right]^{\frac{1}{n}-1} \tag{7.29}$$

Odqvist [21] obtained similar results for principal creep stresses and strains.

7.5 NUMERICAL RESULTS AND DISCUSSION

Wang and Cao [18] determined the elastic constants of lead zirconate titanate [PZT-5H] and characterized it to be a very useful piezoceramic. Elastic constants for Barium Titanate have been given by Royer and Dieulesaint [9]. Graphs are plotted for creep stresses and displacement along the radii ratio $R = r/b$ for the ceramic discs made of lead zirconate titanate [PZT-5H] and Barium Titanate exhibiting transversely isotropic symmetry in Figures 7.1–7.3. The effect of centrifugal forces has been observed by analyzing the model at angular speeds $\Omega^2 = 50$, 100, and 150. The variation in the angular speeds has been shown in proceeding figures. The radial stresses have maximum value at the inner surface of both the rotating disc exhibiting transverse isotropy and circumferential stresses decreases at the inner surface with increase in measure 1/3, 1/5, and 1/7. It is also observed that the centrifugal forces increase the magnitude of radial and circumferential stress at the internal surface. This has been observed for both types of ceramic discs.

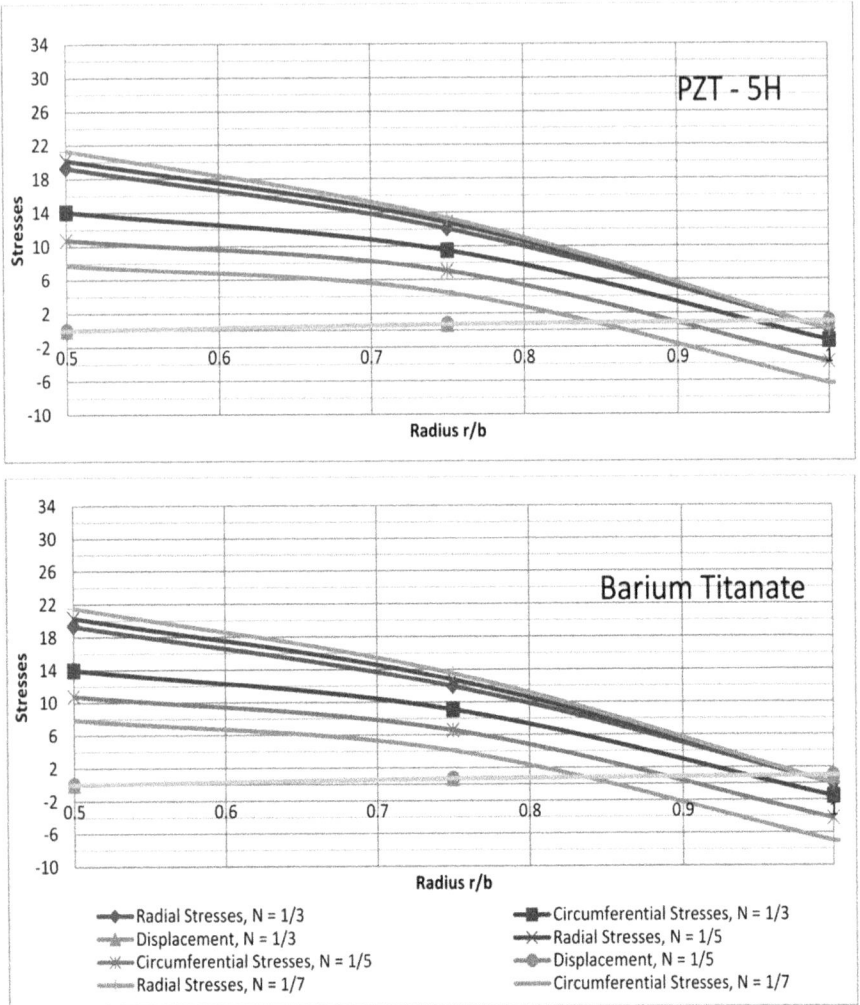

FIGURE 7.1   Principal creep stresses and displacement at $\Omega^2 = 50$.

Similarly, strain rates have been calculated for both types of ceramic discs. The varying rates of strain with respect to changing angular speeds are represented in Figures 7.4–7.6. The radial and circumferential strain rates clearly show that the values of strains are maximum at internal surfaces close to the bore and diminish along the outer surface. Both radial and circumferential strain rates show similar behavior for the ceramic discs. It has also been observed that due to high intensity of stresses and strains close to the inner surface of the discs, the probability of damage to the disc close to the bore is higher.

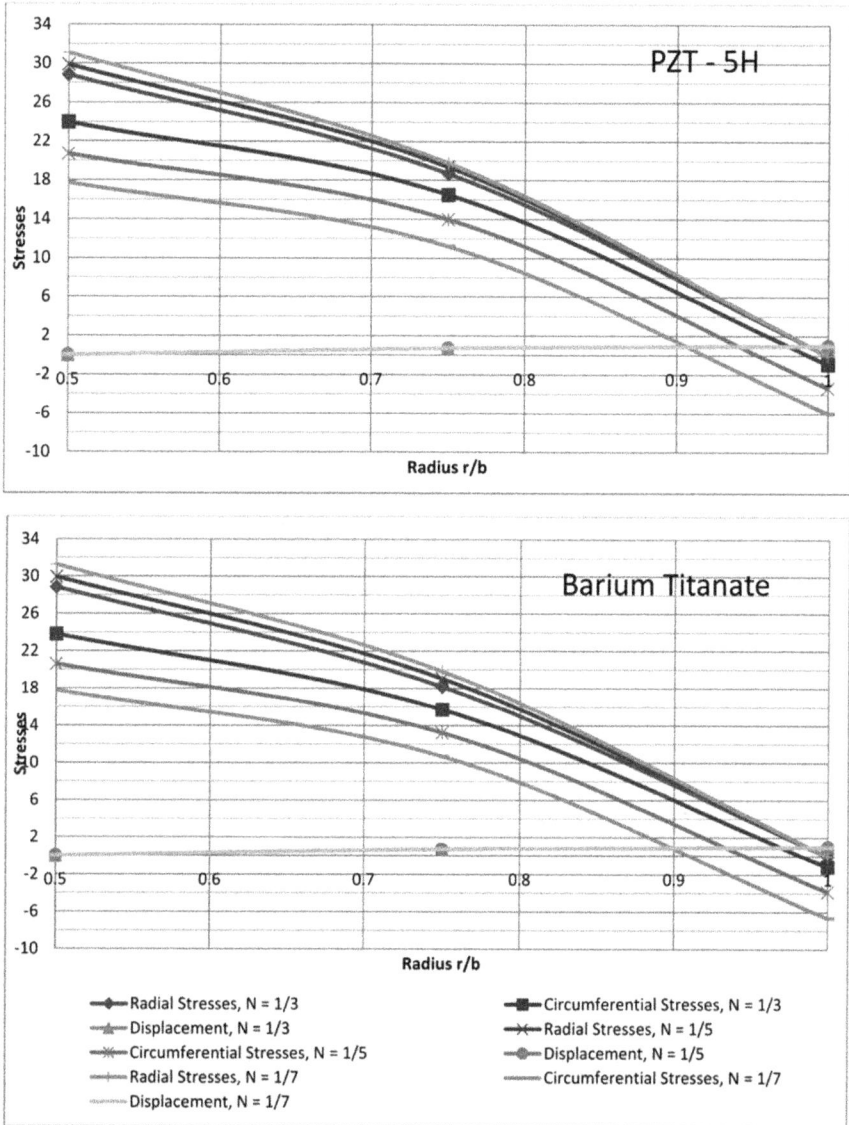

FIGURE 7.2   Principal creep stresses and displacement at $\Omega^2 = 100$.

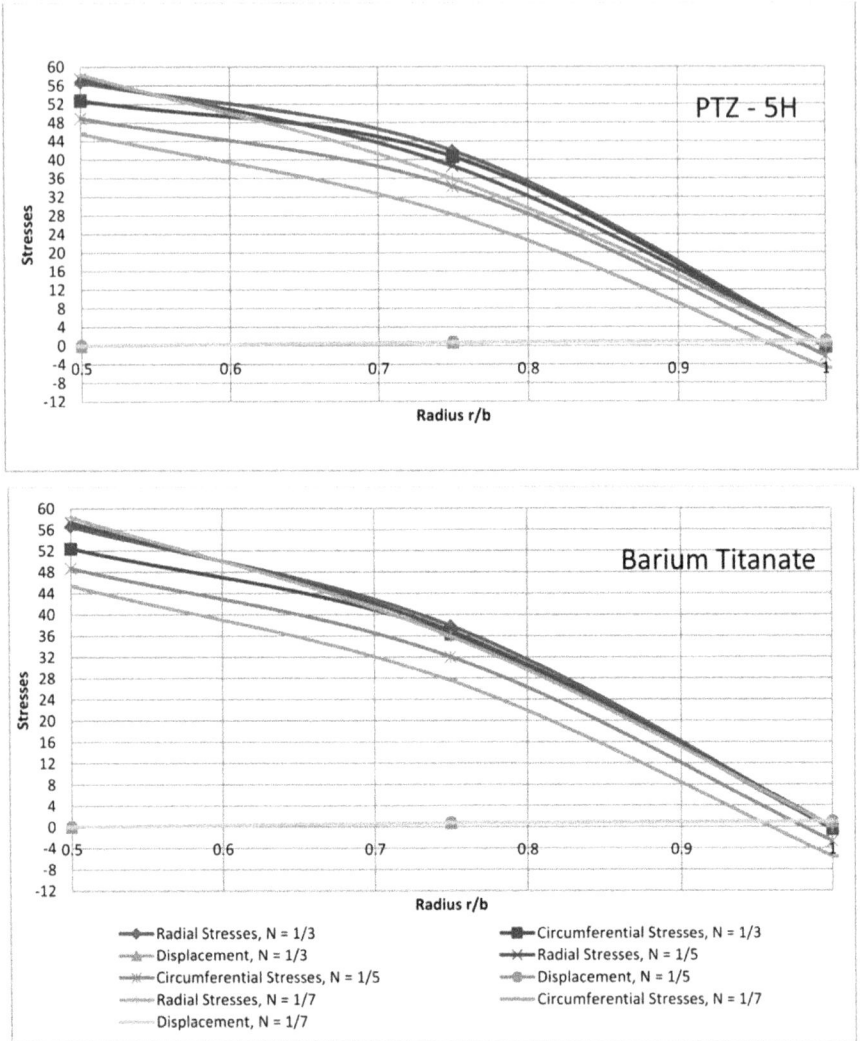

FIGURE 7.3   Principal creep stresses and displacement at $\Omega^2 = 150$.

8.6 CONCLUSIONS

The centrifugal forces increase the values of the principal stresses at the internal surface of the ceramic discs. Strains are maximum at internal surface and diminish toward the outer surface. The rise in strains with increase in centrifugal forces infers to the fact that the disc will tend to facture at the bore adjoining the inclusion when subjected to higher forces. The stresses at various radii ratios have been given in the numerical results to check for the points where the ceramic discs will no longer sustain the centrifugal forces. It is primarily concluded that is mandatory to reinforce such type of ceramic discs with another material in the region close to the bore so as to increase the stress withstanding capacity of the disc thereby leading to a safer design. The paper provides scope for development of functionally graded ceramic discs, reinforced close to the internal radius.

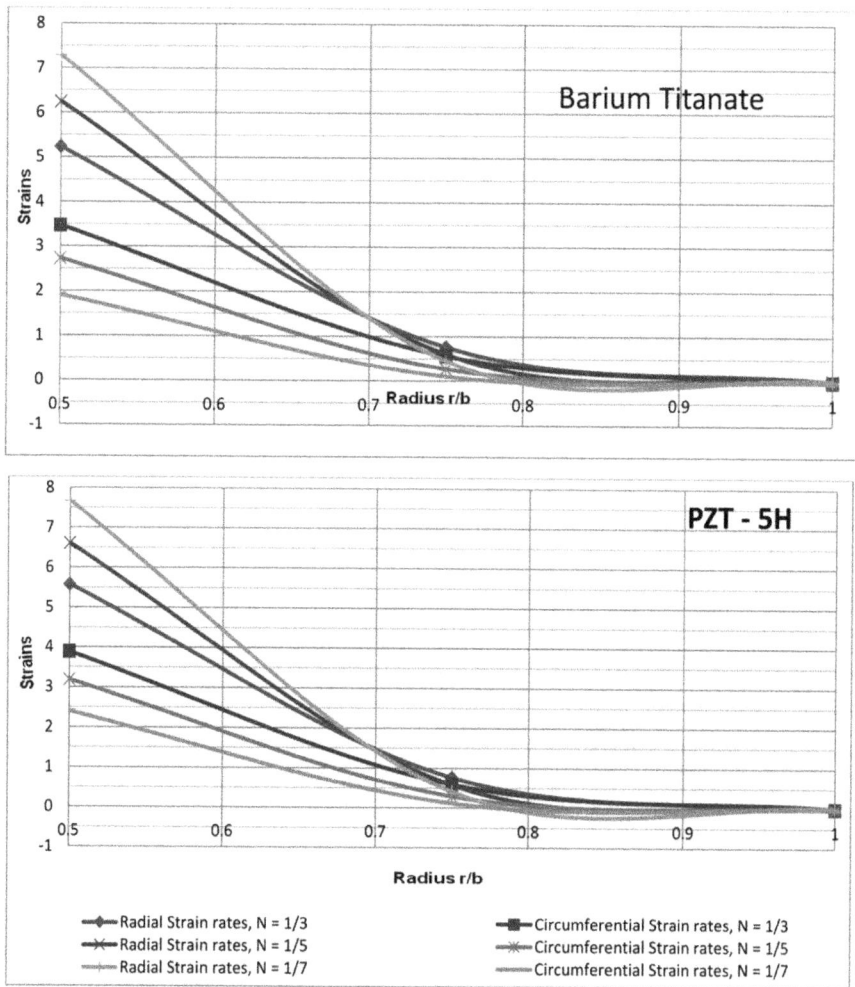

FIGURE 7.4   Creep strain rates at $\Omega^2 = 50$.

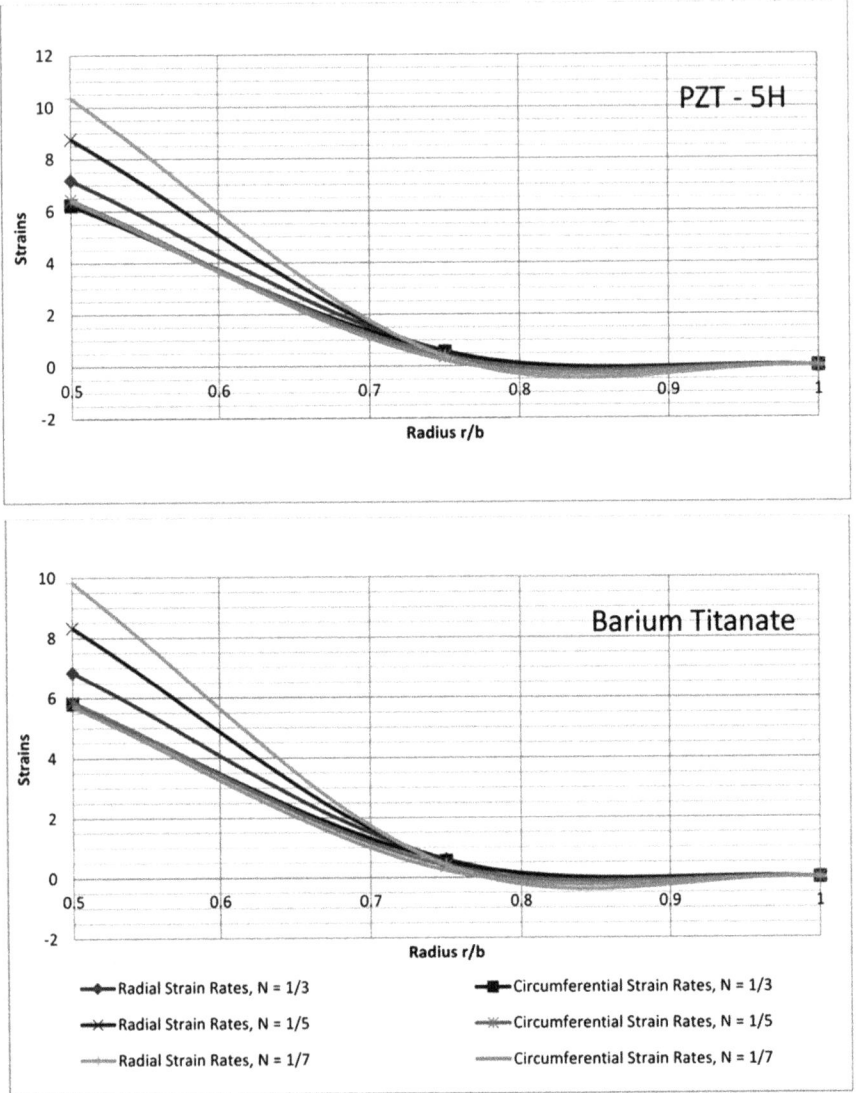

FIGURE 7.5   Creep strain rates at $\Omega^2 = 100$.

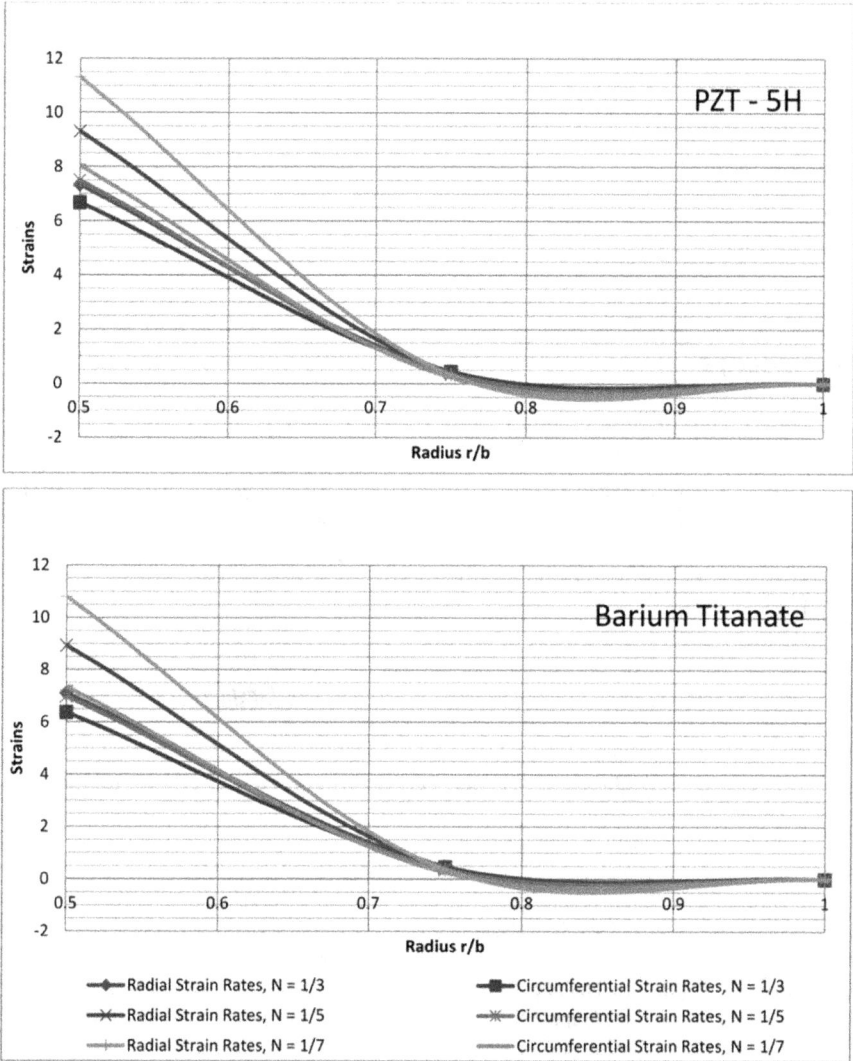

FIGURE 7.6   Creep strain rates at $\Omega^2 = 150$.

KEYWORDS

- **creep**
- **disc**
- **ceramics**
- **stresses**
- **strains**

REFERENCES

1. Sokolinikoff, I.S. *Mathematical Theory of Elasticity*, New York: McGraw-Hill Book Co. (1950).
2. Todhunter, I. and Pearson, K. *History of Plasticity and Strength of Materials*, (1893).
3. Chakrabarty, J., *Theory of Plasticity*, New York: McGraw-Hill Book Coy (1987).
4. Kraus, H., *Creep Analysis*, New York, USA: Wiley (1980) .
5. Odqvist, F.R.N. *Mathematical Theory of Creep and Creep Rupture*, Oxford: Clarendon Press, (1974).
6. Pankaj, T,, Jatinder, K,, Satya, B.S., *Eng. Comp.*, 33, 698–712 (2016).
7. Pankaj, T., Nishi, G., Satya, B.S., *Eng. Comp.*, 34, 1020–1030 (2017).
8. Hetnarski, R. B., Ignaczak, J., *Mathematical Theory of Elasticity*, Taylor and Francis. (2003).
9. Royer, D., Dieulesaint, E., *Elastic Waves in Solids I*, Springer-Verlag, (2000).
10. Timoshenko, S. P., Goodier, J. N. *Theory of Elasticity*, London: McGraw-Hill Book Co., (1951).
11. Gupta, S.K. & Pankaj, Creep, *Defense Sci. J.*, 57(2), 185–195 (2007).
12. Pankaj, S.K.G. *Thermal Sci.*, 11(1), 103–118 (2007).
13. Sanjeev, S., Manoj, S., *Cont. Eng. Sci.* 2(9), 433–440 (2009).
14. Seth, B. R., Nature, 195, 896–897 (1962).
15. Seth, B.R. Int. J. Non-linear Mech 1(1), 35–40 (1966).
16. Shahi, S, Singh, S B, Thakur, P. J. Emerg. Tech. In. Res. 6 (1), 387–395 (2019).
17. Thakur, P., Shahi, S., Gupta, N., Singh, S.B., AIP Conf. Proc., Amer. Inst. of Physics, USA, 1859(1): 020024. (2017).
18. Wang, H, Cao, W. *J. Appl. Phys.* 92(8), 4578–83 (2002).
19. Fung, Y.C. *Foundations of Solid Mechanics*, Engle-Wood Cliffs, N.J. Prentice-Hall. (1965).
20. You, L.H., Zhang, J.J., You, X.Y. *Int. J. Pres. Ves. Piping*, 82, 347–354 (2005).
21. Odqvist, F. K. G., *ASME. J. Appl. Mech*; 43(2), 376 (1976).

CHAPTER 8

ELASTOPLASTIC TRANSITION STRESS BUILDUP IN HOLLOW-SPHERE-SHAPED STRUCTURE OF CARBON NANOTUBE AND GRAPHENE SHEET NANOCOMPOSITE

SHIVDEV SHAHI,[1*] S. B. SINGH,[2], A. K. HAGHI,[3] and S. SARANYA[4]

[1]UIS, Chandigarh University, Gharuan, India

[2]Department of Mathematics, Punjabi University, Patiala, India

[3]Canadian Research and Development Centre of Science and Cultures, Montreal, Canada

[4]Department of Mathematics, Mannar Tirumalai Naicker College, Madurai, India

[*]Corresponding author. shivdevshahi93@gmail.com

ABSTRACT

In this chapter, an analytical solution is being provided for calculation of elastic–plastic transitional stresses in hybrid nanocomposite made of carbon nanotubes (CNT) and graphene sheets (GS) modeled in the form of thin-walled spherical shells, subjected to radial compression at the internal surface. The results are compared with a general use CNT-based nanocomposite. The transition function given by Seth is being used to obtain the pressure at which the initial yielding takes place. The trends for radial and circumferential transitional stresses are obtained using the elastic stiffness constants obtained from young's moduli, shear moduli, and poisson's ratios of the materials at the initial yielding and fully plastic state. The expressions for principal stresses are derived from the transition function considering the material anisotropy.

8.1   INTRODUCTION

Since the discovery of carbon nanotubes in the early 1990s, there has been considerable excitement in the scientific and engineering communities over the remarkable mechanical and physical properties observed for carbon nanotubes. Carbon nanotubes are believed to have elastic moduli of the order of 1 TPa (1000 GPa) with strengths in the range of 30 GPa in addition to exceptionally high electrical and thermal conductivity. These properties combined with recent advances in scaling-up production techniques for carbon nanotubes and the fiber-like structure of nanotubes have generated considerable interest in utilizing carbon nanotubes as nanoscale reinforcement in composites. In composite materials, where we combine distinct phases together for reinforcement, the opportunity exists to design materials for specific properties at various levels of scale. At the microscopic scale, we tailor the local stiffness, strength, toughness, and other properties by controlling the fiber type, loading fraction, and orientation [4, 23]. To realize the potential for carbon nanotubes as reinforcement in composites, we must have a fundamental understanding of how the nanoscale material structure influences the properties of the nanotube as well as how nanotubes interact within a composite [5, 7, 11–14]. Modeling the elastic–plastic behavior of spherical and cylindrical shells and vessels under the influence of internal and external pressure has gained much significance in recent times. Shells made of materials with specific elastic behavior are used in various mechanical components of satellites, submarines, hemispheric dome-shaped antennas, automotives, helmets, etc. Vessels made in shape of shells when used to store fluids at high pressure, experience internal pressure, therefore, it is necessary that the material of the shells should be such that it retains its integrity under such conditions. Modeling of spherical shells made of isotropic materials is available in most standard textbooks [2, 6, 8, 10, 24, 25]. Miller [9] evaluated solutions for stresses and displacements in a thick spherical shell subjected to internal and external pressure loads. You et al. [26] presented a highly precise model to carry out elastic analysis of thick-walled spherical pressure vessels. The researchers in the past have studied the behavior of shells particularly when some assumptions, such as incompressibility of material, creep strain law, and yield condition of Tresca were made. The need for utilization of these specially appointed semi-experimental laws in elastic–plastic transition depends on approach that the transition is a linear phenomenon which is unrealistic. Deformation fields related with irreversible phenomenon, such as elastic–plastic disfigurements, creep relaxation, fatigue, crack, etc., are nonlinear in character. The traditional measures of deformation are not adequate to manage transitions. The concept of generalized strain measures and transition theory given by Seth [15] has been applied to find elastic–plastic stresses in various problems by solving the nonlinear differential equations at the transition points. Thakur [21] successfully analyzed creep transition stresses of a thick isotropic spherical shell by finitesimal deformation under steady state of temperature and internal pressure by using Seth's transition theory. The theory has been used to solve various problems of stress and strain determination in structures modeled in the form of disc [18, 19, 22]. All these problems were based on the recognition of the transition state as a separate state necessitates showing the existence of the used constitutive equation for that state.

In this chapter, an analytical solution is being provided for calculation of elastic–plastic transitional stresses in hybrid nanocomposite made of carbon nanotubes (CNT) and graphene sheets (GS) modeled in the form of thin-walled spherical shells, subjected to radial compression at the internal surface. The results are compared with a general use CNT-based nanocomposite. The transition function given by Seth is being used to obtain the pressure at which the initial yielding takes place. The trends for radial and circumferential stresses are obtained using the elastic stiffness constants obtained from Young's moduli, shear moduli, and poisson's ratios of the materials and the expressions for radial and circumferential stresses derived from the transition function (Figure 8.1).

8.2 GOVERNING EQUATIONS

Consider a spherical shell of constant thickness with internal and external radii a and b, respectively, under internal pressure p.

8.2.1 DISPLACEMENT COORDINATES

The components of displacement in spherical coordinates (r, θ, ϕ) are taken as

$$u = r\left(1 - \beta\right); v = 0; w = 0$$

$$(8.1)$$

where β is position function depending on r.

The generalized components of strain are given by Seth [15, 17] as

$$e_{rr} = \frac{1}{n}\left[1-\left(r\beta'+\beta\right)^{n}\right], e_{\theta\theta} = \frac{1}{n}\left[1-\beta^{n}\right] = e_{\varphi\varphi}; e_{r\theta} = e_{\theta\varphi} = e_{\varphi r} = 0 \quad (8.2)$$

where n is the measure and $\beta' = d\beta/dr$.

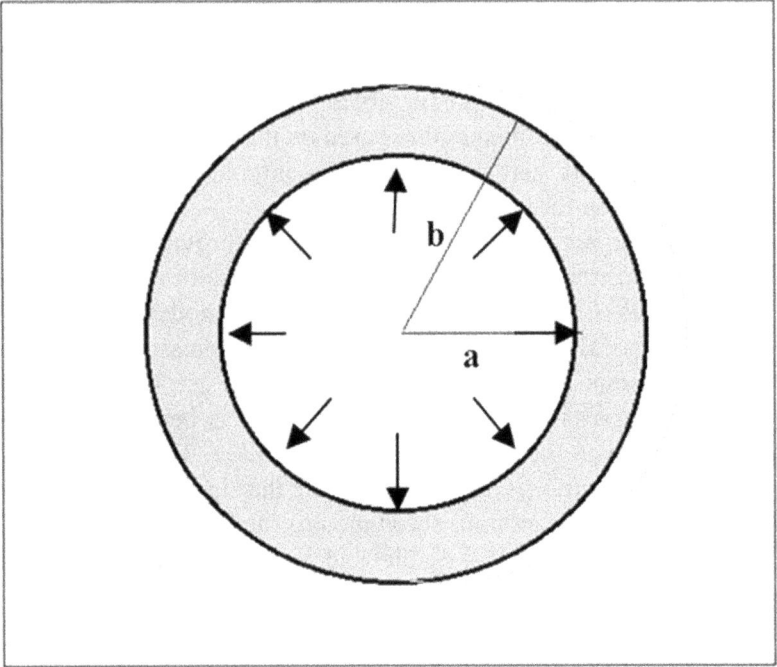

FIGURE 8.1 Cross section of a internally pressurized thin-walled spherical shell.

8.2.2 STRESS–STRAIN RELATION

The stress–strain relations for isotropic material are given by Sokolinokoff [20]

$$T_{ij} = c_{ijkl}e_{kl}, \ (i, j, k, l = 1,2,3),$$

where T_{ij} and e_{kl} are the stress and strain tensors, respectively. These nine equations contain a total of 81 coefficients e_{ijkl}, but not all the coefficients are independent. The symmetry of T_{ij} and e_{kl} reduces the number of independent coefficients to 36. For elastic orthotropic materials that have three mutually orthogonal planes of elastic symmetry, these independent coefficients reduce to 12 and 9 if the coefficients are symmetric. The constitutive equations for orthotropic media are referred from Altenbach et al. [1]:

$$
\begin{bmatrix} T_{11} \\ T_{22} \\ T_{33} \\ T_{23} \\ T_{31} \\ T_{12} \end{bmatrix} = \begin{bmatrix} c_{11} & c_{12} & c_{13} & 0 & 0 & 0 \\ c_{21} & c_{22} & c_{23} & 0 & 0 & 0 \\ c_{31} & c_{32} & c_{33} & 0 & 0 & 0 \\ 0 & 0 & 0 & c_{44} & 0 & 0 \\ 0 & 0 & 0 & 0 & c_{55} & 0 \\ 0 & 0 & 0 & 0 & 0 & c_{66} \end{bmatrix} \begin{bmatrix} e_{11} \\ e_{22} \\ e_{33} \\ e_{23} \\ e_{31} \\ e_{12} \end{bmatrix} \tag{8.3}
$$

Substituting Equation (8.2) in Equation (8.3), we get

$$
T_{rr} = \frac{c_{11}}{n}\left[1-\left(r\beta'+\beta\right)^n\right]+\frac{1}{n}\left(c_{12}+c_{13}\right)\left(1-\beta^n\right); T_{\theta\theta} = \frac{c_{21}}{n}\left[1-\left(r\beta'+\beta\right)^n\right]+\frac{1}{n}\left(c_{22}+c_{23}\right)\left(1-\beta^n\right)
$$

$$
T_{\varphi\varphi} = \frac{c_{31}}{n}\left[1-\left(r\beta'+\beta\right)^n\right]+\frac{1}{n}\left(c_{32}+c_{33}\right)\left(1-\beta^n\right); T_{r\theta} = T_{\theta\varphi} = T_{\varphi r} = 0
$$

$$\tag{8.4}$$

8.2.3 EQUATION OF EQUILIBRIUM

The equations of equilibrium are

$$
\frac{\partial T_{rr}}{\partial r} + \frac{1}{r\sin\theta}\frac{\partial T_{r\varphi}}{\partial\varphi} + \frac{1}{r}\frac{\partial T_{r\theta}}{\partial\theta} + \frac{2T_{rr}-T_{\theta\theta}-T_{\varphi\varphi}+T_{r\theta}\cot\theta}{r} = 0;
$$

$$
\frac{\partial T_{r\theta}}{\partial r} + \frac{1}{r\sin\theta}\frac{\partial T_{\theta\varphi}}{\partial\varphi} + \frac{1}{r}\frac{\partial T_{\theta\theta}}{\partial\theta} + \frac{3T_{r\theta}+\left(T_{\theta\theta}-T_{\varphi\varphi}\right)\cot\theta}{r} = 0; \tag{8.5}
$$

$$
\frac{\partial T_{r\theta}}{\partial r} + \frac{1}{r\sin\theta}\frac{\partial T_{\theta\varphi}}{\partial\varphi} + \frac{1}{r}\frac{\partial T_{\theta\theta}}{\partial\theta} + \frac{3T_{r\theta}+\left(T_{\theta\theta}-T_{\varphi\varphi}\right)\cot\theta}{r} = 0
$$

Substituting Equation (8.4) in Equation (8.5), we see that the equations of equilibrium are all satisfied except:

$$
\frac{\partial T_{rr}}{\partial r} + \frac{2T_{rr}-T_{\theta\theta}-T_{\varphi\varphi}}{r} = 0 \tag{8.6}
$$

$$\frac{T_{\theta\theta} - T_{\varphi\varphi}}{r}\cot\theta = 0 \tag{8.7}$$

From Equation (8.7), the only case of interest is

$$T_{\varphi\varphi} - T_{\theta\theta} = 0 \tag{8.8}$$

Equation (8.8) is satisfied by T_{00} and $T_{\phi\phi}$ as given by Equation (8.2). If $c_{21} = c_{31}$, $c_{22} - c_{33} = c_{32} - c_{23}$, the equation of equilibrium from Equation (8.6) becomes

$$\frac{\partial T_{rr}}{\partial r} + \frac{2(T_{rr} - T_{\theta\theta})}{r} = 0 \tag{8.9}$$

8.2.4 CRITICAL POINTS OR TURNING POINTS

By substituting Equation (8.4) into Equation (8.9), we get a nonlinear differential equation with respect to β:

$$\beta\frac{dP}{d\beta} = \frac{2(c_{11}-c_{21})\{1-\beta^n(P+1)\}}{nc_{11}\beta^nP(P+1)^{n-1}} - \frac{(c_{12}+c_{13})(P+1)^{1-n}}{c_{11}} - (P+1) + \frac{2[c_{12}+c_{13}-(c_{22}+c_{23})](1-\beta^n)}{nc_{11}\beta^nP(P+1)^{n-1}} \tag{8.10}$$

where P is function of β and β is a function of r only.

8.2.5 TRANSITION POINTS

The transition points of β in Equation (8.10) are $P=0$, $P\rightarrow-1$ and $P\rightarrow\pm\infty$.

8.2.6 BOUNDARY CONDITION.

The boundary conditions of the problem are given by

$$r^o = {}^oa; \tau_{rr} = -p$$
$$r^o = {}^ob; \tau_{rr} = 0 \tag{8.11}$$

8.3 PROBLEM SOLUTION

For finding the elastic–plastic stresses, the transition function is taken through the principal stresses at the transition point $P \to \pm\infty$, we define the transition function ζ as:

$$\zeta = 1 - \frac{nT_{rr}}{\left(c_{11} + c_{12} + c_{13}\right)} \cong \frac{1}{\left(c_{11} + c_{12} + c_{13}\right)} \left[\beta^n \left(P+1\right)^n c_{11} + \left(c_{12} + c_{13}\right)\beta^n \right]$$

(8.12)

where ζ be the transition function unction of r only. Taking the logarithmic differentiation of Equation (8.12), with respect to r and using Equation (8.10), we get

$$\frac{d\left(\log\zeta\right)}{dr} = \frac{2}{r\beta^n} \left(\frac{\left[1 - \beta^n\left(P+1\right)^n\right]\left(c_{11} - c_{21}\right) + \left(1 - \beta^n\right)\left[c_{12} + c_{13} - \left(c_{22} + c_{23}\right)\right]}{c_{11}\left(P+1\right)^n + \left(c_{12} + c_{13}\right)} \right)$$

(8.13)

Taking the asymptotic value of Equation (8.13) as $P \to \pm\infty$ and integrating, we get

$$\zeta = A_1 r^{-2K}$$

(8.14)

where A_1 is a constant of integration and $K = c_{11} - c_{21}/c_{11}$. From Equations (8.12) and (8.14), we have

$$T_{rr} = \frac{\left(c_{11} + c_{12} + c_{13}\right)}{n}\left[1 - A_1 r^{-2K}\right]$$

(8.15)

Using boundary condition from Equation (8.11) in Equation (8.15), we get

$$A_1 = b^{2K} \ldots and \ldots p = -\frac{\left(c_{11} + c_{12} + c_{13}\right)}{n}\left[1 - \left(\frac{b}{a}\right)^{2K}\right]$$

(8.16)

Substituting Equation (8.15) in Equation (8.6) and using Equations (8.16) and (8.8), we get

$$T_{\theta\theta} - T_{rr} = \frac{(c_{11} + c_{12} + c_{13})K}{n}\left(\frac{b}{r}\right)^{2K} \qquad (8.17)$$

8.3.1   INITIAL YIELDING

From Equation (8.17), it is seen that $|T_{\theta\theta} - T_{rr}|$ is maximum at the inner surface (i.e., at $r = a$), therefore yielding of the shell will take place at the inner surface of the shell and Equation (8.17) can be written as:

$$\left|T_{\theta\theta} - T_{rr}\right|_{r=a(\text{inner surface})} = \left|\frac{(c_{11} + c_{12} + c_{13})K}{n}\left(\frac{b}{a}\right)^{2K}\right| \equiv Y\,(\text{Yielding})\,(8.18)$$

Using Equation (8.18) in Equations (8.15)–(8.17), we get the orthotropic transitional stresses as in nondimensional components as

$$\sigma_{rr} = -\frac{P_i\left(1 - R^{-2K}\right)}{\left(1 - R_0^{-2K}\right)}; \sigma_{\theta\theta} = -\frac{P_i\left(1 - R^{-2K}\right)}{\left(1 - R_0^{-2K}\right)} + R^{-2K}; P_i = -K^{-1}R_0^{2K}\left(1 - R_0^{-2K}\right)$$

$$(8.19)$$

where $R = r/b$, $R_0 = a/b$, $\alpha_{rr} = T_{rr}/Y$, $\alpha_{\theta\theta} = T_{\theta\theta}/Y$, and $P_i = p/Y$.

8.3.2   FULLY PLASTIC STATE

For fully plastic case [16], $c_{11} = c_{13} = -c_{12}$, $c_{23} = c_{21} = -c_{22}$ stresses and pressure from Equation (8.19) becomes:

$$\sigma_{rr} = -\frac{P_f\left(1 - R^{-2K_1}\right)}{\left(1 - R_0^{-2K_1}\right)}; \sigma_{\theta\theta} = -\frac{P_f\left(1 - R^{-2K_1}\right)}{\left(1 - R_0^{-2K_1}\right)} + R^{-2K_1}; and.P_f = -K_1^{-1}R_0^{2K_1}\left(1 - R_0^{-2K_1}\right)$$

$$(8.20)$$

where $R = r/b$, $R_0 = a/b$, $\alpha_{rr} = T_{rr}/Y^*$, $\alpha_{\theta\theta} = T_{\theta\theta}/Y^*$, $K_1 = (c_{12}-c_{22})/c_{12}$ and $P_f = p_f/Y^*$.

8.4   RESULTS AND DISCUSSION

The above investigations elaborate the initial yielding and fully plastic state of thin spherical shells subjected to internal pressure. The cases of two shells, that is hybrid nanocomposite made of CNT and GS and a general-purpose CNT nanocomposite are considered. In Figure 8.2, curves are plotted for internal pressure at initial yielding state and radii ratio R_0 = a/b for the thin spherical shells. The curves show how the walls of the shells yield when various ratios of radial distances were considered. It is observed that hybrid nanocomposite shells required higher levels of pressure to start initial yielding as compared to shells made of CNT nanocomposite (Table 8.1).

TABLE 8.1   Elastic Constants for the Nanocomposites

Elastic Constants	CNT Nanocomposite	Hybrid CNT + GS Nanocomposite
	Chen and Liu [3]	
E_2	112.5711	118.1274
	0.3672	0.3682

The radial and circumferential stresses that were calculated for the values of internal pressure at initial yielding are plotted in Figure 8.3(a) and (b) and fully plastic state are plotted in Figure 8.4(a) and (b) along the radii ratio R = r/b. For the analysis of thin spherical pressure vessel, the radii ratio R_0 = a/b is considered in the range $0.75 < r < 1.0$. It is observed that the elastic–plastic limits of hybrid (CNT + GS) nanocomposite are greater than that of CNT nanocomposite. The radial stresses required for initial yielding of the shell were greater for hybrid (CNT + GS) nanocomposite. The stresses have greater magnitudes as compared to CNT nanocomposite close to the inner surface of the shell and the stress concentrations diminish as we approach the outer layer of the shell. Similarly, the circumferential stresses were maximum at the internal surface. A significant difference in the stresses was observed at the external surface in both nanocomposites thereby differentiating the principal stress difference at the external surface for hybrid (CNT + GS) nanocomposite. For the fully plastic state, the numerical results in Figure 8.4(a) and (b) infer to the fact that the radial stresses at fully plastic state were greater for hybrid (CNT + GS) nanocomposite but the results were reversed for the circumferential stresses at the fully plastic state.

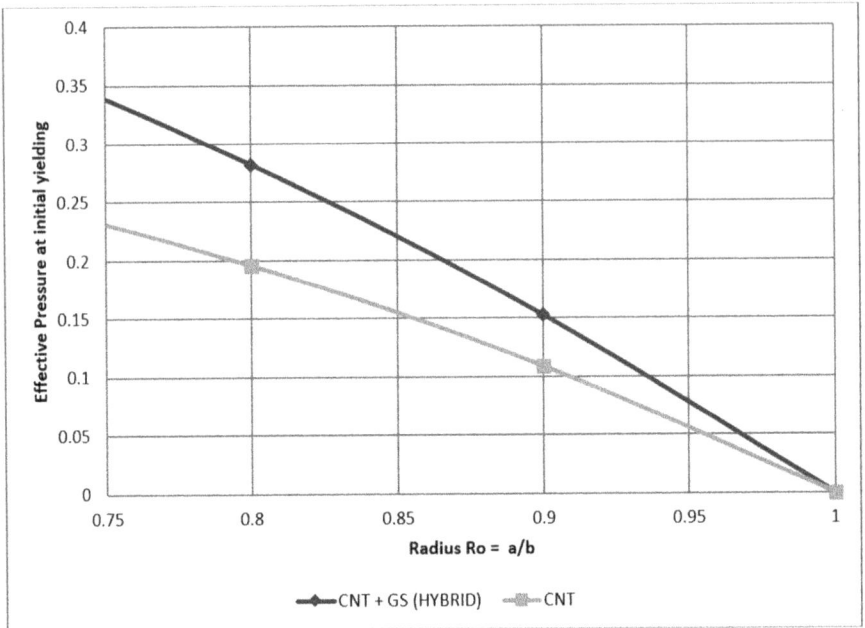

FIGURE 8.2   Effect of pressure in shells along the radius $R_0 = a/b$ at initial yielding.

8.5   CONCLUSION

Elastic–plastic transitional stress concentrations have been determined in nanocomposite spherical shells subjected to internal pressure using Seth's Transition Theory. The pressure required at initial yielding of the shell was greater for the case of hybrid nanocomposite which states better resistance to pressure. The stresses concentrations at fully plastic state in the case of hybrid nanocomposite do not deviate gradually as compared to the CNT nanocomposite. It is sufficient to conclude that Thin spherical shells made of hybrid CNT + GS nanocomposite have a greater radial and circumferential stress-bearing strength and are secure from the design and manufacture perspective.

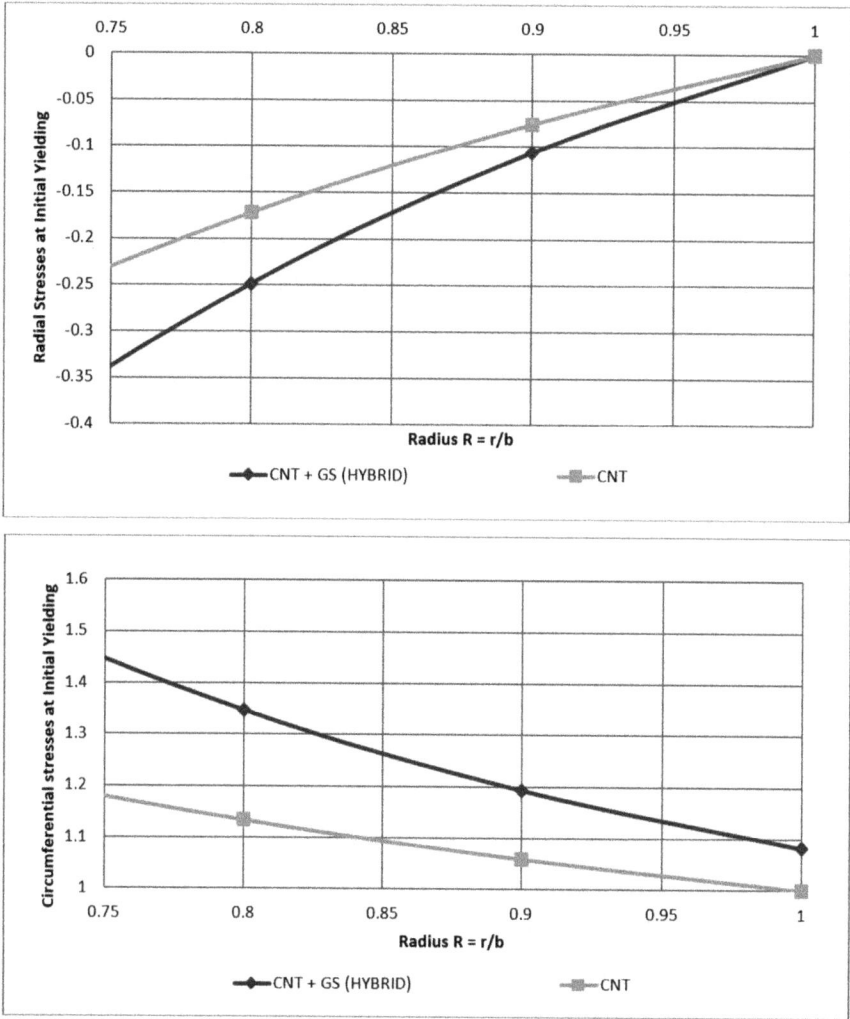

FIGURE 8.3 (a) and (b) Effect of stresses in shells along the radius $R = r/b$ at initial yielding state.

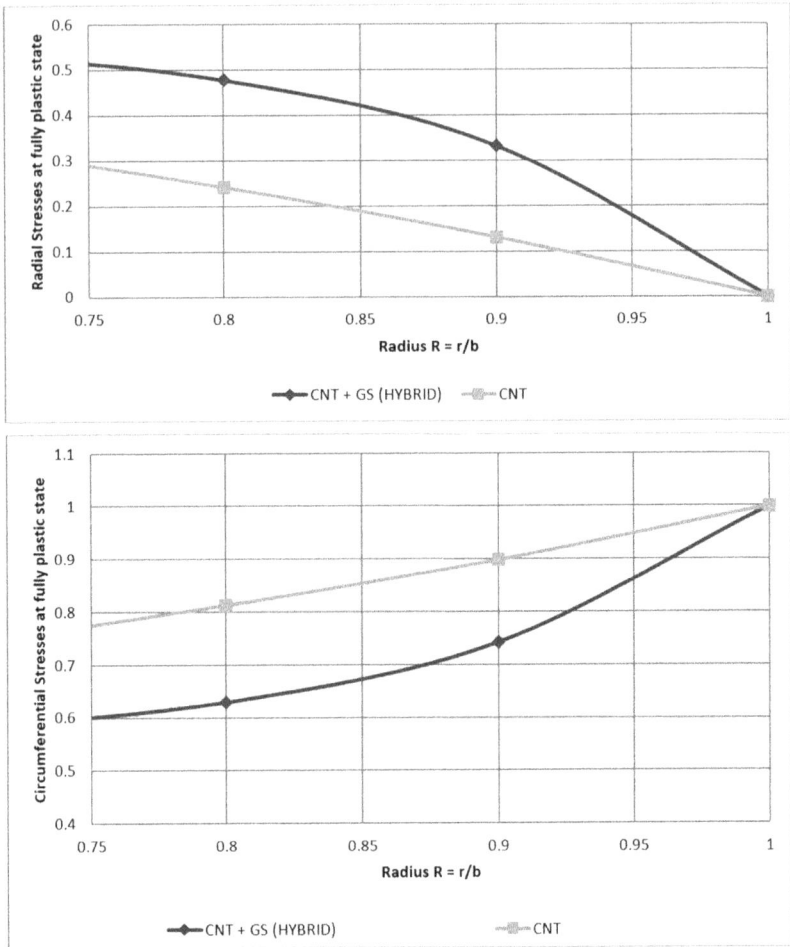

FIGURE 8.4 (a) and (b) Effect of stresses in shells along the radius $R = r/b$ for fully plastic state

KEYWORDS

- transition
- carbon nanotubes
- graphene
- nanocomposite
- pressure vessel

REFERENCES

1.   Altenbach, H., Altenbach, J., Kissing, W. (2004) *Mechanics of Composite Structural Elements*, Springer-Verlag, Berlin.
2.   Boyle, J.T., Spence, J., (1983), *Stress Analysis for Creep*, Butterworths Coy. Ltd. London.
3.   Chen, X. L., Liu, Y. J. (2004) Square representative volume elements for evaluating the effective material properties of carbon nanotube-based composites. *Comput Mater Sci.* 29, 1–11.
4.   Chou, T-W (1992) *Microstructural Design of Fiber Composites*, Cambridge University Press, Cambridge.
5.   Demczyk, B. G., Wang, Y. M., Cumings, J., et al. (2002) Direct mechanical measurement of the tensile strength and elastic modulus of multiwalled carbon nanotubes. *Mater Sci Eng A.* 334, 173–178.
6.   Kraus, H. (1980) *Creep Analysis*, Wiley, New York, USA.
7.   Laurent, C., Flahaut, E., Peigney, A. (2010) The weight and density of carbon nanotubes versus the number of walls and diameter. *Carbon.* 48, 2994–2996.
8.   Lubhan, D., Felger R. P. (1961) *Plasticity and creep of Metals*, Wiley, New York, USA
9.   Miller G.K. (1995), Stresses in a spherical pressure vessel undergoing creep and dimensional changes, *Int J Solids Struc.* 32, 2077–2093
10.   Parkus, H. (1976), *Thermo-Elasticity*, Springer-Verlag, Wien, New York.
11.   Ruoff, R. S., Lorents, D. C. (1995) Mechanical and thermal properties of carbon nanotubes. *Carbon.* 33, 925–930.
12.   Ruoff, R. S., Qian, D., Liu, W. K. (2003) Mechanical properties of carbon nanotubes: theoretical predictions and experimental measurements. *Comptes Rendus Phys.* 4, 993–1008.
13.   Salvetat, J. P., Bonard, J. M., Thomson, N.H., et al. (1999) Mechanical properties of carbon nanotubes. *Appl Phys A Mater Sci Process.* 69, 255–260.
14.   Sears, A., Batra, R. C. (2004) Macroscopic properties of carbon nanotubes from molecular-mechanics simulations. *Phys Rev B.* 69, 235406.
15.   Seth B. R. (1962). Transition theory of elastic-plastic deformation, creep and relaxation, *Nature.* 195, pp. 896–897
16.   Seth, B. R. (1963), Elastic–pastic transition in shells and tubes under pressure. *Z Angew Math Mech.* 43, pp 345–351.
17.   Seth, B. R. (1966) Measure concept in mechanics, *Int J Nonlinear Mech.* 1, 35–40.
18.   Singh, S B, Vakhrushev, A V, Haghi, A K, (2020). *Materials Physics and Chemistry: Applied Mathematics and Chemo-Mechanical Analysis* (1st edn.). Apple Academic Press, Chapter 3 https://doi.org/10.1201/9780367816094
19.   Shahi S, Singh S B, Thakur P, (2019) Modeling creep parameter in rotating discs with rigid shaft exhibiting transversely isotropic and isotropic material behavior, *J Emerg Tech Innov Res.* 6 (1), 387–395.
20.   Sokolinokoff, I.S. (1956), *Mathematical Theory of Elasticity*, 2nd ed.n, McGraw-Hill, New York.
21.   Thakur, P. (2011), Creep transition stresses of a thick isotropic spherical shell by finitesimal deformation under steady-state of temperature and internal pressure, *Thermal Sci.* 15, Suppl. 2, S157–S165

22. Thakur, P., Shahi, S., Gupta, N., Singh, S.B. (2017), Effect of mechanical load and thickness profile on creep in a rotating disc by using Seth's transition theory, AIP Conf. Proc., USA, 1859(1): 020024. doi.org/10.1063/1.4990177

23. Thostenson, E. T., Ren, Z.F., Chou, T.-W. (2001) Advances in the science and technology of carbon nanotubes and their composites: a review, *Compos Sci Technol.* 61, 1899–912,

24. Timoshenko, S. P. and Woinowsky-Krieger, S. (1959). *Theory of Plates and Shells*, 2nd ed. McGraw-Hill, New York.

25. Y.C. Fung, (1965) *Foundations of Solid Mechanics*, Prentice-Hall, Englewood Cliffs, N.J. pp. 13–29.

26. You L.H., Zhang J.J., You X.Y., (2005), Elastic analysis of internally pressurized thick-walled spherical pressure vessels of functionally graded materials, *Int J Pressure Vessels Piping*, 82, 347–354.

CHAPTER 9

DISTURBANCES DUE TO NORMAL HEAT IN A NONLOCAL THERMOELASTIC SOLID WITH TWO TEMPERATURES AND WITHOUT ENERGY DISSIPATION

PARVEEN LATA[1] and SUKHVEER SINGH[2*]

[1]*Department of Basic and Applied Sciences, Punjabi University, Patiala, India*

[2]*Punjabi University APS Neighbourhood Campus, Dehla Seehan, Sangrur, Punjab, India*

Corresponding author. E-mail: sukhveer_17@pbi.ac.in

ABSTRACT

This study is about the disturbances due to normal ramp-type heat in a two-dimensional homogeneous isotropic nonlocal thermoelastic solid with two temperatures in the generalized theory of thermoelasticity. Laplace and Fourier transforms have been used to solve the problem. The analytical expressions of stress components, displacement components, and temperature change have been obtained in the transformed domain and numerical inversion technique has been applied to obtain the results in the physical domain. Numerical-simulated results are depicted graphically to show the effect of two temperatures and nonlocal parameter on the components of displacements, stresses, and conductive temperature. Some special cases have also been deduced from the present investigation.

9.1   INTRODUCTION

The theory of elasticity is concerned about the elastic properties of a material, that is, the regaining of original shape once the deformation forces are removed. Thermoelasticity is the study of the deformations of a material arising due to both mechanical and thermal causes. The nonlocal theory of thermoelasticity is an important theory which states that any physical quantity at a point is not just a function of the values of independent constitutive variables at that point only but a function of their values over the whole body. The theory of thermoelasticity with two temperatures is also one of very important theories in which the heat conduction is considered to be dependent upon two distinct temperatures, namely the conductive temperature and the thermodynamic temperature.

   The concept of nonlocality in elastic materials was developed by Edelen and Laws [6] and Edelen et al. [5]. They proposed a theory of nonlocal interactions and discussed the consequences of global postulate of energy balance to obtain the constitutive equations for the nonlinear theory. Eringen and Edelen [8] took it further and proposed the nonlocal theory of elasticity. It stated that the stress at any single point is affected not just due to the strain at that point but due to the other points of the body too. A unified approach to the development of the basic field equations for nonlocal continuum field theories was given by Eringen [7]. Khurana and Tomar [11] investigated the propagation of plane longitudinal waves through a nonlocal micropolar elastic medium and proved the existence of four dispersive waves and two coupled transverse waves. Singh et al. [25] reviewed the recent developments and theories in nonlocal theory of elastic and thermoelastic materials. A generalized theory of thermoelasticity was given by Lord and Shulman [19], which is known as LS theory.

   Singh et al. [24] studied the propagation of plane harmonic waves and derived the governing relations in nonlocal elastic solid with voids. Marin et al. [20] discussed various results and problems for elastic dipolar materials. Othman and Marin [21] studied the effect of thermal loading due to laser pulse on thermoelastic porous medium under G-N theory. Kaur et al. [10] derived dispersion relation and investigated the propagation of Rayleigh-type surface wave in nonlocal elastic solid. Bachher and Sarkar [1] postulated a new nonlocal theory of thermoelasticity, which is based on Eringen's nonlocal elasticity theory for thermoelastic materials with voids. A material is needed to be classified by its fractional and elastic nonlocality parameter according to this theory. Bellifa et al. [2] presented an efficient zeroth-order nonlocal shear deformation theory for nanobeams. Lata and Singh [14] studied the deformations due to normal force in a nonlocal magneto-thermoelastic solid with the Hall effect.

Chen and Gurtin [3] developed a theory of heat conduction. They suggested that in case of bodies being deformable, the said theory is dependent on two temperatures. Two distinct temperatures are known as the thermodynamic temperature and the conductive temperature. Chen et al. [4] suggested that the heat supplied is directly proportional to the difference between the thermodynamic temperature and the conductive temperature. A generalized two-temperature theory was developed by Youssef [26]. He obtained the uniqueness theorem for equations of two-temperature generalized thermoelasticity. Youssef and Al-Lehaibi [27], after investigating various problems, gave an indication that the two-temperature generalized thermoelasticity is more realistic in describing the state of an elastic body as compared to one temperature. Lata and Singh [16] studied effects of nonlocality and two temperature in a nonlocal thermoelastic solid with memory dependent derivatives. Kumar et al. [12, 13] studied the disturbances in a homogeneous transversely isotropic thermoelastic rotating medium with two-temperatures, in the presence of Hall currents and magnetic field due to thermomechanical sources. Sharma et al. [23] carried the investigation regarding the two-dimensional deformation in a transversely isotropic medium with two temperatures. The disturbances due to inclined load were studied along with graphical representations of the effects of two temperatures. Lata and Singh [15, 17] discussed the disturbances due to two-temperature parameter with angle of inclination in a homogenous isotropic nonlocal thermoelastic solid at different angles of inclination and due to ramp type heat source. Plane wave propagation in a nonlocal magneto-thermoelastic solid with the Hall effect and two-temperature parameter was studied analytically by Lata and Singh [18].

In the present investigation, the effects of nonlocality and two temperatures in a homogeneous isotropic nonlocal thermoelastic solid in the context of generalized theory of thermoelasticity without energy dissipation have been studied. The study can be very helpful in the field of material sciences and new material designing.

9.2 BASIC EQUATIONS

Following Youseff [26] and Eringen [7], the equation of motion for a homogeneous isotropic non local thermoelastic solid with two temperatures is given by

$$\left(\lambda + 2\mu\right)\nabla\left(\nabla\cdot u\right) - \mu\left(\nabla\times\nabla\times u\right) - \beta\nabla T = \left(1 - \varepsilon^2\nabla^2\right)\rho\frac{\partial^2 u}{\partial t^2} \quad (9.1)$$

Heat conduction equation without energy dissipation using Lord–Shulman model [19] is given as

$$K^*\varphi_{,ij} + \rho\left(Q + \tau_0 Q\right) = \beta T_0\left(\dot{e}_{ij} + \tau_0\ddot{e}_{ij}\right) + \rho C_E\left(\dot{T} + \tau_0\ddot{T}\right) \quad (9.2)$$

where $T = \left(1 - a\nabla^2\right)\varphi$ (9.3)

The constitutive relations are given by

$$t_{ij} = \lambda u_{k,k}\delta_{ij} + \mu\left(u_{i,j} + u_{j,i}\right) - \beta T\delta_{ij} \quad (9.4)$$

where λ, μ are material constants, ε is the nonlocal parameter, ρ is the mass density, $u = (u, v, w)$ is the displacement vector, φ is the conductive temperature, a denotes two-temperature parameter, T is absolute temperature and T_0 is reference temperature, K^* is materialistic constant, C_E denotes the specific heat at constant strain, $\beta = (3\lambda + 2\mu)\alpha$ where α is coefficient of linear thermal expansion, e_{ij} are components of strain tensor, δ_{ij} is the Kronecker delta, and t_{ij} are the components of stress tensor.

9.3   FORMULATION OF THE PROBLEM

We consider a homogeneous nonlocal isotropic thermoelastic solid in an initially undeformed state at temperature T_0. For two-dimensional problem, we consider that

$$u = \left(u, 0, w\right) \quad (9.5)$$

Using Equation (9.5) in Equations (9.1) and (9.2), yields

$$\left(\lambda + \mu\right)\frac{\partial e}{\partial x} + \mu\nabla^2 u - \beta\frac{\partial T}{\partial x} = \left(1 - \varepsilon^2\nabla^2\right)\rho\frac{\partial^2 u}{\partial t^2} \quad (9.6)$$

$$\left(\lambda+\mu\right)\frac{\partial_e}{\partial_z}+\mu\nabla^2 w-\beta\frac{\partial T}{\partial z}=\left(1-\varepsilon^2\nabla^2\right)\rho\frac{\partial^2 w}{\partial t^2} \qquad (9.7)$$

$$K^*\left(\frac{\partial^2\varphi}{\partial x^2}+\frac{\partial^2\varphi}{\partial z^2}\right)+\rho\left(Q+\tau_0 Q\right)=\rho C_E\left(\dot{T}+\tau_0\ddot{T}\right)+\beta T_0\frac{\partial}{\partial t}\left\{\left(1+\tau_0\frac{\partial}{\partial t}\right)\left(\frac{\partial u}{\partial x}+\frac{\partial w}{\partial z}\right)\right\}$$

$$(9.8)$$

where $e=\dfrac{\partial u}{\partial x}+\dfrac{\partial w}{\partial z}, \nabla^2=\dfrac{\partial^2}{\partial x^2}+\dfrac{\partial^2}{\partial z^2}$

We define the following dimensionless quantities as

$$\left(x',z'\right)=\frac{\omega_1}{c_1}\left(x,z\right),\left(u',w'\right)=\frac{\omega_1}{c_1}\left(u,w\right),t_{ij}'=\frac{t_{ij}}{\beta T_O},t'=\omega_1 t,a'=\frac{\omega_1^2}{c_1^2}a,K^{*1}=\frac{c_1}{\lambda\omega_1}K^*$$

$$(9.9)$$

where $c_1^2=\dfrac{\mu}{\rho}$ and $\omega_1=\dfrac{\rho c^* c_1^2}{K^*}$

Upon introducing the quantities defined by Equation (9.9) in Equations (9.6)–(9.8), and suppressing the primes, yields

$$\left(1+\delta_1\right)\frac{\partial^2 u}{\partial x^2}+\delta_1\frac{\partial^2 w}{\partial x\partial z}+\frac{\partial^2 u}{\partial z^2}-\delta_2\frac{\partial}{\partial x}\left(1-a\nabla^2\right)\phi=\left(1-\in^2\nabla^2\right)\frac{\partial^2 u}{\partial t^2}$$

$$(9.10)$$

$$\left(1+\delta_1\right)\frac{\partial^2 u_3}{\partial x_3^2}+\delta_1\frac{\partial^2 w}{\partial x\partial z}+\frac{\partial^2 w}{\partial x^2}-\delta_2\frac{\partial}{\partial z}\left(1-a\nabla^2\right)\phi=\left(1-\in^2\nabla^2\right)\frac{\partial^2 w}{\partial t^2}$$

$$(9.11)$$

$$K^*\left(\frac{\partial^2\phi}{\partial x^2}+\frac{\partial^2\phi}{\partial z^2}\right)+\rho\left(1+c_1\tau_0\frac{\partial}{\partial t}\right)Q=\rho C_E\frac{\partial}{\partial t}\left(1+c_1\tau_0\frac{\partial}{\partial t}\right)T+\beta T_0\frac{\partial}{\partial t}\left\{\left(1+c_1\tau_0\frac{\partial}{\partial t}\right)\left(\frac{\partial u}{\partial x}+\frac{\partial w}{\partial z}\right)\right\}$$

$$(9.12)$$

where

$$\delta_1 = \frac{\lambda + \mu}{\mu}, \delta_1 = \beta \frac{T_0}{\mu}$$

We assume the medium to be at rest initially. Then the initial and regularity conditions are given by

$$u(x,z,0) = 0 = \dot{u}(x,z,0)$$

$$w(x,z,0) = 0 = \dot{w}(x,z,0)$$

$$\varphi(x,z,0) = 0 = \dot{\varphi}(x,z,0)..for..z \geq 0, \infty < x < \infty$$

$$u(x,z,t) = w(x,z,t) = \varphi(x,z,t) = 0\, for..t \geq 0..when..z \to \infty \quad (9.13)$$

Laplace and Fourier Transforms are defined by

$$\overline{f}(x_1,x_3,s) = \int_0^\infty f(x_1,x_3,t)e^{-st}dt \quad (9.14)$$

$$\overline{f}(\xi,x_3,s) = \int_{-\infty}^\infty f(x_1,x_3,t)e^{i\xi x_1}dx_1 \quad (9.15)$$

Using Laplace and Fourier transforms defined by Equations (9.14)–(9.15), upon Equations (10)–(12), we obtain a system of equations,

$$\left[-\xi^2\left(1+\delta_1\right)+\left(1+\epsilon^2\,s^2\right)\frac{d^2}{dz^2}-\left(1+\epsilon^2\,\xi^2\right)s^2\right]\hat{\overline{u}}+\left[i\delta_1\xi\frac{d}{dz}\right]\hat{\overline{w}}-\left[1+a\xi^2-a\frac{d^2}{dz^2}\right]\hat{\overline{\varphi}}. \quad (9.16)$$

$$\left[i\delta_1\xi\frac{d}{dz}\right]\hat{\overline{u}}+\left[\left(1+\delta_1++\epsilon^2\,s^2\right)\frac{d^2}{dz^2}-\xi^2-\left(1+\epsilon^2\,\xi^2\right)s^2\right]\hat{\overline{w}}-i\delta_2\xi\left[1+a\xi^2\right] \quad (9.17)$$

$$\left[i\delta_4\delta_5\xi\right]\hat{\overline{u}}+\left[\delta_4\delta_5\frac{d}{dz}\right]\hat{\overline{w}}+\left[K^*(-\xi^2-\delta_3\delta_4+\frac{d^2}{dz^2}\right]\hat{\overline{\varphi}}=0 \quad (9.18)$$

where $\delta_3 = \rho C_E s, \delta_4 = 1 + c_1 \tau_0 s, \delta_5 = \beta T_0 s$.

Without considering internal heat source and setting, from Equations (9.16), (9.17), and (9.18), we yield a set of homogeneous equations which will have a nontrivial solution if determinant of coefficient vanishes so as to give a characteristic equation as

$$\left[A\frac{d^6}{dx_3^6} + B\frac{d^4}{dx_3^4} + C\frac{d^2}{dx_3^2} + D \right]\left(\hat{\bar{u}}, \hat{\bar{w}}, \hat{\bar{\varphi}} \right) = 0 \tag{9.19}$$

where $A = K^* \zeta_1 \zeta_2$,

$B = -\zeta_1 \zeta_2 \zeta_3 - K^* \left(\zeta_4 - \zeta_{10} \right) + a \zeta_5 \zeta_6$

$C = \left(\zeta_1 \zeta_4 + \zeta_2 \zeta_9 - \zeta_{10} \right) \zeta_3 + K^* \zeta_4 \zeta_9 - \zeta_5 \zeta_7$,

$D = \zeta_5 \zeta_8 - \zeta_3 \zeta_4 \zeta_9$

where $\zeta_1 = 1 + \epsilon^2 s^2$, $\zeta_2 = 1 + \zeta_1 + \epsilon^2 s^2$, $\zeta_3 = K^* \xi^2 + \zeta_3 \zeta_4$, $\zeta_4 = \xi^2 + s^2 (1 + \epsilon^2 \xi^2)$, $\zeta_5 = \xi^2 \delta_2 \delta_4 \delta_5$, $\zeta_6 - 1 + \epsilon^2 s^2$, $\zeta_7 = 1 - \epsilon^2 s^2 (1 + a\xi^2)$, $\zeta_8 = \xi^2 (1 + a\xi^2)$, $\zeta_9 = \xi^2 (1 + \delta_1) + s^2 (1 + \epsilon^2 \xi^2)$, $\zeta_{10} = \xi^2 \delta_1^2$, $\zeta_{11} = 1 + a\xi^2$, $\zeta_{12} = \iota \xi \delta_2 \delta_4 \delta_5$

The roots are $\pm\lambda_i (i = 1,2,3)$, making use of the radiation conditions that as the solutions of equation can be written as

$$\tilde{\bar{u}} = A_1 e^{-\lambda_1 z} + A_2 e^{-\lambda_2 z} + A_3 e^{-\lambda_3 z} \tag{9.20}$$

$$\tilde{w} = d_1 A_1 e^{-\lambda_1 z} + d_2 A_2 e^{-\lambda_2 z} + d_3 A_3 e^{-\lambda_3 z} \tag{9.21}$$

$$\tilde{\bar{\varphi}} = l_1 A_1 e^{-\lambda_1 z} + l_2 A_2 e^{-\lambda_2 z} + l_3 A_3 e^{-\lambda_3 z} \tag{9.22}$$

where,

$$d_i = \frac{P^* \lambda_i^4 + Q^* \lambda_i^2 + R^*}{P\lambda_i^4 + Q\lambda_i^3 + R\lambda_i^2 + S\lambda_i + T}; i = 1,2,3 \tag{9.23}$$

$$l_i = \frac{P^{**} \lambda_i^4 + Q^{**} \lambda_i^2 + R^{**}}{P\lambda_i^4 + Q\lambda_i^3 + R\lambda_i^2 + S\lambda_i + T}; i = 1,2,3 \tag{9.24}$$

$P^* = K^* \zeta_1,\ Q^* = -K^*(\zeta_9 + \zeta_1),\ R^* = \zeta_3\zeta_9 - \zeta_5\zeta_{11},\ P^{**} = \zeta_1\zeta_2,\ Q^{**} = \zeta_{10} - \zeta_1\zeta_4 - \zeta_2\zeta_9,$
$R^{**} = \zeta_4\zeta_9,\ P = K^*\zeta_2, Q = a\zeta_{12},\ R = \zeta_2\zeta_3 - K^*\zeta_4,\ S = \zeta_{11}\zeta_{12},\ T = \zeta_3\zeta_4$

9.3.1 BOUNDARY CONDITIONS

We consider that normal force is applied on the half-surface . The boundary conditions are as follows:

(1) $\qquad\qquad 1. \rightarrow t_{zz}(x,z,t) = G(t)\delta(x)$ $\qquad\qquad$ (9.25)

(2) $\qquad\qquad 2. \rightarrow t_{xz}(x,z,t) = 0$ $\qquad\qquad$ (9.26)

(3) $\qquad\qquad 3. \rightarrow \varphi(x,z,t) = 0$ $\qquad\qquad$ (9.27)

where, $\delta(x)$ is dirac delta function of and is a function defined as

$$G(t) = \left\{ T_1 \frac{t^{0; t \le 0}}{t_{0_{T_1; t > t_0}}}; 0 < t \le t_0 \right\}$$ (9.28)

where t_0 indicates the length of the time to raise the heat and T_1 is a constant, this means that the boundary of the half-space, which is initially at rest and has a fixed temperature t_0, is suddenly raised to a temperature equal to function $G(t)\delta(x)$ and maintained at this temperature afterward.

Applying Laplace and Fourier transform to Equation (9.25), we get

$$\bar{\varphi}(\zeta,0,s) = \bar{G}(s), \text{ where } \bar{G}(s) = T_1 \frac{\left(1 - e^{-st_0}\right)}{t_0 s^2}$$ (9.29)

Applying the Laplace and Fourier transform defined by Equations (9.14) and (9.15) on the boundary conditions (9.25)–(9.27) and then using the dimensionless quantities defined by Equation (9.9) and using Equations (9.2) and (9.4) and substituting values of from Equations (9.28)–(9.30), and solving, we obtain the components of displacement, stress, and conductive temperature as

$$\hat{\bar{u}} = T_1 \frac{\left(1 - e^{-st_0}\right)}{\Delta t_0 s^2} \left\{ \sum\nolimits_{i=1}^{3} M_{1i} e^{-\lambda_i z} \right\} \tag{9.30}$$

$$\hat{\bar{w}} = T_1 \frac{\left(1 - e^{-st_0}\right)}{\Delta t_0 s^2} \left\{ \sum\nolimits_{i=1}^{3} d_i M_{1i} e^{-\lambda_i z} \right\} \tag{9.31}$$

$$\hat{\bar{\varphi}} = T_1 \frac{\left(1 - e^{-st_0}\right)}{\Delta t_0 s^2} \left\{ \sum\nolimits_{i=1}^{3} l_i M_{1i} e^{-\lambda_i z} \right\}. \tag{9.32}$$

$$\tilde{\bar{t}}_{zz} = T_1 \frac{\left(1 - e^{-st_0}\right)}{\Delta t_0 s^2} \left\{ \sum\nolimits_{i=1}^{3} R_i M_{1i} e^{-\lambda_i z} \right\} \tag{9.33}$$

$$\tilde{\bar{t}}_{zx} = T_1 \frac{\left(1 - e^{-st_0}\right)}{\Delta t_0 s^2} \left\{ \sum\nolimits_{i=1}^{3} \Delta_{2i} M_{1i} e^{-\lambda_i z} \right\} \tag{9.34}$$

$$\tilde{\bar{t}}_{xx} = T_1 \frac{\left(1 - e^{-st_0}\right)}{\Delta t_0 s^2} \left\{ \sum\nolimits_{i=1}^{3} S_i M_{1i} e^{-\lambda_i z} \right\} \tag{9.35}$$

where $\Delta = \sum_{i=1}^{3} M_{3i} N_i$

$M_{11} = \Delta_{22}\Delta_{33} - \Delta_{32}\Delta_{23}, M_{12} = \Delta_{21}\Delta_{33} - \Delta_{31}\Delta_{23}, M_{13} = \Delta_{32}\Delta_{22} - \Delta_{31}\Delta_{22},$

$M_{31} = \Delta_{22}\Delta_{33} + \Delta_{32}\Delta_{23}, M_{32} = \Delta_{21}\Delta_{33} + \Delta_{31}\Delta_{23}, M_{33} = \Delta_{32}\Delta_{21} - \Delta_{31}\Delta_{22},$

$\Delta_{2i} = \iota\xi d_i - \lambda_i, \Delta_{3i} = l_i\lambda_i, N_i = \lambda_i d_i(\lambda + 2\mu) + \beta l_i, R_i = \lambda_i d_i(\lambda + 2\mu) + \beta\theta_0 l_i,$

$S_i = \iota\xi(\lambda + 2\mu) - \beta\theta_0 l_i; i - 1, 2, 3$

9.4   PARTICULAR CASES

1. If $a = 0$, then from Equations (9.30)–(9.35), we obtain the corresponding expressions for displacement components, stress components, and conductive temperature for nonlocal isotropic solid without two temperatures.
2. If $=0$, then from Equations (9.30)–(9.35), we obtain the corresponding expressions for displacement components, stress components, and conductive temperature for isotropic solid without nonlocal effects and with two temperatures.

9.5 INVERSION OF THE TRANSFORMATION

To obtain the solution of the problem in physical domain, we must invert the transforms in Equations (9.30)–(9.35). Here the displacement components, normal and tangential stresses, and conductive temperature are functions of , the parameters of Laplace and Fourier transforms and respectively, and hence are of the form . To obtain the function in the physical domain, we first invert the Fourier transform using

$$\bar{f}(x,z,s) = \frac{1}{2\pi}\int_{-\infty}^{\infty} e^{-i\xi x}\hat{f}(\xi,z,s)\,d\xi = \frac{1}{2\pi}\int_{-\infty}^{\infty}\left|\cos(\xi x)f_e - i\sin(\xi x)f_0\right|d\xi \quad (9.36)$$

where, f_e and f_0 are, respectively, the even and odd parts of $f(\xi,z,s)$. Thus, Equation (9.36) gives the Laplace transform $f(x,z,s)$ of the function $f(x,z,t)$. Following Honig and Hirdes [9], the Laplace transform function $f(x,z,s)$ can be inverted to $f(x,z,t)$. The last step is to calculate the integral in Equation (9.36) which has been done as per the method described in Press et al. [22]. The method of evaluating this integral involves the use of Romberg's integration with adequate step size. The results from successive refinements of the extended trapezoidal rule followed by extrapolation of the results to the limit when the step size tends to zero are also used.

9.6 NUMERICAL RESULTS AND DISCUSSION

Magnesium material is chosen for the purpose of numerical calculation which is isotropic and physical data for which is given as

$$\lambda = 9.4\times10^{10}\,\mathrm{Nm^{-2}},\ \mu = 3.278\times10^{10}\,\mathrm{Nm^{-2}},\ K^* = 1.7\times10^{10}\,\mathrm{Wm^{-1}K^{-1}},$$
$$\rho = 1.74\times10^3\,\mathrm{kgm^{-3}},\ T_0 = 298K,\ C^* = 10.4\times10^2\,\mathrm{Jkg^{-1}\,deg^{-1}},\ \omega_1 = 3.58, a = 0.05$$

A comparison of values of displacement components, u and w stress components, t_{zz}, t_{xx} and t_{zx}, and conductive temperature φ for a transversely isotropic thermoelasic solid with distance x has been made for local and nonlocal parameter $\varepsilon = 1.0$ and is presented graphically for $a = 0$ and $a = 0.1$ in Figures 9.1–9.6.

1. The black colored solid line with squares as symbols and the red colored solid line with circles as symbols, respectively, corresponds to $\varepsilon = 0$ and $a = 0$ and $a = 0.1$.

2. The blue colored solid line with squares as upward triangles and
    the purple colored solid line with downward triangles as symbols,
    respectively, corresponds to $\varepsilon = 0$ and $a = 0$ and $a = 0.1$.

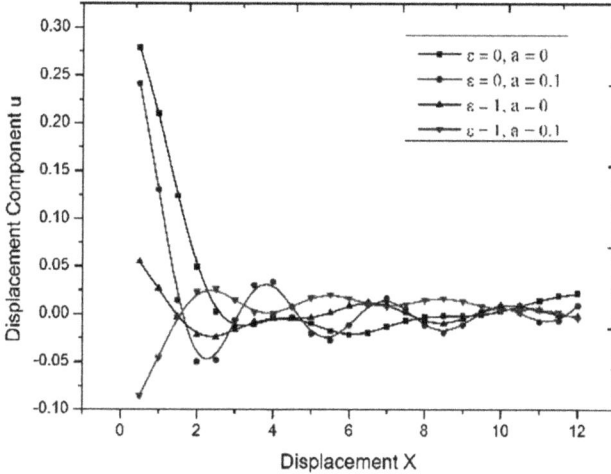

FIGURE 9.1 Variation of displacement component u with displacement x.

From Figures 9.1 to 9.6, it is clear that there is oscillatory motion followed
by the variations of displacement components, stress components, and
conductive temperature. The graphs have been plotted for different values of
$\varepsilon = 0$; $a = 0$; $\varepsilon = 0$, $a = 0.1$; $\varepsilon = 1$, $a = 0$ and $\varepsilon = 1$, $a = 0.1$. From the graphs,
we can conclude that

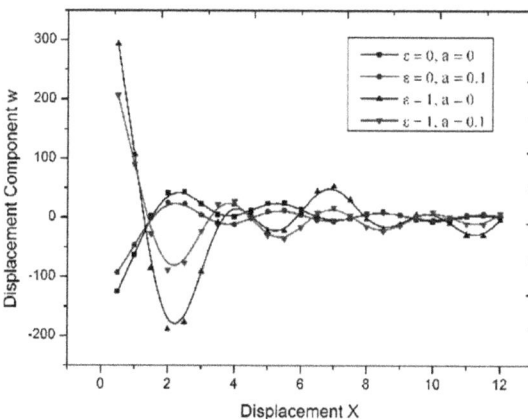

FIGURE 9.2 Variation of displacement component w with displacement x.

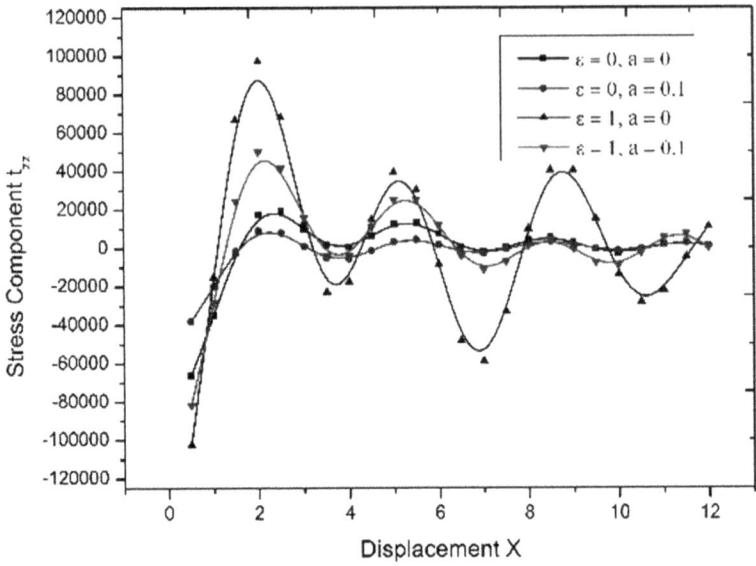

FIGURE 9.3   Variation of stress component t_{zz} with displacement x.

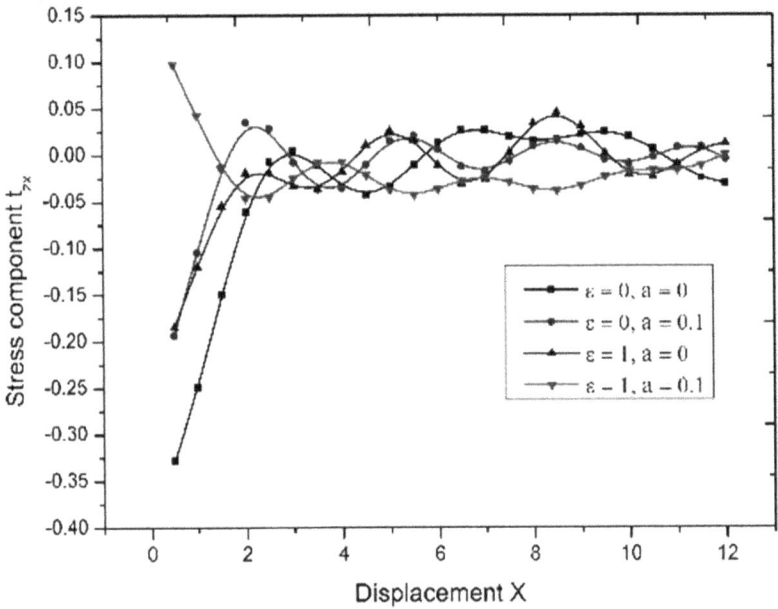

FIGURE 9.4   Variation of stress component t_{zx} with displacement x.

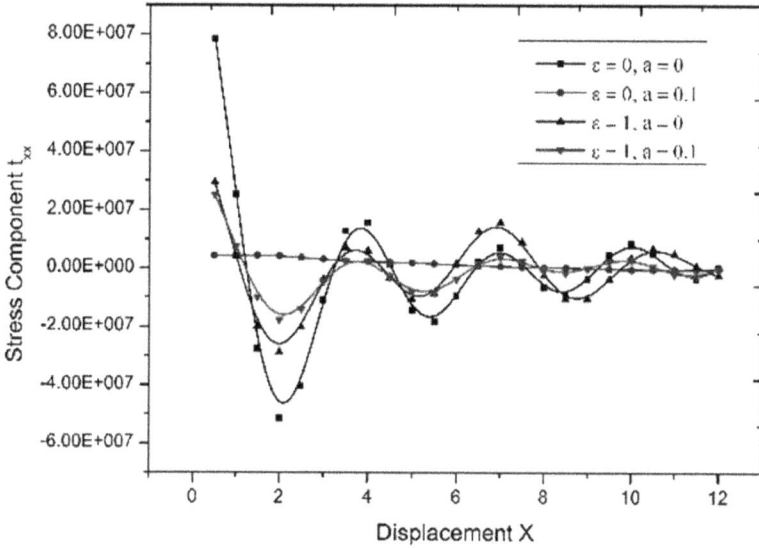

FIGURE 9.5   Variation of stress component t_{xx} with displacement x.

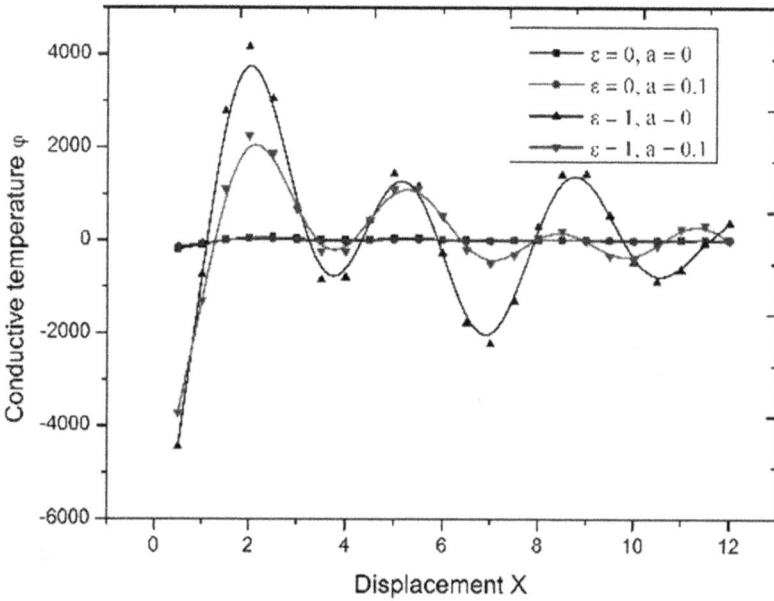

FIGURE 9.6   Variation of conductive temperature φ with displacement x.

1. The nonlocal parameter is affecting the variations of all the components.
2. Two-temperature parameter is also varying the variations of different components.

9.7 CONCLUSION

In the present discussion, the numerical results have been depicted graphically showing the effects of nonlocal parameter and two temperatures on the components of displacements, stresses, and conductive temperature. From above investigation, it is observed that there is a significant impact on displacement components, stress components, and conductive temperature due to effects of nonlocality and two temperatures. The amplitude of all the physical quantities (discussed above) either increase or decrease with nonlocal parameters as well as with the two temperatures. In the presence and the absence of nonlocal parameters, the stress and displacement components follow an oscillatory path with respect to x. The normal force plays a significant role in the distribution of all the physical quantities. The results of this chapter give an inspiration to study nonlocal parameter effects further in thermoelastic bodies. These results will be very useful for the researchers working in the field of material science, geophysics, acoustics, and so forth.

KEYWORDS

- **thermoelasticity**
- **nonlocality**
- **nonlocal theory of thermoelasticity**
- **Eringen model of nonlocal theories**
- **two temperatures**
- **normal heat**

REFERENCES

1. Bachher, M. and Sarkar, N. (2018), "Nonlocal theory of thermoelastic materials with voids and fractional derivative heat transfer," *Waves in Random and Complex Media*, https://10.1080/17455030.2018.1457230.

2. Bellifa, H., Benrahou, K.H., Bousahla, A.A., Tounsi, A. and Mahmoud, S.R. (2017), "A nonlocal zeroth-order shear deformation theory for nonlinear postbuckling of nanobeams," *Structural Engineering and Mechanics*, **62**(6), 695–702.

3. Chen, P.J. and Gurtin, M.E. (1968), "On a theory of heat conduction involving two temperatures," *ZAMP*, **19**, 614–627.

4. Chen, P.J., Gurtin, M.E. and Willams, W.O. (1969), "On the thermodynamics of non-simple elastic material with two temperatures," *ZAMP*, **20**, 107–112.

5. Edelen, D.G.B, Green, A.E. and Laws, N. (1971), "Nonlocal continuum mechanics," *Archive for Rational Mechanics and Analysis*, **43**, 36–44.

6. Edelen, D.G.B. and Laws, N. (1971), "On the thermodynamics of systems with nonlocality," *Archive for Rational Mechanics and Analysis*, **43**, 24–35.

7. Eringen, A.C. (2002), *Nonlocal Continuum Field Theories*, Springer, New York.

8. Eringen, A.C. and Edelen, D.G.B. (1972), "On nonlocal elasticity," *International Journal of Engineering Science*, **10**, 233–248.

9. Honig, G. and Hirdes, U. (1984), "A method for the numerical inversion of Laplace transform," *Journal of Computational and Applied Mathematics*, **10**, 113–132.

10. Kaur, G., Singh, D. and Tomar, S.K. (2018), "Rayleigh type wave in nonlocal elastic solid with voids," *European Journal of Mechanics*, https:/10.1016/j.euromechsol.2018.03.015.

11. Khurana, A. and Tomar, S.K. (2013), "Reflection of plane longitudinal waves from the stress-free boundary of a nonlocal, micropolar solid half-space," *Journal of Mechanics of Materials and Structures*, **8**(1), 95–107.

12. Kumar, R., Sharma, N. and Lata, P. (2016), "Effects of Hall current in a transversely isotropic magnetothermoelastic two temperature medium with rotation and with and without energy dissipation due to normal force," *Structural Engineering and Mechanics*, **57**(1), 91–103.

13. Kumar, R., Sharma, N. and Lata, P. (2016), "Thermomechanical interactions in the transversely isotropic magnetothermoelastic medium with vacuum and with and without energy dissipation with combined effects of rotation, vacuum and two temperatures," *Applied Mathematical Modelling*, **40**, 6560–6575.

14. Lata, P. and Singh, S. (2020), "Deformation in a nonlocal magneto-thermoelastic solid with hall current due to normal force," *Geomechanics and Engineering*, **22** (2), 109-117. https://doi.org/10.12989/gae.2020.22.2.109

15. Lata, P. and Singh, S. (2019), "Effect of nonlocal parameter on nonlocal thermoelastic solid due to inclined load," *Steel and Composite Structures*, **33**(1), 123–131.

16. Lata, P. and Singh, S. (2020), "Thermomechanical interactions in a nonlocal thermoelastic model with two temperature and memory dependent derivatives," *Coupled Systems Mechanics*, **9**(5), 397–410. https://doi.org/10.12989/CSM.2020.9.5.397

17. Lata, P. and Singh, S. (2020), "Effects of nonlocality and two temperature in a nonlocal thermoelastic solid due to ramp type heat source," *Arab Journal of Basic and Applied Sciences*, **27**(1), 358–364. https://doi.org/10.1080/25765299.2020.1825157

18. Lata, P. and Singh, S. (2020), "Plane wave propagation in a nonlocal magneto-thermoelastic solid with two temperature and Hall current," *Waves in Random and Complex Media*. https://doi.org/10.1080/17455030.2020.1838667

19. Lord, H.W. and Shulman, Y. (1967), "A generalized dynamical theory of thermoelasticity," *Journal of the Mechanics and Physics of Solids*, **15**(5), 299–309. https://doi.org/10.1016/0022-5096(67)90024-5.

20. Marin, M., Ellahi, R. and Chirila, A. (2017), "On solutions of Saint–Venant's problem for elastic dipolar bodies with voids," *Carpathian Journal of Mathematics*, **33**(2), 219–232.

21. Othman, M.I.A and Marin, M. (2017), "Effect of thermal loading due to laser pulse on thermoelastic porous medium under G-N theory," *Results in Physics*, **7**, 3863–3872.

22. Press, W.H., Teukolshy, S.A., Vellerling, W.T. and Flannery, B.P. (1986), *Numerical Recipes in Fortran: The Art of Scientific Computing*, Cambridge University Press, Cambridge.

23. Sharma, N., Kumar, R. and Lata, P. (2015), "Disturbance due to inclined load in the transversely isotropic thermoelastic medium with two temperatures and without energy dissipation," *Material Physics and Mechanics*, **22**, 107–117.

24. Singh, D., Kaur, G. and Tomar, S.K. (2017), "Waves in nonlocal elastic solid with voids," *Journal of Elasticity*, https:/10.1007/s10659–016-9618-x.

25. Singh, S., Lata, P. and Singh, SBIR. (2020), "Recent developments in the theory of nonlocality in elastic and thermoelastic mediums," *Applied Mechatronics and Mechanics: System Integration and Design,* CRC Press.

26. Youssef, H.M. (2005), "Theory of two-temperature-generalized thermoelasticity," *IMA Journal of Applied Mathematics*, **71**, 383–390.

27. Youssef, H.M. and Al-Lehaibi, E.A. (2007), "State space approach of two-temperature generalized thermoelasticity of one-dimensional problem," *International Journal of Solids and Structures*, **44**, 1550–1562.

CHAPTER 10

OPTIMIZATION OF SPARK GAP IN POWDER MIXED WIRE ELECTRIC DISCHARGE MACHINING THROUGH GENETIC ALGORITHM APPROACH

SWARUP S. DESHMUKH,[1] ARJYAJYOTI GOSWAMI,[1*] RAMAKANT SHRIVASTAVA,[2] and VIJAY S. JADHAV[2]

[1]*Department of Mechanical Engineering, National Institute of Technology Durgapur, Durgapur, West Bengal, India*

[2]*Mechanical Engineering Department, Government College of Engineering Karad, Karad, Maharashtra, India*

Corresponding author. E-mail: arjyajyoti.goswami@me.nitdgp.ac.in

ABSTRACT

Focus of industry is continuously shifting toward miniaturized components which cannot be fabricated through conventional manufacturing methods, which has led to the development of newer methods of manufacturing. Powder mixed wire electric discharge machining (pwEDM) is a route of machining which modifies the conventional wire Electric Discharge Machining (w-EDM). In pwEDM, powder is added to the dielectric fluid which influences the properties of the finally fabricated part. Spark gap is an important parameter in pwEDM which must be closely controlled. The spark gap should ideally be equal to the wire diameter but in practice it is always slightly more. This slight deviation in spark gap results in inaccuracy in the fabricated part and as such makes the pwEDM process unsuitable for precise manufacturing. This study is focused on reducing the average deviation in spark gap through control of the input process parameters such as pulse on time, pulse off time, servo voltage (SV), wire feed (WF), and powder concentration. The experiments were conducted as per Taguchi L27 array and the spark gap was measured using an inverted microscope. The data collected through Taguchi Array was used to develop a mathematical model through regression analysis. The mathematical model, this developed, was optimized through genetic algorithm. After optimization, the average deviation in the spark gap corresponding to a particular level of parameters was found to be 23.185 µm, which was minimum for the given set of conditions. Such optimization is useful for precise machining of parts using pwEDM.

10.1   INTRODUCTION

10.1.1   STEP-BY-STEP PROCESS OF WEDM

Wire electric discharge machine (w-EDM) is the modified version of Electric Discharge Machine. It is used to cut complex shapes from high hardness; high strength and conductive materials which are not easily cut by the conventional means. The electrode used in w-EDM is in the form of wire. This wire electrode is surrounded by dielectric fluid, that is, deionized water. A potential difference is applied between the electrode and workpiece. The negative terminal is connected to a wire electrode and a positive terminal is connected to the workpiece because of which the wire electrode acts as cathode while the workpiece acts as an anode. A small gap is maintained between them because of which the electron from the cathode terminal tries to move toward the anode. This electron continuously strikes the molecules of the deionized water and knocks out electrons, thus forming an ionized channel during its passage. As more and more electrons join in on the path more number of electrons are liberated, each trying to move toward the anode. Thus, a spark is generated in the gap between the cathode and the anode.

Spark generation in WEDM is shown in Figure 10.1.

Due to this spark energy, high temperature is generated at the location where the spark strikes, which causes melting and vaporization of the workpiece. During pulse off time (POFFT), the workpiece is rapidly cooled with the help of deionized water and debris particle produced during machining is flushed out. The debris particle or eroded particles present in the deionized water is removed through filtration and the cycle is repeated.

Eroded particles removals by filtration are shown in Figure 10.2.

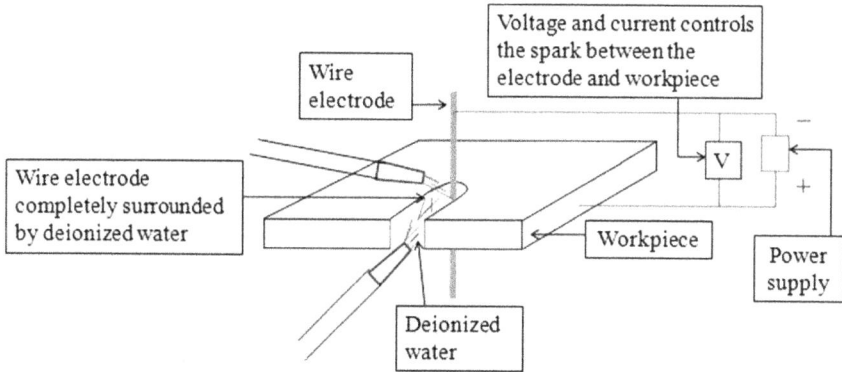

FIGURE 10.1   Spark generations in WEDM.

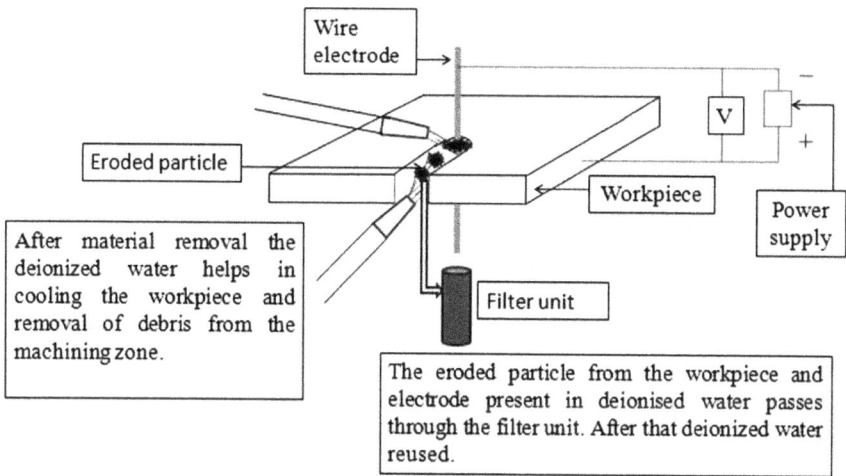

FIGURE 10.2   Removals of eroded particles by filtration.

10.1.2   POWDER MIXED WIRE EDM

After the addition of powder, the electrical conductivity of the dielectric fluid increases, its resisting capacity decreases and the spark needs to travel greater distances due to which the spark gap increases. After addition of powder in deionized water, distribution of the spark energy takes place.

The working principle of powder mixed WEDM (pwEDM) is shown in Figure 10.3.

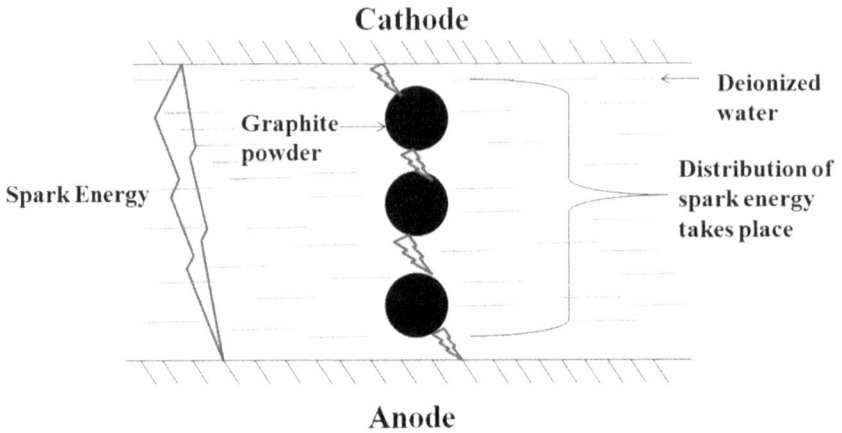

FIGURE 10.3   Principle of pwEDM.

Due to distribution of spark energy, the crater size in pwEDM is smaller as compared to wEDM, which ultimately improves the surface finish of the machined part.

The spark gap totally depends on the powder concentration (PC) and electrical conductivity of powder. A powder having high electrical conductivity loses its electron at high temperature, thus increasing the conductivity of the de-ionized water. This results in spark travelling greater distance than previous and consequently, the spark gap is more.

When the powder is added initially, it increases the conductivity of dielectric fluid but at the initial lower concentrations the powder particles are insufficient to distribute spark energy efficiently. Due to this, the spark travels greater distance and higher spark gap is observed.

After a particular value of PC, clogging in the spark gap area is observed. At this concentration, the spark energy is distributed efficiently and is unable to travel long distance. So the spark gap value at a particular concentration is lesser as compared to the case with no powder added.

10.1.3   GENETIC ALGORITHM

Any engineering solution has a large number of feasible solutions, which when taken together constitutes the Feasible Design Space. The aim of genetic algorithm (GA) is to search through this Feasible Design Space to find the best solution which fits the problem. GA is suitable for finding the optimum combinations of parameters or solutions from a large and potentially huge search space. The conventional method of optimization scans through only a small set of data and derives the optimum solutions from them.

The process starts with a set of solutions (a symbolic version of chromosomes) called populations. The solutions of one population are taken to form a new population, just like chromosomes of one population are combined to form the new generation of the population. The whole process is motivated by the prognosis that the new population will be better than the previous one to fit the evolution. It basically means the set of solutions of the next generation will fit the objective of optimization in a better manner than the set of solutions of the previous population.

The solutions (or chromosomes) which are selected to form a new population are selected on the basis of their fitness to the objective. Just like in natural world, only those traits are passed on to the next generation which is helpful in surviving in a particular set of conditions (analogous to optimization objective). The more suitable a particular set of solutions (chromosomes) is, the better chances it has for reproduction. This is repeated till some conditions for the improvement of the best solution are satisfied.

The flowchart of GA is shown in Figure 10.4.

1. *Objective function (f(x))*: In this experimentation, objective function is to minimize the spark gap by setting process parameters at an optimum level.
2. *Fitness function (F(x))*: Generally, GA deals with the maximization problem. So, it is necessary to convert the minimization problem into maximization problem. This conversion does not change the location of minimum point after transformation of the maximization problem into minimization problem. This new function is called as the fitness function. In maximization problem, this fitness function is same as the objective function. But in minimization problem, it converts with the help of Equation (10.1).

$$F(x) = \frac{1}{1 + f(x)} \qquad (10.1)$$

where $F(x)$ = fitness function and $f(x)$ = objective function.
3. *Initial population*: In order to implement the GA random population, that is, random binary coded strings consisting of 0's and 1's are considered. The length of binary coded string depends on the accuracy of required solution.

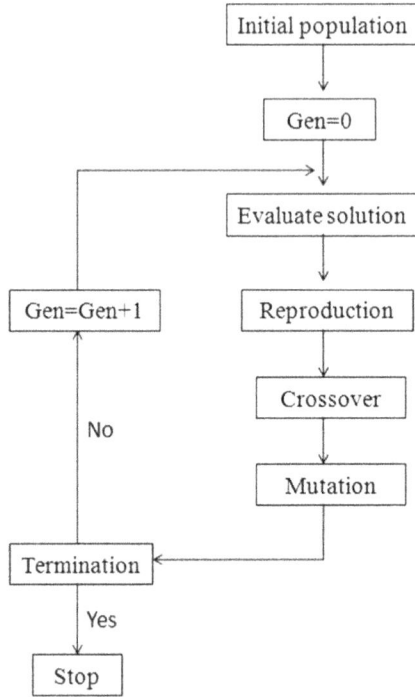

FIGURE 10.4   Flowchart of genetic algorithm.

4. *Reproduction*: Reproduction is the first step applied after choosing the initial population. In this step, the only those strings which have higher fitness value than the predefined criteria can be copied into the mating pool. The strings having higher fitness value have a higher probability to contribute better in upcoming generation. There are various reproduction methods, among them roulette wheel is the easiest and most commonly used. In this method, slot size of roulette wheel is proportional to fitness value. The group of strings in the mating pool made new population for the generations.

5. *Crossover*: After reproduction, crossover operation is carried out. Crossover is performed on the strings in the mating pool having better fitness value. The string which participates in mating depends on crossover probability. Crossover probability is the ratio of the number of strings participating in mating to the population size. In crossover, pair of strings in the mating pool interchange their bits and develop new strings. There are various crossover techniques, among

all of them single-point crossover and multi-point crossover are most commonly used. Example of a single-point crossover are:

Consider the two strings of 6 bits. Apply single point crossover at 4th position and interchange bits after the 4th position. After interchanging new strings are developed.

Single-point crossover (Crossover probability = 0.9)

1. String 1—100110; (1) New string 1—100111
2. String 2—110011; (2) New string 2—110010

6. *Mutation*: If there are a total of 60 bits in a string, mutation probability of each and every bit is to be checked. Mutation probability is generally in the range of 0001–0.05. The bits having mutation probability less than a predefined value can be changed, that is, if bit is 0 it can be made 1 and vice versa. The need of mutation is to keep diversity in the population.

Example: (Mutation probability =0.001)

1. String 1—100111; (2) New string 1—100101
2. String 2—110010; (2) New string 2—110011

After mutation new population is generated, then it is decoded and calculates objective function value. It completes one generation of the process and this process continues until the termination criterion is achieved.

10.2   LITERATURE REVIEW

Electric discharge machining is characterized by the low material removal rate; high electrode wear rate and somewhat poor surface finish, which is not as per the requirement. For improvement in the material removal rate, electrode wear rate and surface finish researchers added different type of conductive powders, that is, silicon, graphite, silicon carbide, tungsten in dielectric fluid, thus arriving at powder assisted electric discharge machining. Thermophysical properties of these powder particles have a significant effect on the response variable of EDM. There is scope as well as a need to optimize the process parameters of pwEDM to minimize the spark gap with help of nature inspired optimization technique like GA. Literature review is given in Table 10.1.

TABLE 10.1 Literature Review

Reference	Type; Response Variables	Workpiece and Powder Added	Conclusion
[1]	*Type:* PMEDM *Response variables:* Spark gap, surface roughness, Surface topography	*Workpiece:* SKH-54 *Powder:* Graphite, Silicon, silicon carbide, molybdenum sulphide	Among all the powders, only graphite has semiconducting property, and also high thermal conductivity which distributes the spark energy. Due to this the size of craters could be controlled, that is, flat and shallow craters were observed. Graphite powder has excellent lubricity compared to other powders which enabled it to achieve the mirror surface finish in electric discharge machining.
[2]	*Type:* PMEDM *Response variables:* Spark gap, MRR and TWR	*Workpiece:* Mould steel SKD-11 *Powder:* Aluminium (Al) Chromium (Cr), Silicon carbide (Sic),Copper (Cu)	Higher spark gap was observed for aluminium powder compared to Cr and SiC due to less electrical resistivity as compared to Cr and SiC. In case of Cu powder no major difference was observed in spark gap as compared to EDM without powder since owing to high density of copper powder it completely settles down at the bottom tank. Lesser spark gap with Cr power resulted in greater MRR. Also, higher spark density and greater spark explosion pressure were observed in case of Cr powder.
[3]	*Type:* PMEDM *Response variables:* Machining efficiency and surface roughness	*Workpiece:* Steel workpiece *Powder:* Aluminium	The conducting powder particle in the machining area causes series discharge responsible for the enlargement in electric discharge gap. At higher value of pulse current increment in the spark energy causes greater etching away of material from the steel. In PMEDM loss of spark energy takes place due to enlargement in electric discharge gap and widened the electric discharge passage. This is responsible for the improvement in the surface roughness value.

TABLE 10.1   *(Continued)*

Reference	Type; Response Variables	Workpiece and Powder Added	Conclusion
[4]	*Type:* PMEDM *Response variables:* MRR, Ra	*Workpiece:* EN-31 *Powder:* silicon powder	Mathematical model was developed using response surface methodology and ANOVA was performed to find out the significant process parameters. The powder concentration (PC) and Ip (Pulse current) has significant effect on the MRR. Higher value of PC promotes the bridging effect (series discharge). This series discharge helps in improving the MRR. Higher concentration of powder distributes spark energy in the machining area which helps in improving the surface roughness.
[5]	*Type:* PMEDM and powder mixed milling μ-EDM *Response variables:* spark gap, MRR, SR (Ra), peak to valley SR (Rmax),	*Workpiece:* WC-CO *Powder:* Nano Graphite powder	The rotary motion of electrode easily flushed out the debris particle in the spark gap area, while it is difficult to flush out the debris particle in the spark gap due to stationary electrode in powder mixed die sinking EDM. More spark gap was observed in the powder mixed milling μ-EDM compared to powder mixed die sinking EDM. As powder concentration increases it settles down in the spark gap area if it is small in size resulting in secondary sparking. This is also a reason for increment in the MRR and EWR. The low density of graphite powder allows it to mix properly in dielectric fluid. Due to its high thermal conductivity it dissipates more spark energy and smaller crater size was observed which helps in improving the Ra and peak to valley SR.

TABLE 10.1   *(Continued)*

Reference	Type; Response Variables	Workpiece and Powder Added	Conclusion
[6]	*Type:* PMEDM *Response variables:* MRR	*Workpiece:* Composite material (AA6061/10%SiC) *Powder:* Tungsten	Response surface methodology was used to optimize the process parameters of powder mixed EDM to increase the MRR. It was observed that MRR increases with the increases in the current and pulse on time value. At higher value of current and pulse on time, maximum spark energy was generated which eroded more material from the workpiece. At optimum value, 48% increment in MRR was observed.
[7]	*Type:* PM-WEDM *Response variables:* MRR, Recast layer thickness	*Workpiece:* Tungsten carbide-cobalt (WC-5.3%Co) *Powder:* Aluminum and silicon powder	Bridging effect decreases the insulting strength of dielectric fluid and increases the conductivity which increases the sparking rate. Increased sparking rate results in faster erosion rat from the material. The white layer thickness of material with use of aluminum and Silicon powder is 6 and 4 μm, respectively, compared to 14 μm when no powder is added in the dielectric fluid.
[8]	*Type:* PM-μWEDM *Response variables:* Spark gap, MRR	*Workpiece:* Gold coated silicon *Powder:* Nano carbon powder	In this study MRR and spark gap was measured while varying the powder concentration as 0.1 g/L, 1 g/L ,2 g/L. It was concluded that MRR increases from 1% to 33%. The variation is initial increase followed by a decrease and spark gap increases from 2% to 159% for different powder concentration.

10.3   EXPERIMENTAL SETUP

For the present work, Electronica Ecocut WEDM, Model-ELPULS-15 is used, consolidated with the diffused Brass wire of 0.25 mm diameter as an electrode. The machine comprises of a worktable for holding the workpiece during machining, wire feeding mechanism, computer control system and filtration arrangement for recirculation of deionized water. The workpiece is held on the worktable with the help of fasteners and the wire is continuously fed from a spool. Least possible movement of the worktable is 1 micron. The machine used for the present work was a 5-axis machine having X–Y–Z–U–V axes. For normal cutting, the table is moved in X and Y axes while for tape cutting it is moved in the U and V axes. The conductivity of dielectric fluid, that is, deionized water was checked by conductivity meter before adding it into a clean tank. This dielectric fluid is pumped through the upper and lower nozzle during the machining with help of pump 1 (shown in Figure 10.5) and it flushes out the debris particle generated after the machining.

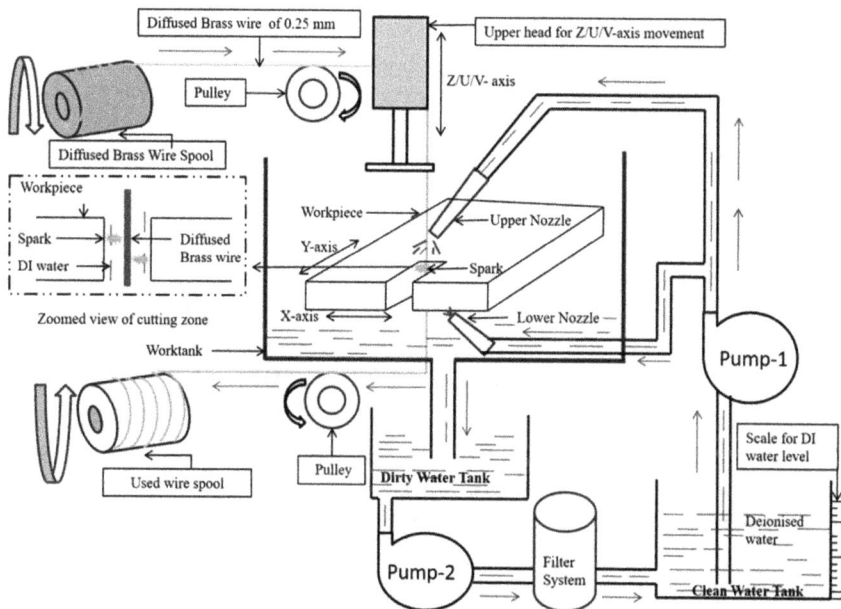

FIGURE 10.5   Schematic of wire electric discharge machining.

This impure dielectric fluid (i.e., debris particle + deionized water) is collected in dirty tank and pumped through the filter unit. This filter unit filters debris particle and produces clean deionized water to be utilized in the next machining cycle. This water is continuously recirculated until the completion of the machining operation. Schematic diagram and actual photograph of w-EDM is shown in Figure 10.5 and Figure 10.6, respectively.

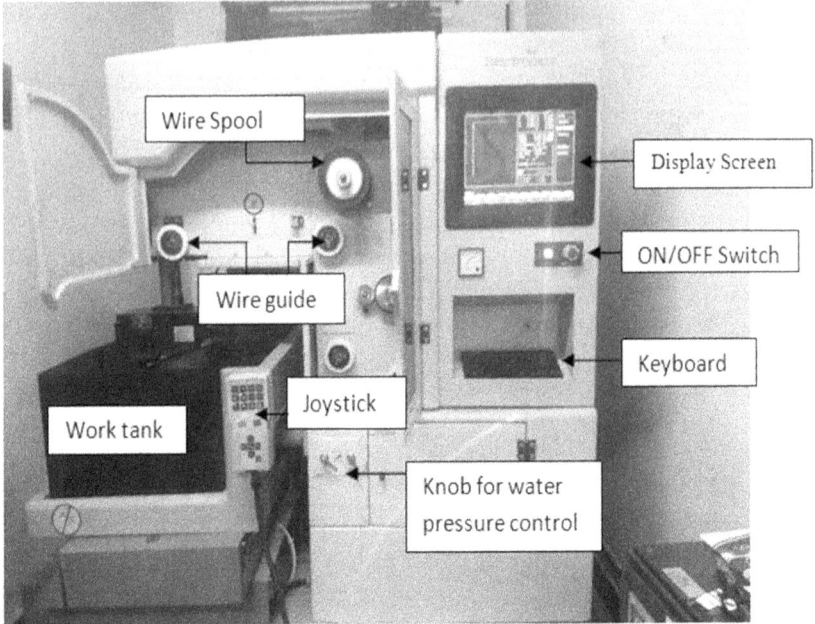

FIGURE 10.6   Actual set up for wire electric discharge machine.

The original w-EDM setup has been modified for the powder mixed wire electric discharge machining (pwEDM) setup. The graphite powder of 550 mesh size, that is, 25 µm was added in clean deionized water and pumped through the upper and lower nozzle of the w-EDM. The powder mixed dielectric fluid enters at the gap between the electrode and workpiece in the machining area. This graphite powder distributes the spark energy in the machining area which helps in reducing the spark gap.

This powder mixed dielectric fluid in working tank is pumped through the clean tank with help of pump 2 (shown in Figure 10.7). The collected powder mixed dielectric fluid is flushed out after the completion of first machining cycle and the tank was cleaned using a cotton cloth. The flushing was done to maintain the PC in the upcoming machining cycle. In the next cycle, fresh deionized water of 50 L was filled in the clean tank and graphite powder was added. This is repeated until the completion of the experimentation. To deal with the problem of sedimentation, that is, settling down of graphite powder at the bottom of clean tank, mechanical stirrer was attached to the clean water tank. This stirrer performed the stirring operation continuously throughout the machining operation. Schematic of pwEDM is shown in Figure 10.7.

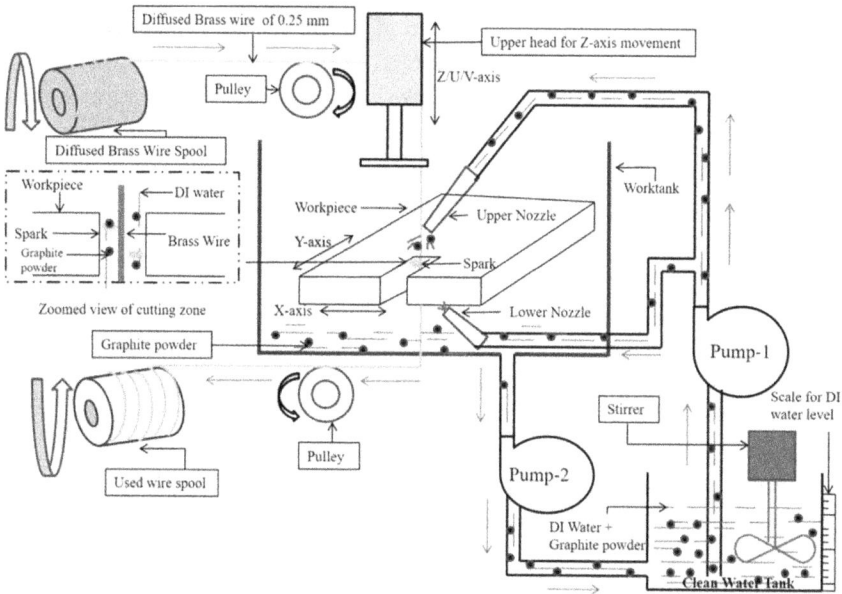

FIGURE 10.7 Schematic of powder assisted electric discharge machining (pwEDM) (dotted block shows the schematic of close up view of the machining zone).

10.3.1 WORKPIECE

In this experimentation, AISI 4140 was chosen as a workpiece material. It is in the category of high strength, high hardness and difficult to cut material. This material has wide applications in the moulding and forging industry. The common applications of this material are shaft, nut, bolt, crankshaft, etc. The chemical composition of this material is shown in Table 10.2.

TABLE 10.2 Chemical Composition of AISI 4140

C%	Cr%	Fe%	Mn%	Mo%	Si%	P%	S%
0.38–0.43	0.8–1.10	96.7–97.8	0.75–1.00	0.15–0.25	0.15–0.35	≤0.035	≤0.040

10.3.2 DIELECTRIC FLUID

For this experimentation work, de-ionized water was selected as dielectric fluid. The conductivity of deionized water was maintained below 50 micro-Siemens throughout the experimentation. Kerosene was also used as dielectric fluid. Kerosene is a compound of carbon and hydrogen. During machining, it decomposes to produce carbon, due to high temperature at cutting zone area. Therefore in the case of kerosene, formation of carbide takes place and it sticks to electrode reduces discharge efficiency. Because of that deionized water is preferred to use as dielectric fluid. Deionized water mostly used than kerosene because it has a lower viscosity than kerosene due to this it can flows into the smaller gap easily.

10.3.3 DESIGN MATRIX

DOE (Design of Experiment) technique provides a better combination of input process parameters to obtain the optimum value of response variables. pwEDM involves many input parameters to be considered during the machining process. The pilot experimentation was performed by varying one process parameter at a time and measuring the spark gap value. From this pilot experimentation, process factors such as PONT (μs), POFFT (μs), SV (volt), WF (m/min) and PC (g/L) and process parameter levels are considered. These factors are considered to obtain the best value for spark gap when cutting material is like AISI 4140. The process parameter levels selections and their respective values for further experimentation are shown in Table 10.3.

TABLE 10.3   Summary Table of Parameter Levels

Factors / Levels	PONT (μs) (A)	POFFT (μs) (B)	SV (V) (C)	WF (m/min) (D)	PC (gm/lit) (E)
1	115	45	10	2	0
2	120	50	20	4	0.2
3	125	55	30	8	0.3

10.3.4 ORTHOGONAL ARRAY (L27)

On the basis of the selection of process parameters and its level, the orthogonal array is generated. In this case, all input parameters are considered at three levels. Observations for L27 experimentation is shown in Table 10.4.

TABLE 10.4 Observations for L27 Experimentation

Sr. No.	PONT	POFFT	SV	WF	PC	SG (µm)
1	115	45	10	2	0	30.245
2	115	45	10	2	0.2	36.320
3	115	45	10	2	0.3	28.545
4	115	50	20	4	0	29.325
5	115	50	20	4	0.2	55.270
6	115	50	20	4	0.3	53.570
7	115	55	30	8	0	33.650
8	115	55	30	8	0.2	54.540
9	115	55	30	8	0.3	52.600
10	120	45	20	8	0	29.315
11	120	45	20	8	0.2	46.525
12	120	45	20	8	0.3	46.265
13	120	50	30	2	0	28.545
14	120	50	30	2	0.2	55.025
15	120	50	30	2	0.3	50.410
16	120	55	10	4	0	35.145
17	120	55	10	4	0.2	48.465
18	120	55	10	4	0.3	39.720
19	125	45	30	4	0	29.760
20	125	45	30	4	0.2	55.515
21	125	45	30	4	0.3	53.085
22	125	50	10	8	0	28.300
23	125	50	10	8	0.2	38.265
24	125	50	10	8	0.3	37.780
25	125	55	20	2	0	38.505
26	125	55	20	2	0.2	40.935
27	125	55	20	2	0.3	40.285

10.3.5   SAMPLE PHOTOGRAPH

Rectangular jobs of sizes 10 mm × 10 mm × 10 mm were cut from the work material as test specimens for every experimental run and sample photograph of AISI 4140 steel Work Material are shown in Figure 10.8.

FIGURE 10.8   Photograph of the sample work material.

10.3.6   SPARK GAP MEASUREMENT

The spark gap is calculated using Equation (10.2).

$$\text{Spark Gap} = \frac{\text{Kerfwidth - Wirediameter}}{2}(\mu m) \qquad (10.2)$$

The concept of the spark gap is shown in Figure 10.9.

Kerf width is measured with the help of SuXma Met-I metallurgical inverted microscope, generally it is in micrometre or micron. This metallurgical inverted microscope is shown in Figure 10.10.

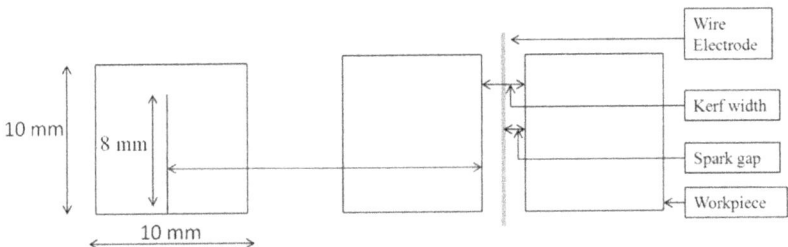

FIGURE 10.9   Concept of spark gap.

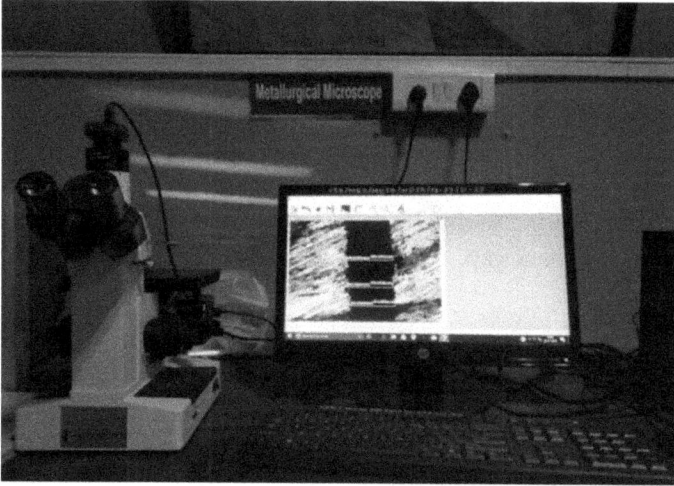

FIGURE 10.10   SuXma Met-I metallurgical inverted microscope.

10.4   RESULTS AND DISCUSSIONS

10.4.1   TAGUCHI ANALYSIS FOR SPARK GAP

Figure 10.11 shows the average spark gap values at various levels of pulse on time (PONT), POFFT, servo voltage (SV), wire fed rate and PC. It indicates that the level 1 of a PONT, that is, 115 μs gives the maximum average spark gap value = 41.56 μm whereas the level 2, that is, 120 μs gives the lowest average spark gap value = 39.94 μm. The level 3, that is, 125 μs gives average spark gap value = 40.27 μm. Hence, it is clear that PONT should be kept at level 2 in order to have lower average spark gap values. In the case of POFFT, level 1 of POFFT, that is, 45 μs gives lowest average spark gap value = 38.40 μm whereas the level 2, that is, 50 μs gives maximum average spark gap value = 41.83 μm and level 3, that is, 55 μs gives average spark gap value = 41.54 μm, so for minimum spark gap POFFT must be set to a level 1.

In case of SV, level 1, that is, 10 V gives minimum average spark gap value, that is, 34.75 μm, level 2 of SV, that is, 20 V gives average spark gap value = 41.11 μm and level 3 of SV, that is, 30 V gives maximum average spark gap value, that is, 45.90 μm. Therefore, to have a lower average value of spark gap, SV should be at level 1. In the case of wire feed (WF) rate, level 1 of WF, that is, 2 m/min gives minimum average spark gap value =

38.76 μm and level 2, that is, 4 m/min gives maximum spark gap value = 43.32 μm, level 3, that is, 8 m/min gives average spark gap value 39.69 μm. So the WF rate should keep at level 1 to have a lower average value of spark gap. In case of PC, level 1 of PC, that is, 0.0 g/L gives lowest average spark gap value = 29.20 μm and level 2, that is, 0.2 g/L gives maximum average spark gap value= 47.87 μm, level 3, that is, 0.3 g/L gives the average value of spark gap = 44.70 μm. So that PC rate should keep at level 1 to have a lower average value of spark gap.

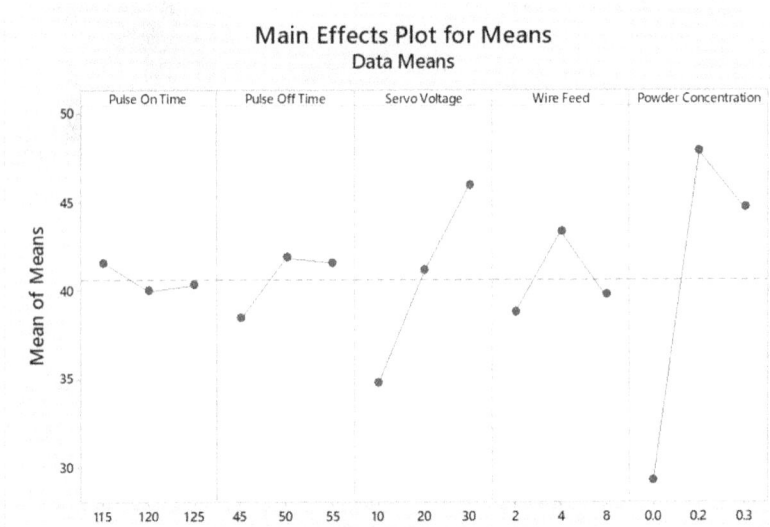

FIGURE 10.11 Main effect plot for spark gap data means vs. pulse on time, pulse off time, servo voltage, wire feed, and powder concentration.

At higher value of PONT and lower value of POFFT, more spark energy is developed as compared to low value of PONT and high value of POFFT. This high spark energy is efficiently distributed by powder particles so it is unable to travel greater distance. As a result, lower spark gap was observed at PONT= 120 μs, POFFT = 45 μs as compared to PONT = 115 μs, POFFT = 50 μs.

From ANOVA, it is clear that SV is the second significant process parameter in pwEDM. As the SV increases, it increases the spark discharge intensity. Due to its high significance as SV increases from 10 to 20 V and then from 20 to 30 V, the spark discharge intensity also increases. Powder particles are incapable to distribute this high spark discharge intensity. Due to this the spark, energy travels greater distance and results in higher spark gap. Therefore, minimum spark gap is obtained at SV = 10 V.

WF rate is the third most significant parameter in pwEDM. As WF rate increases, new portion of wire quickly comes in the spark gap area. The fresh wire discharges the spark more efficiently compared to eroded portion of wire. Due to this, the spark travels greater distance and more spark gap was observed at higher value of WF rate. As a result of this, greater spark gap was observed at 4 m/min compared to 2 m/min. At higher value of feed rate wire consumption is more, also it increases the cost of manufacturing.

PC is the most significant parameter as compared to other parameters. Minimum spark gap was observed at no powder condition. But as PC increases, it increases the conductivity of dielectric fluid; due to this, spark travels more distance and it increases the spark gap. At initial level of PC, insufficient powder particles are present and hence they are unable to distribute the spark energy efficiently. So as the PC increases from a particular value the powder particles clogs the spark gap. At this concentration, the powder particles are sufficient to distribute the spark energy efficiently. At this particular PC, lesser spark gap was observed compared to the case without addition of powder.

10.4.1.1   RESPONSE TABLE FOR MEAN

Table 10.5 enumerates all the ranking results for various process parameters. Rank 1 of PC and rank 2 of SV indicates that it is the most dominant parameter among four parameters.

TABLE 10.5   Response Table for Means of Spark Gap by Taguchi

Level	PONT	POFFT	SV	WF	PC
1	41.56	38.40*	34.75*	38.76*	29.20*
2	39.94*	41.83	41.11	43.32	47.87
3	40.27	41.54	45.90	39.69	44.70
Delta	1.63	3.44	11.15	4.56	18.67
Rank	5	4	2	3	1

10.4.1.2   ANALYSIS OF VARIANCE FOR SPARK GAP

Analysis of variance (ANOVA) for data means is shown in Table 10.6. It indicates that the PC (g/L) is most affecting parameter on spark gap with 57.88% followed by SV (V) 18.14% and WF rate (m/min) 3.36% contribution and POFFT (μs) 2.10% contribution and PONT (μs) 0.43% contribution, respectively. Figure 10.12 shows the percentage contribution of each process parameter affecting spark gap.

TABLE 10.6 Response Table for Means of Spark Gap by ANOVA

Source	DF	Seq SS	Adj SS	Adj MS	F-Value	P-Value	Contribution
PONT	2	13.30	13.30	6.650	0.19	0.829	0.43%
POFFT	2	65.26	65.26	32.628	0.93	0.415	2.10%
SV	2	563.06	563.06	281.532	8.02	0.004	18.14%
WF	2	104.41	104.41	52.204	1.49	0.256	3.36%
PC	2	1796.94	1796.94	898.470	25.60	0.000	57.88%
Error	16	561.59	561.59	35.099			18.09%
Total	26	3104.56					100.00%

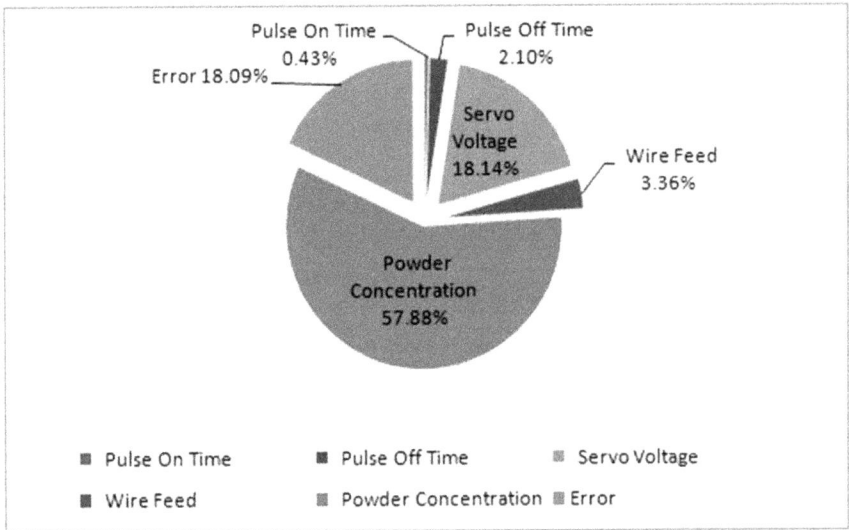

FIGURE 10.12   Effect of process parameters on spark gap.

10.4.1.3   ANALYSIS BY S/N RATIO

To obtain optimal cutting performance of spark gap, the lower the better quality characteristic must be taken.

The main effect plots of S/N ratio of SG are shown in Figure 10.13 the optimum levels for pwEDM parameters can be predicted from the main effect plots for S/N ratio of the spark gap. It shows the level number 2 for PONT, level 1 for POFF, level 1 for SV, level 1 for WF rate and level 1 for PC, gives maximum value of S/N ratio. For the lower value of the SG S/N ratio should be higher as shown in the main effect plot for S/N (SG). At the last confirmation test has been performed. Table 10.7 shows spark gap result at initial level of process parameters and after optimization, respectively. Microscopic image of kerf width at initial condition and optimum condition are shown in Figure 10.14a and 10.14b, respectively.

TABLE 10.7   Validation Test

Powder Mixed Wire EDM Process Parameters	Initial Level and Its Value	Optimum Process Parameter Level and Its Value		
		Taguchi Method Prediction	Genetic algorithm	Experiment
PONT	1-(115)	2-(120)	3-(125)	3-(125)
POFFT	1-(45)	1-(45)	1-(45)	1-(45)
SV	1-(10)	1-(10)	1-(10)	1-(10)
WF	1-(2)	1-(2)	1-(2)	1-(2)
PC	1-(0.0)	1-(0.0)	1-(0.0)	1-(0.0)
Spark gap (μm)	30.245	26.470	23.185	24.326
% Decrement in spark gap *(Taguchi method)*		12.48%		
% Decrement in spark gap *(Genetic Algorithm)*		19.57%		
%Error between predicted and experimental value *(Genetic Algorithm)*		4.69%		

10.4.2   OPTIMIZATION BY GENETIC ALGORITHM

The optimal selection of process parameters of pwEDM helps to reduce spark gap which avoids the inaccuracy in fabricated part and makes it suitable for precise machining.

In this section, the process parameters of pwEDM are optimized with the help of GA. The fitness function is derived from regression model. Regression model reveals the relationship between the powder-mixed w-EDM process parameters and the response. The regression model using Taguchi method is obtained by Minitab software.

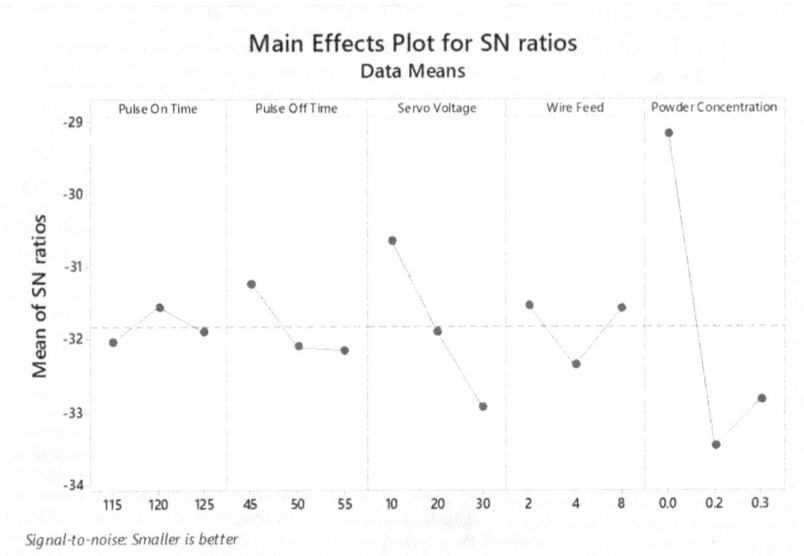

FIGURE 10.13   Main effect plots for S/N ratio of spark gap (L27array).

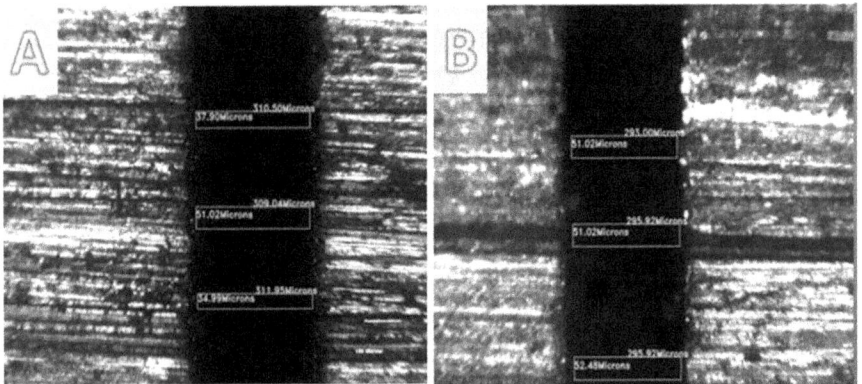

FIGURE 10.14   Kerf Width at (a) initial condition, (b) Taguchi optimum condition.

An empirical regression model is chosen and it is given in Equation 10.3.

$$Y = a + bX_1 + cX_2 + dX_3 + eX_4 + fX_5 \qquad (10.3)$$

where, Y is the response, that is, Spark gap; X_1 is the process parameters and "a, b, c, d, e and f" are the coefficients.

The equation of spark gap getting after the regression analysis is shown in Equation 4 and its constraint is given below:

Find out of optimum value: [PONT (X_1), POFFT (X_2), SV (X_3), WF (X_4), PC (X_5)]

Minimize: spark gap $f\{X_1, X_2, X_3, X_4, X_5\}$

Spark gap $(f) = 19.6 - 0.129\,X_1$

$$+ 0.314X_2 + 0.557X_3 + 0.004X_4 + 57.6X_5 \qquad (10.4)$$

Within the range: 115 µs < PONT < 125 µs, 45 µs < POFFT < 55 µs, 10 V < SV < 30 V, 2 m/min < WF < 8 m/min, 0.0 g/L < PC < 0.3 g/L;

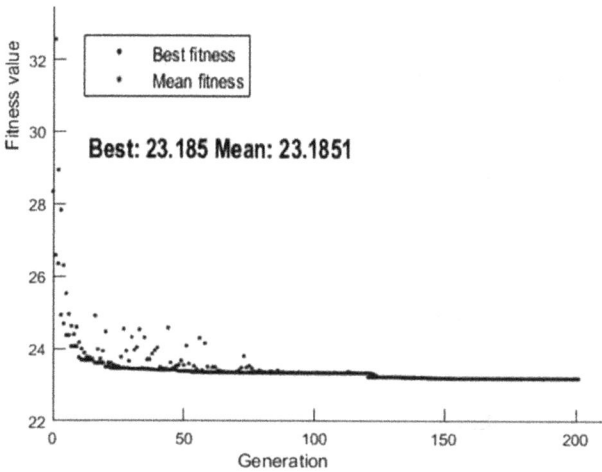

FIGURE 10.15   Genetic algorithm fitness value of spark gap.

GA tool box in MATLAB (R2017a) is utilized to find out the optimum level of process parameters which minimizes the spark gap. In this experimentation, population size was taken as 50, fitness scaling done by proportional function and a Tournament selective criterion was used. Set crossover fraction of 0.8 with single point crossover function and adaptive feasible mutation function was utilized for optimization. The optimum value of spark gap was obtained at PONT of 125 µs, POFFT of 45 µs, SV of 10 V, WF rate of 2 m/min and PC of 2 g/L. The optimum value of spark gap predicted by the GA approach is equal to 23.185 µm, it is shown in Figure 10.15.

Validation tests were performed, results for which are shown in Table 10.7.

10.5   CONCLUSIONS

1. Taguchi, ANOVA and S/N ratio analysis for spark gap gives the levels of the optimum parameters as level 2 of PONT, that is, 120 µs, level 1 of POFFT, that is, 45 µs, level 1 of SV, that is, 10 V, level 1 of WF rate, that is, 2 m/min and level 1 of PC, that is, 0.0 g/L. At this level, the spark gap is 26.470 (µm).
2. Confirmation test shows that after single objective optimization using Taguchi technique, spark gap decreases from 30.245 (µm) to 26.470 (µm), that is, percentage decrement in spark gap = 12.48%.
3. ANOVA results reveal that PC is most significant process parameter among all the considered parameters. The percentage contribution of PC is equal to 57.88%.
4. The GA gives optimum value of spark gap which is equal to 23.185 µm and process parameters values at optimum condition are PONT = 125 µs, POFFT = 45 µs, SV = 10 V, WF = 2 m/min and PC = 0.0 g/L.
5. After implementation of GA, validation test showed a 19.57% decrement in spark gap also the % error between predicted and experimental value of spark gap is only 4.69%.

ACKNOWLEDGEMENT

The authors would like to acknowledge Government Engineering College, Karad for providing the research facilities.

KEYWORDS

- powder mixed WEDM
- graphite powder
- spark gap
- Taguchi method
- genetic algorithm

REFERENCES

1. Wong, Y.S.; Lim, L.C.; Rahuman, I.; Tee, W.M. Near-mirror-finish phenomenon in EDM using powder-mixed dielectric. *Journal of Materials Processing Technology.* 1998, 79, 30–40.
2. Tzeng, Y.F.; Lee, C.Y. Effects of powder characteristics on electrodischarge machining efficiency. *International Journal of Advanced Manufacturing Technology.* 2001, 17, 586–592.
3. Zhao, W.S; Meng, Q.G.; Wang Z.L. The application of research on powder mixed EDM in rough machining. *Journal of Materials Processing Technology.* 2002, 129, 30–33.
4. Kansal, H.K.; Singh, S.; Kumar, P.; Parametric optimization of powder mixed electrical discharge machining by response surface methodology. *Journal of Materials Processing Technology.* 2005, 169, 427–436.
5. Jahan,M.P.; Rahman,M.; Wong,Y.S. Study on the nano-powder-mixed sinking and milling micro-EDM of WC-Co. *International Journal of Advanced Manufacturing Technology.* 2011, 53, 167–180.
6. Singh, B.; Kumar, J.; Kumar S.; Influences of process parameters on MRR improvement in simple and powder-mixed EDM of AA6061/10%SiC composite. *Materials and Manufacturing Processes.*2015, 30, 303–312.
7. Kumar, V.; Sharma, N.; Kumar, K,; Khanna, R. Surface modification of WC-Co alloy using Al and Si powder through WEDM: A thermal erosion process. *Particulate Science and Technology.* 2018, 36, 7, 878–886.
8. Jarin, S.; Saleh, T.; Rana, M.; Muthalif, A. G. A.; Ali, M. Y. An experimental investigation on the effect of nanopowder for micro-wire electro discharge machining of gold coated silicon. *Procedia Engineering.* 2017, 184, 171–177.

CHAPTER 11

STUDY OF OCEAN WAVE FLOW AROUND A VERTICALLY SUBMERGED RECTANGULAR PLATE IN INTERMEDIATE DEPTH OF WATER

PRADIP DEB ROY

Department of Mechanical Engineering, National Institute of Technology Silchar, Assam, India

E-mail: pdroy02@yahoo.co.in

ABSTRACT

The chapter presented is a study of ocean wave flow in the intermediate depth of water around a vertically submerged rectangular plate of various plate thicknesses under two different conditions: (1) a surface-piercing rectangular thin plate and (2) a bottom-standing rectangular thin plate. In this chapter, an analytical solution of linear wave theory is solved. The solution method is confined under the different relative depth of water and different time-period conditions. The results are validated through Simpson's 1/3 rule. The solution of horizontal wave forces on the plate with respect to the wave amplitudes is obtained in the intermediate depth of water. It is observed that the wave force on the plate at a surface-piercing position is maximum for $d/L = 0.48$ compared to a bottom standing position. It is also observed that the wave forces on the plate at Type I and also at Type II are both gradually converging with the increase of the time period. Wave forces also increase with the thickness of the plate. The linear wave is studied numerically and validates its results with the analytically, showing good agreement and their percentage errors approximately less than to 3.419%.

11.1   INTRODUCTION

The interaction between linear wave and maritime structures is of fundamental importance in coastal and ocean engineering. An attempt is made to design a new-type wave energy converter, wherein kinetic energy obtainable from the ocean wave can be easily converted into the form of potential energy and subsequently can be used for any other purpose. A vertically submerged rectangular plate of different thickness is considered in the flow field, which acts as an energy absorber of a wave energy converter and moves in the surge direction due to the action of an ocean wave. The plate is an element of the entire design of the wave energy converter. The aim of this paper is to study the wave flow around a vertical rectangular plate in the intermediate depth of water ($0.05 \leq d/L \leq 0.5$). This is the initial step of the research for the development and the design of the wave energy converter. Linear wave theory (LWT) is used here to study the wave force and flow parameter on the plate. Finally, the analytical results are validated with numerical results. Simpson's 1/3 rule is used here for the validation. Flow parameters are studied around the plate in the intermediate depth of water by ANSYS FLUENT software and using the volume of fluid (VOF) method. Two different configurations are used here for the study.

- Type I: Surface-piercing rectangular plate.
- Type II: Bottom-standing rectangular plate.

Various types of studies on the wave force were carried out at different times by many authors on different types of submerged structures, like a vertical barrier of different geometrical shapes, etc. Hanssen and Torum [1] experimentally studied the breaking wave forces on tripod concrete structures on shoal using a Morison's equation. The main purpose of their study was to investigate both horizontal and vertical forces as well as overturning moments due to wave acting on the tripod.

Meylan [2] presented a variational equation for the wave force on floating thin plates in his article "A variational equation for the wave forcing of floating thin plates". He had used free surface Green's function for the solution. Solutions of the variational equation were presented for some simple thin geometric plates using polynomial-based functions. Maiti and Sen [3] investigated a numerical time-simulation algorithm for analyzing highly nonlinear solitary waves interacting with plane gentle and steep slopes, employing mixed Eulerian–Lagrangian method and also found out the pressures and forces on the inclined wall by Bernoulli's equation.

Sundaravadivelu et al. [4] experimentally investigated the measurement of wave force and moments on an intake well due to regular wave using Linear Diffraction Theory.

Tsai and Jeng [5] studied the forces on vertical walls due to obliquely incident waves employing the Fourier series. For the calculation of the short-crested wave system, Prabhakar and Sundar [6] numerically computed the variations of pressure on vertical walls at a constant depth of water by the Fourier series approximation method and their results obtained from this method were compared with the experimental results conducted by Nagai [7]. Prabhakar and Sundar also compared their results with the results of Fenton [8] for the pressure curves at still water level (SWL) and seabed. Mallayachari and Sundar [9] investigated the wave pressure exerted on vertical walls due to regular and random waves, using Fourier series approach.

Most of the other authors investigated the breaking wave forces exerted on a vertical circular cylinder (e.g., Sawaragi and Nochino [10]; Apelt and Piorewiez [11]; Hovden and Torum [12]). Neelamani et al. [13] experimentally investigated wave forces on a seawater intake structure consisting of a perforated square caisson in a regular and random wave.

Deb Roy and Ghosh [14] investigated the wave force on a vertical submerged circular thin plate in shallow water by the use of the Morison's equation. The main purpose of their study was to investigate both horizontal and overturning moments due to regular wave acting on the plate.

Deb Roy and Ghosh [15] investigated the wave force on a vertical submerged circular thin plate in shallow water due to oblique waves at a different incident angle of the wave.

Deb Roy and Ranjan [16] theoretically investigated the wave force on a vertical submerged rectangular thin plate in shallow water by the use of the Fourier series technique.

Teo [17] theoretically investigated wave pressure on a vertical wall due to short-crested waves: fifth-order approximation.

Jeng [18] theoretically investigated wave kinematics of partial reflection from a vertical wall at a different angle due to short-crested waves: third-order approximation.

Machado et al. [19] simulated the generation and propagation of regular waves by CFD software, ANSYS CFX. The authors mainly focus on different methods to generate the waves. The authors also analyzed to prevent the wave reflection from the end of the numerical wave tank (NWT) using the numerical beach.

Wu Yun-Ta et al. [20] investigated the solitary wave propagation by two approaches. The main aim of this work was to generate stable and accurate solitary waves.

Finnegan and Goggins [21] numerically simulated LWT with the wave–body interaction in deep water using the ANSYS CFX program. The simulating results were matched with the LWT and the results were showing good agreement.

Kim et al.'s [22] ANSYS FLUENT codes were used to simulate the wave–structure interactions at the offshore substructure. To prevent the reflection of waves, a damping domain was created at the end of the tank. Overtopping wave-energy converter stored the water of incoming waves during high tide in a reservoir at a higher level than the mean sea water level in the form of potential energy. In its natural way the water back to the sea, the water passes through turbines generating electricity.

11.2   NUMERICAL METHOD

The numerical simulation of the LWT is analyzed in a two-dimensional (2D) NWT by VOF method in the intermediate depth of water ($0.05 \leq d/L \leq 0.5$). The domain of the computation is a rectangle ($l_x \times l_z$), with one end is wave generating zone (l_{x1}) and the other end is damping zone (l_{x2}). The calm water depth (d) is 5 m and x (=0, 20, 30 m) is the distance from the wavemaker where the entire analysis was carried out. The Cartesian coordinates are employed in the x–z plane, vertical to the fluid surface. The z-axis is directed vertically upward from the SWL, which is measured positive upward and the x-axis is measured along the direction of propagation of waves. The wave is generated in an NWT at $x = 0$ and flow in the x-direction. Consider the free surface of oscillating is in the z-direction. The effect due to flow in the y-direction is neglected. The top boundary of the NWT opens to the atmosphere. The bottom wall and the right wall of the computational domain are modeled as solid walls where the no-slip boundary condition is applied. The geometry of the NWT corresponding dimensions and location of boundaries are shown in Figure 11.1.

For the computational domain discretization, the rectangular grid was used. The user define function was used to generate the linear wave to obtain the highest quality results. Grid with approximately 120,451 nodes was tested for investigating the numerical results. To analyze the free surface more accurately, mesh refinement was needed near the vicinity of the mean water level as shown in Figure 11.2. The finer meshes were observed near the free surface and near the wavemaker, while coarser meshes were found towards the right wall and the bottom of the domain. The size of the mesh increased gradually towards the end of the NWT to provide a numerical damping effect. Linear waves of wave height $H = 0.12$ m and 0.2 m are generated in an NWT, under the time periods $T = 3.29$ s and 5 s, respectively.

11.2.1   GOVERNING EQUATION

Consider unsteady, incompressible, inviscid, irrotational flow without surface tension and with atmospheric pressure $p_a = 0$. Continuity equation and Navier–Stokes equation are used as the governing equations of the present problem

$$\frac{\partial u}{\partial x} + \frac{\partial w}{\partial z} = 0 \tag{11.1}$$

$$\frac{\partial u}{\partial t} + u\frac{\partial u}{\partial x} + w\frac{\partial u}{\partial z} = -\frac{1}{\rho}\frac{\partial p}{\partial x} + v\left[\frac{\partial^2 u}{\partial x^2} + \frac{\partial^2 u}{\partial z^2}\right] \tag{11.2}$$

$$\frac{\partial w}{\partial t} + u\frac{\partial w}{\partial x} + w\frac{\partial w}{\partial z} = -\frac{1}{\rho}\frac{\partial p}{\partial z} + v\left[\frac{\partial^2 w}{\partial x^2} + \frac{\partial^2 w}{\partial z^2}\right] \tag{11.3}$$

where u and w are the velocity components in the x and z directions, respectively; t is the time; p is the pressure; ρ is the density which varies across the air and water interface, and $v(=0)$ is the kinematic viscosity of the inviscid fluid.

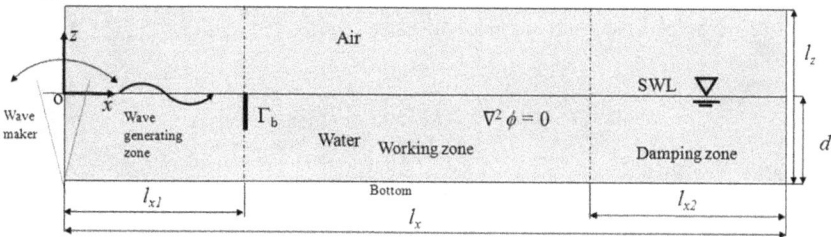

FIGURE 11.1   Definition sketch of numerical wave tank.

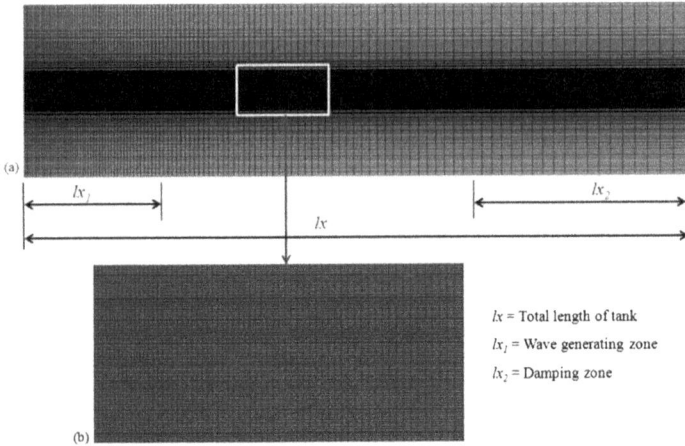

FIGURE 11.2   Mesh refinement near the free surface.

In the present study, the interface of the two fluids is modelled by the VOF method [24]. The volume fraction function α_q is defined as the ratio of the volume occupied by the *qth* phase in a cell to the total volume of the cell. Accordingly, $\alpha_q = 1$ means that the cell is filled with the fluid of the *qth* phase, $\alpha_q = 0$ means that the cell does not contain any fluid of the *qth* phase, when $0 < \alpha_q < 1$, the cell is called an interface cell. For water wave problems, $q = 1$ or 2 since there are only two phases: air and water. α_q is determined by the following equations:

$$\frac{\partial \alpha_q}{\partial t} + \nabla \cdot \left(\alpha_q \overline{V} \right) = 0 \tag{11.4}$$

$$\sum_{q=1}^{2} \alpha_q = 1 \tag{11.5}$$

where $\overline{V}$ is the velocity vector. The density of the air–water mixture is computed by volume fraction, as follows:

$$\rho = \alpha_q \rho_w + \left(1 - \alpha_q \right) \rho_a \tag{11.6}$$

where ρ_w and ρ_a are the water density and the air density, respectively. In the present work, material properties of air and water are defined as $\rho_a = 1.225$ kg/m³, $\rho_w = 998.2$ kg/m³.

The specification of the system at which simulation was run is an Intel XEON processor and a 64 GB RAM. An additional graphics card of 2 GB was also installed. A simulation was performed based on RANSE with the k–ε turbulence model. The time discretization of the equations was achieved with the first-order implicit scheme and a second-order upwind scheme. The time step size was set at 0.01 s, the number of time steps was selected 4000 and the maximum iteration was 20. Based on these parameters and the computational capabilities, the simulation took 20 h to solve.

11.2.2 GEOMETRY OF DAMPING

To prevent the reflection of waves from the boundary, damping is essential at the right end of the NWT. There are different methods proposed by different authors in their literature for the absorption of incident waves, namely:

- Controlling the total time of simulation to ensure that the first wave generated does not reach the end of the tank [25].
- Mesh size increases at the end of the tank to dissipate waves [26].
- Implementing a beach [27].
- To prevent the reflection of waves from the affecting area, tank length must increase [28].
- Creating a damping zone [29].

In this work, we have used three methods to dissipate the wave energy: (1) mesh size increases at the end of the tank to dissipate waves, (2) to prevent the reflection of waves, tank length must increase, and (3) creating a damping zone. Furthermore, the height of the NWT above the SWL is increased to 5 m to prevent air circulation at the top boundary.

11.3 MATHEMATICAL FORMULATION

Consider a rectangular plate Γ_b of dimensions l_1, l_2 and l_3 submerged vertically in an intermediate depth of water. Here, l_1 is the thickness of the plate, l_2 is the width of the plate and l_3 is the height of the plate. The system is idealized as a 2D Cartesian coordinate system and hence l_2 is considered a unity. The plate is subjected to an incoming wave which is assumed to be traveling in the x-direction. The fluid domain occupies the region $-d \leq z \leq 0$,

$0 \le x \le l_x$ except for the plate in the fluid region. The plate Γ_b of negligible draft occupies the following two positions, as shown in Figure 11.3.

- Type I: $x = 0$, 30, and 20, $-z_2 \le z \le 0$.
- Type II: $x = 0$, 30, and 20, $-d \le z \le -z_1$.

11.3.1 GOVERNING EQUATION

The governing equation and the boundary conditions are nondimensional-ized by the following nondimensional parameters:

$$\left(x^*, z^*, \eta^*, d^*\right) = k\left(x, z, \eta, d\right), \ \phi^* = \sqrt{\frac{k^3}{g}}\phi, \ t^* = \omega t, \ \omega^* = \frac{\omega}{\sqrt{gk}} \quad (11.7a)$$

$$P^* = \frac{kP}{\rho g}, \ F_x^* = \frac{F_x}{0.5\rho g d^2} \quad (11.7b)$$

where $k = 2\pi/L$ (L = wavelength) and $\omega = 2\pi/T$ (T = time period) are the wave number and wave frequency, respectively. Dimensionless quantities are denoted by the symbol star (*) and will be omitted in Sections 11.3 and 11.4 for the sake of simplicity.

Let $\phi(\overline{x},t)$ denote velocity potential and the governing differential equations for irrotational wave motion are given by the Laplace equation:

$$\nabla^2 \varphi\left(\overline{x},t\right) = 0 \quad (11.8)$$

11.3.2 BOUNDARY CONDITIONS

To solve the governing equation (11.8), a set of boundary conditions is required. They are summarized below:

(a) The dynamic boundary condition at the free surface (DFSBC):

$$\omega\varphi_t + \eta = 0, \ at \ z = 0 \quad (11.9)$$

(b) The kinetic boundary condition at the free surface:

$$\varphi_z = \omega\eta_t, \ at \ z = 0 \tag{11.10}$$

(c) Combined free surface boundary condition:

$$\varphi_z = \omega\eta_t, \ at \ z = 0 \tag{11.11}$$

(d) Bottom boundary condition (BBC):

$$\varphi_z = 0, \ at \ z = -d \tag{11.12}$$

Where η is the wave profile elevation, g is the acceleration due to gravity, t is the time and d is the depth of water.

11.4   METHOD OF SOLUTIONS

Employing the Laplace equation (11.8) into the BBC and the linearized DFSBC, the desire velocity potential for small amplitude linear waves can be derived by the method of separation of variables. A useful form of this velocity potential is

$$\varphi = \frac{1}{\omega}\frac{\cosh(z+d)}{\cosh(d)}\sin(x-t) \tag{11.13}$$

Integrating the velocity potential into the linearized DFSBC and letting $z = 0$ yields the equation for the wave surface profile.

$$\eta = \cos(x-t) \tag{11.14}$$

The dispersion relation

$$\omega^2 = \tanh d \tag{11.15}$$

11.4.1 PRESSURE AND FORCE ON THE PLATE

Evaluation of pressure induced by the wave on the plate can be determined from Bernoulli's equation:

$$P = -z - \omega\varphi_t - \frac{1}{2}\left[\varphi_x^2 + \varphi_z^2\right] \tag{11.16}$$

The horizontal and vertical force F_x per unit width of the plate can be determined by integrating the above expression over the wetted contour.

$$F_x = \int_{\Gamma_b} Pn_x ds \tag{11.17}$$

Here ρ is the water density and n_x is the x-component of the exterior unit normal vector $\bar{n}$ on Γ_b. As the plate is submerged vertically, so $n_x = 1$. Determination of the time derivative ϕ_t and velocities ϕ_x are done by first-order difference rules. In the configuration shown, there is no variation of pressure in the y-direction, as the waves are assumed to be long crested and propagation in the x-direction. Referring to Figure 11.3, wave-induced pressure on the plate in the x-direction and z-direction are to be calculated.

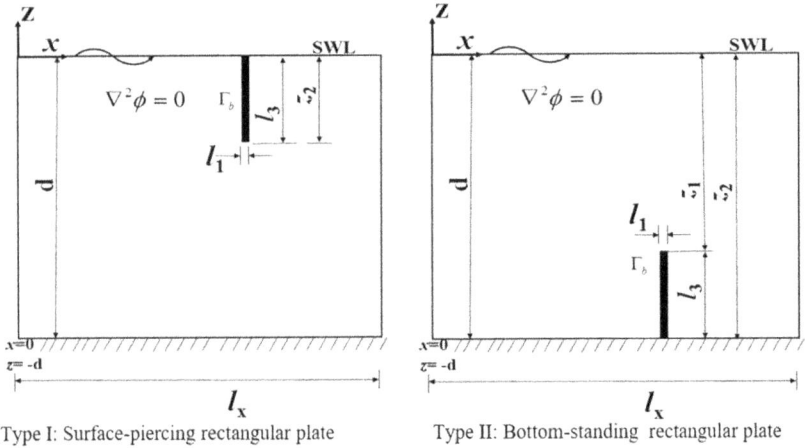

Type I: Surface-piercing rectangular plate   Type II: Bottom-standing rectangular plate

FIGURE 11.3 Definition sketch of the plate positions.

11.5 RESULT AND DISCUSSION

The aim of this chapter is to study wave flow around a vertical rectangular plate in the intermediate depth of water ($0.05 \leq d/L \leq 0.5$) and also to investigate the effect of ocean wave force on the plate by LWT as shown in Figure 11.3. For each type, results have been estimated up to four decimal places in order to achieve accuracy. Finally, the analytical results are validated with the numerical results. Simpson's 1/3 rule is used here for the validation. Flow parameters are studied around the plate in the intermediate depth of water by ANSYS FLUENT software and using the VOF method.

To demonstrate the effects of wave force around the plate, representative sets of analytical results with the numerical results for horizontal force (F_x) for the two types of a vertically submerged rectangular thin plate in an intermediate depth of water between the ranges $0.16 \leq d/L \leq 0.48$ are shown in Tables 11.1–11.3. It is observed from Tables 11.1–11.3 that analytical results are in good agreement with the numerical results and their percentage errors are very less in each case. As shown in Tables 11.1–11.3, wave forces on the plate are more of a relative depth of water ($d/L = 0.48$) compared to the other relative depth of water ($d/L = 0.31, 0.19, 0.16$). It is also observed that the wave force increases with the thickness of the plate. From Tables 11.1 to 11.3, it is clear that on a/d (= 0.002, 0.004, 0.006, 0.008, 0.01, 0.012) and d/L (= 0.48), the agreement of analytical results with the numerical results is less than equal to 3.419%. This concludes that wave force on the plate is important at a relative depth of water ($d/L = 0.48$) compared to the other relative depth of water.

Figures 11.4–11.6 show the plot of the horizontal force (F_x) on the plate versus dimensionless wave amplitude ($a/d = 0.002$ to 0.012) for various relative depths of water $d/L = 0.48, 0.31, 0.19, 0.16$ and for various plate thickness ($l_1 = 0.001$ m, 0.01 m and 0.02 m). Forces are nondimensionalized by the factor ($1/2\rho gd^2$). It is observed that the horizontal force on the plate is very high for higher relative depth of water ($d/L = 0.48$) compared to the lower relative depth of water ($d/L = 0.31, 0.19, 0.16$) for an (a/d) ratio. Also, it is seen that the horizontal force gradually decreases with the decreasing z/d (= 0 to −1) for an (a/d) ratio. Figure 11.4 also shows that for the horizontal force of the two types, which gradually converges to the decreasing value of the relative depth of water and when $d/L = 0.16$, the convergence is very close. Figures are also showing that the results obtained from the analytical solutions are in good agreement with the results obtained from numerical solutions. It is also observed that the effect of plate thickness takes an important role in the horizontal force. It is observed that Figure 11.6 for the plate thickness $l_1 = 0.02$ m experienced more horizontal force on the plate compared to the lower plate thickness ($l_1 = 0.001$ m, 0.01 m). Figure 11.4 shows that the horizontal force on the plate is very low for the plate thickness ($l_1 = 0.001$ m).

TABLE 11.1 Comparison of Horizontal Force between Analytical Values (F_x) and the Numerical Values (F_n) in Various Relative Depths of Water for $d = 5$ m, $l_1 = 0.001$ m

a/d	$T = 2.6$ s $d/L = 0.48$			$T = 3.29$ s $d/L = 0.31$			$T = 4.5$ s $d/L = 0.18$			$T = 5$ s $d/L = 0.16$		
	$F_x \times 10^{-9}$ Analytical	$F_n \times 10^{-9}$ Numerical	Error (%)	$F_x \times 10^{-9}$ Analytical	$F_n \times 10^{-9}$ Numerical	Error (%)	$F_x \times 10^{-9}$ Analytical	$F_n \times 10^{-9}$ Numerical	Error (%)	$F_x \times 10^{-9}$ Analytical	$F_n \times 10^{-9}$ Numerical	Error (%)
Type I ($z/d = 0$)												
0.002	0.1087	0.1050	3.418	0.0507	0.0490	3.369	0.0207	0.0200	3.346	0.0158	0.0153	3.343
0.004	0.2173	0.2099	3.418	0.1014	0.0980	3.369	0.0414	0.0400	3.346	0.0317	0.0306	3.343
0.006	0.3260	0.3148	3.418	0.1520	0.1469	3.369	0.0621	0.0600	3.346	0.0475	0.0459	3.343
0.008	0.4346	0.4197	3.418	0.2026	0.1958	3.369	0.0827	0.0799	3.346	0.0632	0.0611	3.343
0.010	0.5432	0.5246	3.418	0.2532	0.2446	3.369	0.1033	0.0998	3.346	0.0789	0.0763	3.343
0.012	0.6518	0.6295	3.418	0.3037	0.2934	3.369	0.1238	0.1197	3.346	0.0946	0.0915	3.343
Type II ($z/d = -1$)												
0.002	0.0149	0.0144	3.418	0.0171	0.0166	3.369	0.0128	0.0123	3.369	0.0109	0.0105	3.348
0.004	0.0299	0.0288	3.418	0.0342	0.0331	3.370	0.0255	0.0246	3.370	0.0217	0.0210	3.347
0.006	0.0448	0.0432	3.418	0.0513	0.0495	3.369	0.0382	0.0369	3.369	0.0326	0.0315	3.346
0.008	0.0597	0.0576	3.419	0.0683	0.0660	3.369	0.0508	0.0491	3.369	0.0434	0.0419	3.346
0.010	0.0745	0.072	3.418	0.0853	0.0824	3.369	0.0634	0.0613	3.369	0.0542	0.0523	3.347
0.012	0.0894	0.0863	3.419	0.1022	0.0988	3.370	0.0760	0.0735	3.370	0.0649	0.0627	3.346

TABLE 11.2 Comparison of Horizontal Force Between Analytical Values (F_x) and the Numerical Values (F_n) in Various Relative Depths of Water for $d = 5$ m, $l_1 = 0.01$ m

a/d	T = 2.6 s d/L = 0.48			T = 3.29 s d/L = 0.31			T = 4.5 s d/L = 0.18			T = 5 s d/L = 0.16		
	$F_x \times 10^{-9}$ Analytical	$F_n \times 10^{-9}$ Numerical	Error (%)	$F_x \times 10^{-9}$ Analytical	$F_n \times 10^{-9}$ Numerical	Error (%)	$F_x \times 10^{-9}$ Analytical	$F_n \times 10^{-9}$ Numerical	Error (%)	$F_x \times 10^{-9}$ Analytical	$F_n \times 10^{-9}$ Numerical	Error (%)
Type I (z/d = 0)												
0.002	10.87	10.50	3.418	5.070	4.900	3.369	2.070	2.000	3.346	1.587	1.534	3.343
0.004	21.73	20.99	3.418	10.14	9.800	3.369	4.140	4.000	3.346	3.170	3.064	3.343
0.006	32.60	31.48	3.418	15.20	14.69	3.369	6.210	6.000	3.346	4.750	4.591	3.343
0.008	43.46	41.97	3.418	20.26	19.58	3.369	8.270	7.990	3.346	6.326	6.115	3.343
0.010	54.32	52.46	3.418	25.32	24.46	3.369	10.33	9.980	3.346	7.899	7.634	3.343
0.012	65.18	62.95	3.418	30.37	29.34	3.369	12.38	11.97	3.346	9.467	9.151	3.343
Type II (z/d = −1)												
0.002	1.490	1.440	3.418	1.710	10.66	3.369	1.280	1.230	3.346	1.091	1.055	3.343
0.004	2.990	2.880	3.418	3.420	3.310	3.369	2.550	2.460	3.346	2.179	2.106	3.343
0.006	4.480	4.320	3.419	5.130	4.950	3.369	3.820	3.690	3.346	3.263	3.154	3.343
0.008	5.970	5.760	3.419	6.830	6.600	3.369	5.080	4.910	3.346	4.343	4.198	3.343
0.010	7.450	7.200	3.419	8.530	8.240	3.369	6.340	6.130	3.346	5.420	5.239	3.343
0.012	8.900	8.630	3.419	10.22	9.880	3.369	7.600	7.350	3.346	6.493	6.276	3.343

TABLE 11.3 Comparison of Horizontal Force between Analytical Values (F_x) and the Numerical Values (F_n) in Various Relative Depths of Water for $d = 5$ m, $l_1 = 0.02$ m

a/d	$T = 2.6$ s $d/L = 0.48$			$T = 3.29$ s $d/L = 0.31$			$T = 4.5$ s $d/L = 0.18$			$T = 5$s $d/L = 0.16$		
	$F_x \times 10^{-9}$ Analytical	$F_n \times 10^{-9}$ Numerical	Error (%)	$F_x \times 10^{-9}$ Analytical	$F_n \times 10^{-9}$ Numerical	Error (%)	$F_x \times 10^{-9}$ Analytical	$F_n \times 10^{-9}$ Numerical	Error (%)	$F_x \times 10^{-9}$ Analytical	$F_n \times 10^{-9}$ Numerical	Error (%)
Type I ($z/d = 0$)												
0.002	43.50	42.00	3.418	20.30	19.60	3.369	8.300	8.020	3.346	6.350	6.140	3.343
0.004	86.90	84.00	3.418	40.50	39.20	3.369	16.57	16.02	3.346	12.68	12.26	3.343
0.006	130.4	125.9	3.418	60.80	58.80	3.369	24.84	24.01	3.346	19.00	18.37	3.343
0.008	173.8	167.9	3.418	81.00	78.30	3.369	33.08	31.98	3.346	25.30	24.46	3.343
0.010	217.3	209.8	3.418	101.3	97.80	3.369	41.31	39.93	3.346	31.59	30.54	3.343
0.012	260.7	251.8	3.418	121.5	117.4	3.369	49.53	47.87	3.346	37.87	36.60	3.343
Type II ($z/d = -1$)												
0.002	6.000	5.800	3.418	6.900	6.600	3.369	5.110	4.940	3.346	4.370	4.220	3.343
0.004	11.90	11.50	3.418	13.70	13.20	3.369	10.20	9.860	3.346	8.720	8.420	3.343
0.006	17.90	17.30	3.419	20.50	19.80	3.369	15.27	14.76	3.346	13.05	12.62	3.343
0.008	23.90	23.00	3.419	27.30	26.40	3.369	20.33	19.65	3.346	17.37	16.79	3.343
0.010	29.80	28.80	3.419	34.10	33.00	3.369	25.38	24.53	3.346	21.68	20.96	3.343
0.012	35.80	34.50	3.419	40.90	39.50	3.369	30.40	29.39	3.346	25.97	25.10	3.343

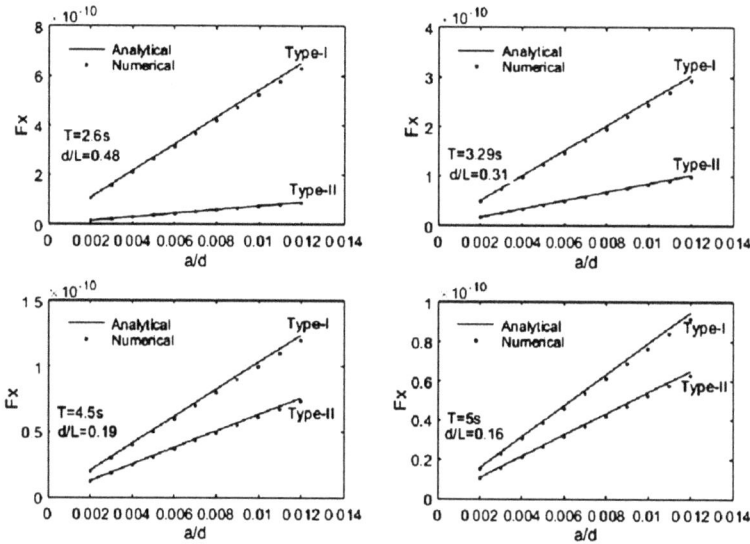

FIGURE 11.4 Dimensionless horizontal force (F_x) versus dimensionless wave amplitude (a/d) for the case of a vertically submerged rectangular plate under two different configurations. Here, $t = 0$, $x = 0$, $d = 5$ m, $l_1 = 0.001$ m, $l_3 = 1$ m.

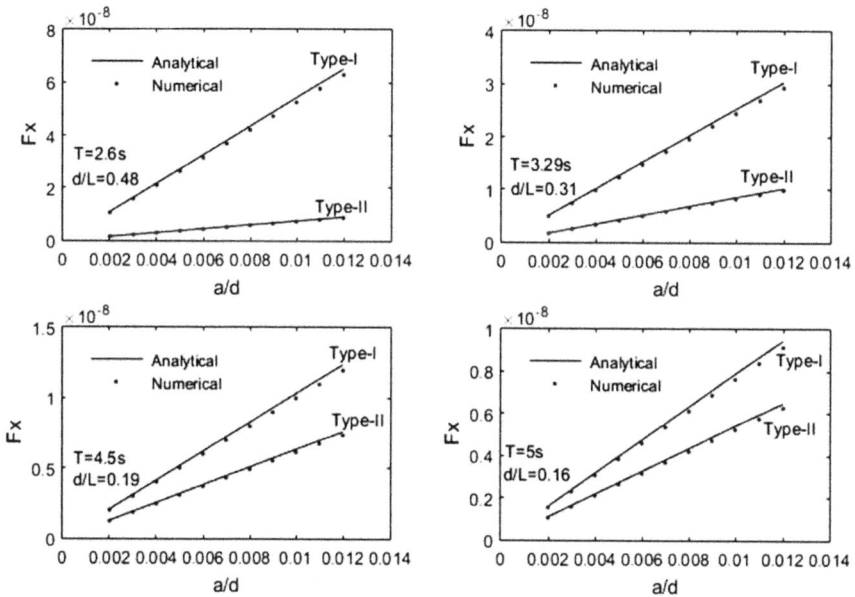

FIGURE 11.5 Dimensionless horizontal force (F_x) versus dimensionless wave amplitude (a/d) for the case of a vertically submerged rectangular plate under two different configurations. Here, $t = 0$, $x = 0$, $d = 5$ m, $l_1 = 0.01$ m, $l_3 = 1$ m.

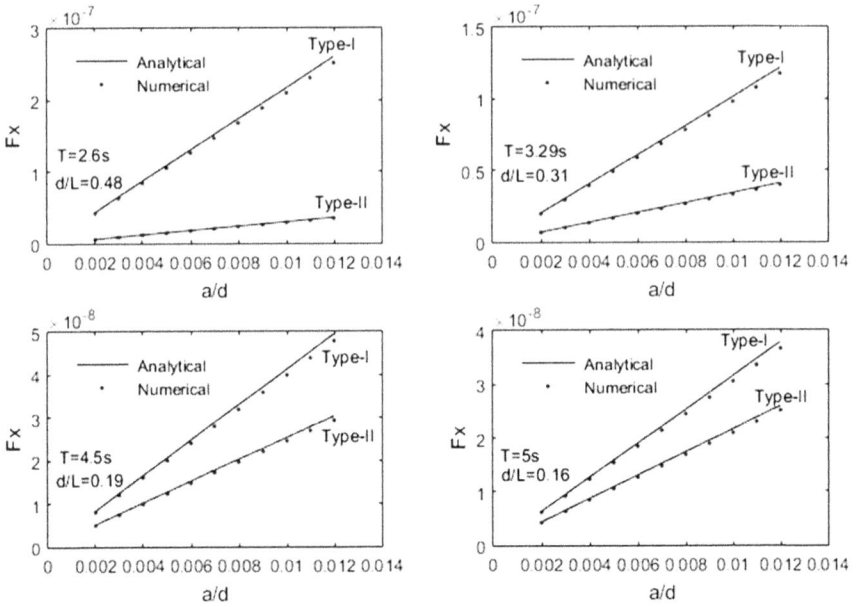

FIGURE 11.6 Dimensionless horizontal force (F_x) versus dimensionless wave amplitude (a/d) for the case of a vertically submerged rectangular plate under two different configurations. Here, $t = 0$, $x = 0$, $d = 5$ m $l_1 = 0.02$ m, $l_3 = 1$ m.

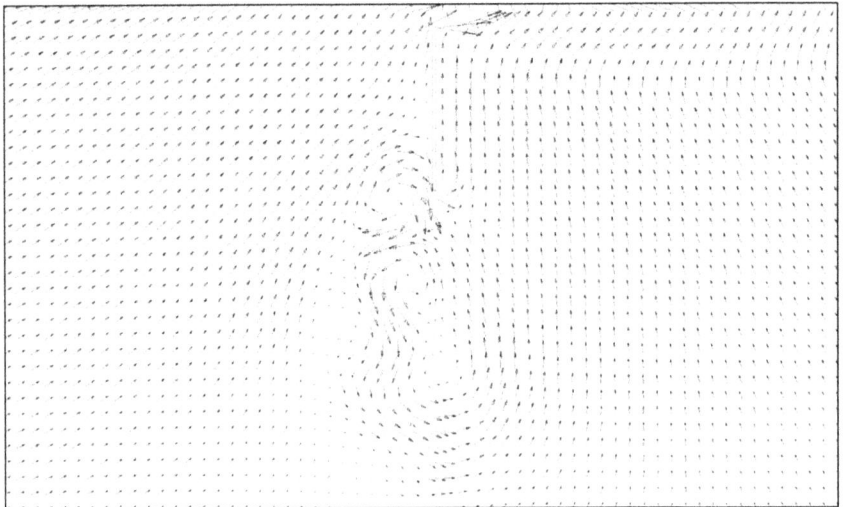

FIGURE 11.7 Velocity vector of the particle motion around a vertically submerged rectangular plate for $l_1 = 0.02$ m, $H = 0.2$ m and $T = 5$ s.

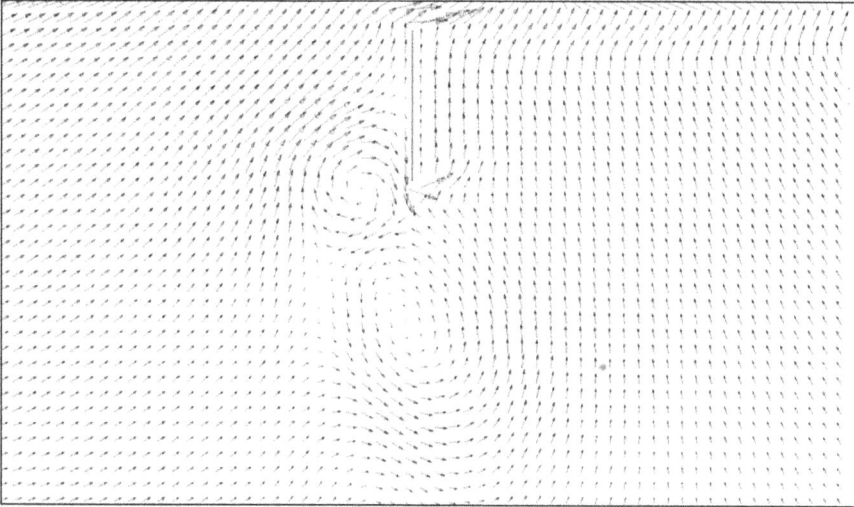

FIGURE 11.8 Velocity vector of the particle motion around a vertically submerged rectangular plate for $l_1 = 0.01$ m, $H = 0.2$ m and $T = 5$ s.

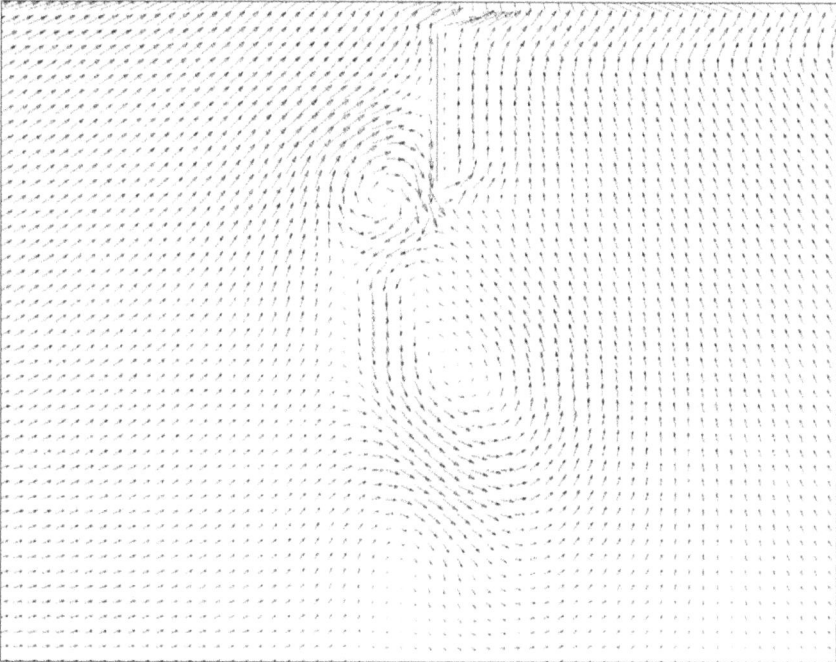

FIGURE 11.9 Velocity vector of the particle motion around a vertically submerged rectangular plate for $l_1 = 0.001$ m, $H = 0.2$ m and $T = 5$ s.

Figures 11.7–11.12 show the velocity vector of the particle motion around a vertically submerged rectangular plate of various plate thicknesses and for the wave height $H = 0.2$ m and 0.12 m and time period $T = 5$ s and $T = 3.29$ s, respectively. The thicknesses of the plate are considered 0.001 m, 0.01 m, and 0.02 m. Arrow lines indicate the direction of the motion of the particles. It is observed that the distribution of particle motion on the plate at the left side indicates that pressure distributions on the plate are not uniform in all cases, which indicates that moment generated on the plate due to the motion of waves.

Figure 11.7 shows that the particle motion around the plate for the wave height $H = 0.2$ m, time period $T = 5$ s and the plate thickness $l_1 = 0.02$ m. the right side of the plate shows that the velocity of the particle motion, moving upwards direction from the bottom of the tank and left side shows that the particle motion normally exerted on the plate which indicates that wave force act on the plate. Just left-bottom of the plate, there are three vortices generated. This indicates the wave force exerted on the plate is not uniform.

Figure 11.8 shows that the particle motion around the plate for the wave height $H = 0.2$ m, time period $T = 5$ s and the plate thickness $l_1 = 0.01$ m. Just left-bottom of the plate, there are two vortices generated. This indicates the wave forces exerted on the plate is not uniform.

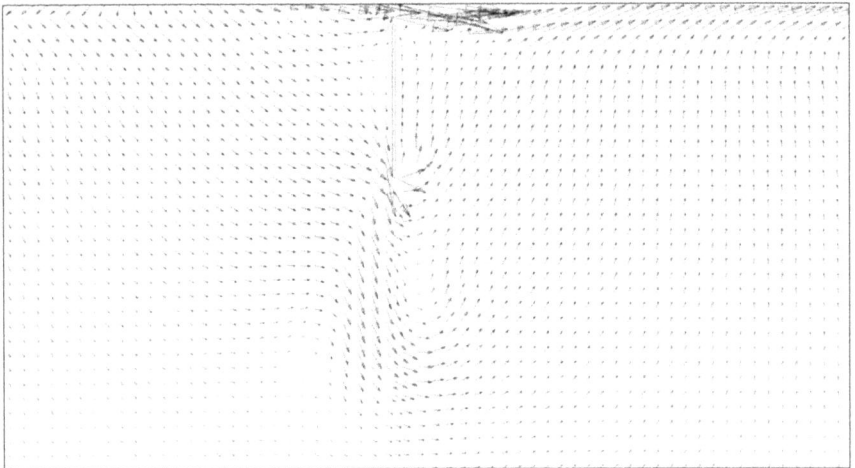

FIGURE 11.10 Velocity vector of the particle motion around a vertically submerged rectangular plate for $l_1 = 0.02$ m, $H = 0.12$ m and $T = 3.29$ s.

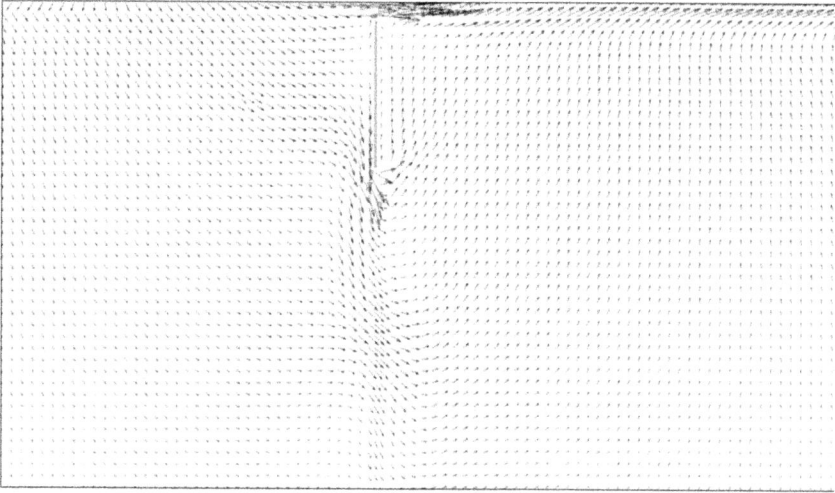

FIGURE 11.11   Velocity vector of the particle motion around a vertically submerged rectangular plate for $l_1 = 0.01$ m, $H = 0.12$ m and $T = 3.29$ s.

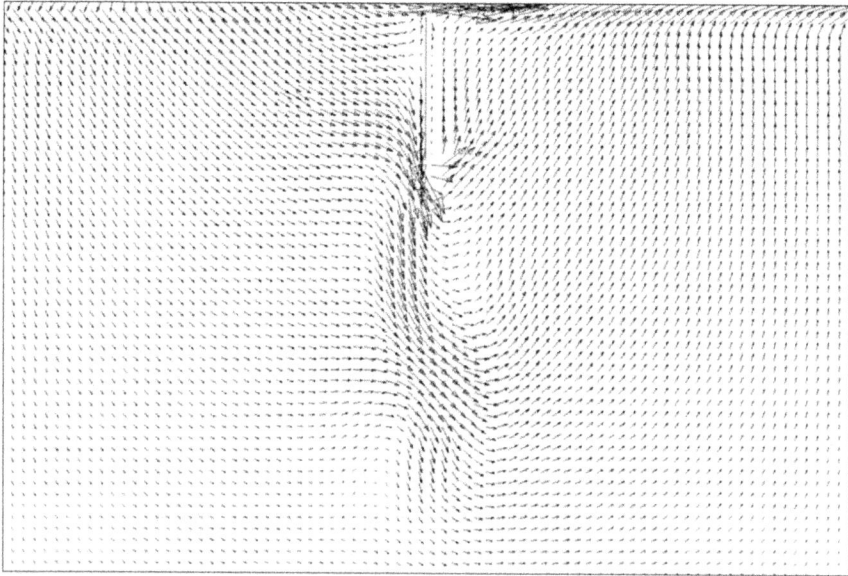

FIGURE 11.12   Velocity vector of the particle motion around a vertically submerged rectangular plate for $l_1 = 0.001$ m, $H = 0.12$ m and $T = 3.29$ s.

Figure 11.9 shows that the particle motion around the plate for the wave height $H = 0.2$ m, time period $T = 5$ s and the plate thickness $l_1 = 0.001$ mm. Just left-bottom of the plate, there are two vortices generated and another vortex is also generated at the bottom of the tank. This indicates that the wave force exerted on the plate is not uniform.

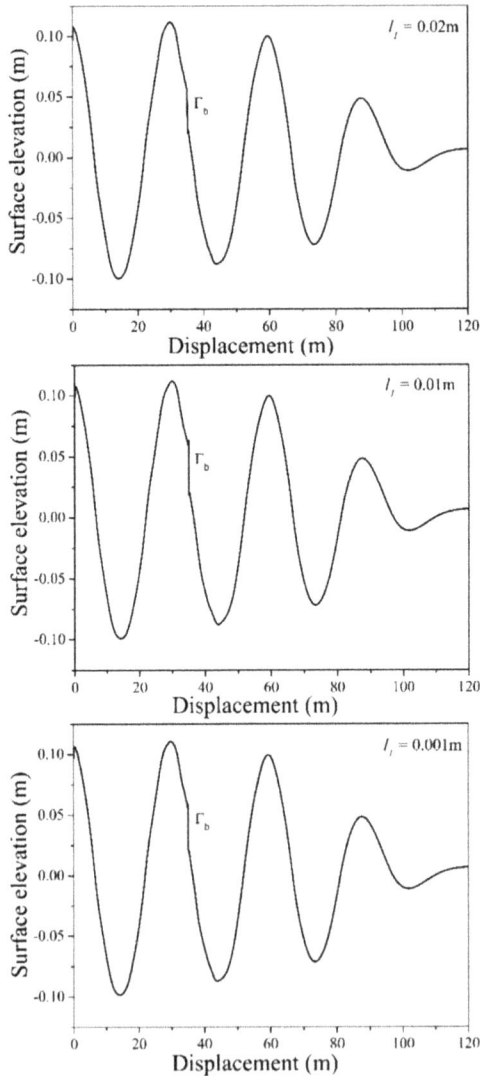

FIGURE 11.13 Surface elevation versus the displacement curves for $H = 0.2$ m, $d = 5$ m, $L = 30.5062$, $T = 5$ s, $x = 35$ m and different plate thickness ($l_1 = 0.02$ m, 0.01 m and 0.001 m).

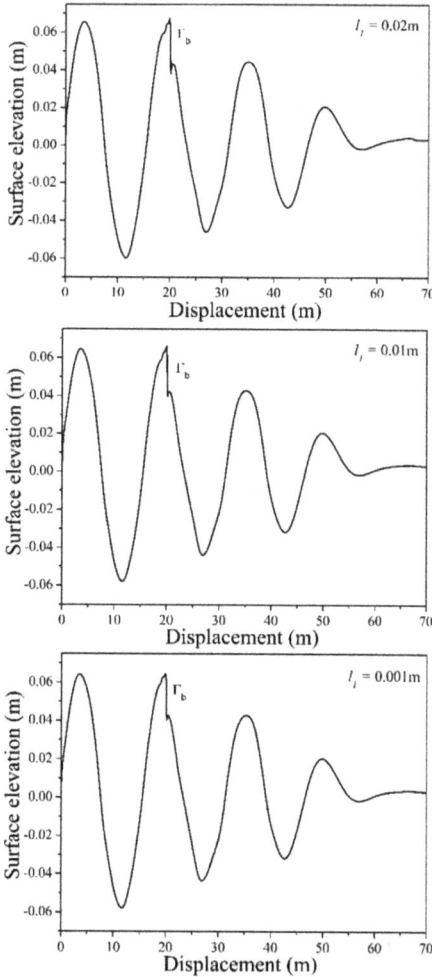

FIGURE 11.14   Surface elevation versus the displacement curves for $H = 0.12$ m, $d = 5$ m, $L = 15.9913$, $T = 3.29$ s, $x = 20$ m and different plate thickness ($l_1 = 0.02$ m, 0.01 m and 0.001 m).

Figure 11.10 shows that the particle motion around the plate for the wave height $H = 0.12$ m and time period $T = 3.29$ s and the plate thickness $l_1 = 0.02$ m. It is observed from Figure 11.10 that no vortex is generated as Figures 11.7– 11.9 and the particle start to move in the upward direction from the right-bottom of the plate. This indicates that the wave forces exerted on the plate is not enough.

Figure 11.11 shows that the particle motion around the plate for the wave height $H = 0.12$ m and time period $T = 3.29$ s and the plate thickness $l_1 = 0.01$ m.

Figure 11.12 shows that the particle motion around the plate for the wave height $H = 0.12$ m and time period $T = 3.29$ s and the plate thickness $l_1 = 0.001$ m.

Figures 11.13 and 11.14 show the surface elevation versus the displacement curves for various wave heights ($H = 0.12$ m and 0.2 m) and different plate thickness ($l_1 = 0.001$ m, 0.01 m and 0.02 m). The plate Γ_b is at the position $x = 35$ m for the time period $T = 5$ s and at the position $x = 20$ m for the time period $T = 3.29$ s. It is observed from the figures that due to presence of obstacle flow parameters are not smooth. Flow parameters are more disturbed for large plate thickness and disturbance reduces for the lower plate thickness.

11.6   CONCLUDING REMARKS

The present work is to study of ocean wave flow around a vertically submerged rectangular plate in the intermediate depth of water ($0.16 \leq d/L \leq 0.48$) under two different configurations. It is observed that the wave force on the plate at the mean free surface ($z/d = 0$) is maximum for $d/L = 0.48$ and compared to the other relative depth of water. It is also shown that for the horizontal force of the two types, which gradually converges to the decreasing value of the relative depth of water and when $d/L = 0.16$, the convergence is very close. This finding provides a useful guide to the designer, in the design of a vertical submerged rectangular plate in an intermediate depth of water. The horizontal wave force also depends on the thickness of the plate. It is observed that wave force is more for plate thickness 0.02 m and low for plate thickness 0.001 m. It is observed that the velocity vector of the particle motion on the plate is not uniform. The velocity vector depends on the plate thickness. Just left-bottom of the plate, there are three vortices generated of the plate thickness 0.02 m. This indicates the wave force exerted on the plate is not uniform.

KEYWORDS

- linear wave theory
- Laplace's equation
- rectangular plate
- horizontal wave force
- ANSYS FLUENT
- Simpson's 1/3 rule

REFERENCES

1. Hanssen, A.G., Torum, A., Breaking wave forces on tripod concrete structure on shoal. Journal of Water Way, Port, Coast and Ocean Engineering, 125 (1999) 304–310.
2. Meylan, M.H., A variational equation for the wave forcing of floating thin plates. Applied Ocean Research, 23 (2001) 195–206.
3. Maiti, S., Sen, D., Computation of solitary waves during propagation and runup on a slope. Ocean Engineering, 26 (1999) 1063–1083.
4. Sundaravadivelu, R., Sundar, V., Rao, T.S., Technical note: wave forces and moments on an intake well. Ocean Engineering, 26 (1997) 363–380.
5. Tsai, C.P., Jeng, D.S., Forces on vertical walls due to obliquely incident waves. In: Coastal Engineering. Elsevier, Amsterdam, The Netherlands, 1990.
6. Prabhakar, V., Sundar, V., Standing wave pressures on walls. Ocean Engineering, 28 (2001) 439–455.
7. Nagai, S., Pressure of standing waves on vertical wall. Journal of the Waterways and Harbors Division, ASCE, 95 (1969) 53–76.
8. Fenton, J.D., Wave force on vertical walls. Journal Waterway, Port, Coastal and Ocean Engineering, ASCE, 111 (4) (1985) 693–719.
9. Mallayachari, V., Sundar, V., Pressure exerted on vertical walls due to regular and random waves. In: *Proceedings of the International Symposium: Waves-Physical and Numerical Modeling*, Vancouver, Canada, August (1994) 21–24.
10. Sawaragi, T., Nochino, M., Impact forces of nearly breaking waves on a vertical circular cylinder. Coastal Engineering in Japan, Tokyo. (1984) p. 27.
11. Apelt, C.J., Piorewiez, J., Laboratory studies of breaking wave forces acting on vertical cylinders in shallow water. Coastal Engineering, 11 (3) (1987) 241–262.
12. Hovden, S.I., Torum, A., Wave forces on vertical cylinder on a reef. In: *Proceedings of the Third International Conference on Port Coastal Engineering for Developing Countries* (COPEDEC), Mombasa, Kenya, 1991.
13. Neelamani, S., Bhaskar, N.V., Vijayalokshmi, K., Technical note: wave force on a seawater intake caisson. Ocean Engineering, 29 (2002) 1247–1263.
14. Deb, Roy, P., Ghosh, S., Wave force on vertically submerged circular thin plate in shallow water. Ocean Engineering, 33 (2006) 1935–1953.
15. Deb, Roy, P., Ghosh, S., Force on vertically submerged circular thin plate in shallow water due to oblique wave. Indian Journal of Marine Sciences, 38(4) (2009) 411–417.
16. Deb, Roy, P., Ranjan, R., Aquatic Procedia, Elsevier. 4 (2015) 95–102.
17. Teo, H, T., Technical note: Ocean Engineering, 30 (2003) 2157–2166.
18. Jeng, D, S., Ocean Engineering, 29 (2002) 1711–1724.
19. Machado, M, M, Fábio., Lopes, A.M.G, Ferreira, A.D., Numerical simulation of regular waves: Optimization of a numerical wave tank. Ocean Engineering, 170(2018) 89–99.
20. Wu, Y.T., Hsiao, S.C., Generation of stable and accurate solitary waves in a viscous numerical wave tank. Ocean Engineering, 167 (2018) 102–113.
21. Finnegan, W., Goggins, J., Numerical simulation of linear water waves and wave structure interaction. Ocean Engineering, 43 (2012) 23–31.
22. Kim, S.Y., Kim, K.M., Park, J.C., Jeon, G.M., Chun, H.H., Numerical simulation of wave and current interaction with a fixed offshore substructure. International Journal of Naval Architecture and Ocean Engineering, 8 (2016) 188–197.

23. Han, Z., Liu, Z., Shi, H., Numerical study on overtopping performance of a multi-level breakwater for wave energy conversion. Ocean Engineering, 150 (2018) 94–101.
24. Hirt, C.W., Nichols, B.D., Volume of Fluid (VOF) method for the dynamics of free boundaries. Journal of Computational Physics, 39(1) (1981) 201–225.
25. Finnegan, W., Goggins, J., Numerical simulation of linear water waves and wave–structure interaction. Ocean Engineering, 43 (2012) 23–31.
26. Silva, M.C., Vitola, M.D.A., Pinto, W.T., Levi, C.A., Numerical simulation of monochromatic wave generated in laboratory: Validation of a CFD code. In *23rd Congresso Nacional De Transporte Aquaviário, Construção Naval e Offshore*, Rio de Janeiro, Brazil. 2010, pp. 25–29.
27. Saincher, S., Banerjeea, J., Design of a numerical wave tank and wave flume for low steepness waves in deep and intermediate water. Procedia Engineering, 116 (2015) 221–228.
28. Prasad, D.D., Ahmed, M.R., Lee, Y.H., Sharma, R.N., Validation of a piston-type wavemaker using Numerical Wave Tank. Ocean Engineering, 131 (2017) 57–67.
29. Westphalen, J., Greaves, D.M., Williams, C.J.K., Hunt-Raby, A.C., Zang, J., Focused waves and wave–structure interaction in a numerical wave tank. Ocean Engineering, 45 (2012) 9–21.

CHAPTER 12

INVESTIGATIONS OF FLOW BEHAVIOR IN CYLINDER AND DISC MADE OF MONOLITHIC AND COMPOSITE MATERIALS

SAVITA BANSAL,[1] S. B. SINGH,[1*] and A. K. HAGHI[2]

[1]Department of Mathematics, Punjabi University, Patiala, India

[2]Canadian Research and Development Center of Sciences and Cultures, Montreal, Canada

*Corresponding author. E-mail: sbsingh69@yahoo.com

NOTATIONS

ε_r = radial strain
ε_θ = tangential strain
ε_z = axial strain
ε = effective strain
$\dot{\varepsilon}_r$ = radial strain rate
$\dot{\varepsilon}_\theta$ = tangential strain rate
$\dot{\varepsilon}_z$ = axial strain rate
$\dot{\varepsilon}$ = strain rate (or creep rate)
σ_r = radial stress
σ_θ = tangential stress
σ_z = axial stress
σ = effective stress
F, G, and H = anisotropic constants

ABSTRACT

Most of the engineering or mechanical components that undergo fast wear and tear resemble the shape of a rotating disk, cylinder, or shell. This chapter helps the various researchers to know about the analysis of creep in cylinder and disk done by many authors. These two shapes have been actively studied by various researchers over the past years so that they can develop the materials which are stronger, stiffer, and are less prone to creep. A good amount of phenomenological study has been conducted in the field of creep analysis of rotating disks and rotating cylinders. Most of the researchers observed that monolithic materials such as aluminum and steel are not as strong as composite materials. In today's time, scientists of material sciences are looking to create composites with more strength and having less impact of variation of temperature and pressure at a much less cost. This has led to a huge amount of research in functionally graded material that is designed for specific function and applications to show a graded variation in the materials properties, at an optimum cost. The properties of functionally graded material have been found to be more appealing than of conventional composites and are ideal for creation of future high speed space crafts which can withstand extreme temperature and pressure conditions of space tours. With the advent of nanotechnology, use of nanomaterials fast started picking up for making of engineering components from nanotech-based functionally graded materials. This chapter is split into two parts. The first section focuses on creep analysis in cylinder and the second part highlights the creep in disk conducted by various researchers during 1954 and 2020. The expressions of strains, stresses, and strain rates were calculated using constitutive equations, equilibrium equations, associated flow rule, yield criterion, and various assumptions.

12.1   INTRODUCTION

Solid mechanics is the branch of science used to study the behavior of solid materials and their dynamics. It is the study of the behavior of materials like wood, steel, granite, plastics, alloys, metals, biological materials, engineering material, and so on, when are in motion or/and are under applied internal or external pressure resulting in deformation, thermal changes, chemical interactions, and other electromagnetic properties. Different materials have different physical properties and the usage of the materials for a particular purpose is typically based on their properties. The properties of a solid material include various aspects like strength, toughness, hardness, malleability, brittleness, ductility, creep, etc. One of the very important and common properties that is looked into while choosing a solid material is the material's strength and its creep.

Mechanics of solids are fundamental to many branches of physics, including material science. Through solid mechanics principles, we try to find answers to some of the questions like:

How to give further strength to a long bridge or large structures?

How to save the gears and other mechanical and engineering components from deformation?

What material should be used to replace organic/human body parts?

Till what temperature a certain physical behavior of a solid material will not change?

The science of solid mechanics is also used to study deformable materials and structures (disk, cylinder, shell). The use of composite materials for making such materials and structures are in practice for many decades and is an active field of study in material sciences.

Composite material is a combination of two or more materials with improved and desired properties, such as strength, rigidity, and stiffness. The science of making composite material involves the requirement of combining the good properties of one class of material with the others. Here, two materials that have different physical and chemical properties are placed together either at a microscopic or at a macroscopic scale, to get a composite material with desired characteristics. For example, hard cardboard is formed by bonding several thin layers of paper together to make a hard and strong composite. A concrete pillar or a beam is made by placing iron mesh with cement, gravel, and sand to give the final composite material in the shape of a pillar or a beam, extra strength. Some naturally occurring composites are wood (cellulose fibers embedded in lignin) and bone (stiff material in a soft organic matrix with holes filled with fluid). Some man-made composites are concrete, plywood fiberglass, etc.

With respect to the physical behavior of materials, including composites, they are divided into two types, isotropic material and anisotropic material. An isotropic material is material having same mechanical, physical, thermal, and electrical properties in all directions. Anisotropic material has different physical properties in different directions, namely, wood which is stronger along the grain than across it. Isotropic materials are in general mathematical approximations to the real value. However, for small deformations, the equilibrium equations, compatibility equations, and the strain displacement relations remain the same whether the material is isotropic or anisotropic.

Orthotropic materials like timber rolled metals, sandwich structures, and laminated structures have three mutually orthogonal planes of symmetry. Intersecting lines of these planes are known as the axes of anisotropy.

Giving strength to material or forming a new material as composite was an ancient art. Man knew of the art of creating stronger composite material by way of mixing two different materials since time immemorial. In ancient history, Egyptians made pyramids with bricks mixed into chopped straws and Japanese warriors used laminated metals. The ancient art has now become a field of scientific study in all our cutting-edge experiments and endeavors, like space missions.

These days, steel bars are used in cement to increase its tensile strength. Modern technologies demand materials having multifunctional characteristics and having properties significantly better than the existing material. Composite materials are used in commercial and military aircraft, automobiles, buses, trucks, naval vessels, boats and ships, infrastructure (bridges, highways, and buildings), coastal use, sports, agriculture, armored vehicles, musical instruments, robotics, and many more.

When a certain force (physical or chemical) is applied to any material, including a composite, it gets deformed. To increase the tensile strength and ductility of a material, particularly a composite, many types of fibers, or metals, namely, steel, titanium, etc., or alloys are used as reinforcement in composite laminates. Fiber reinforcement composites are classified as continuous fiber composite, chopped fiber composites, hybrid composites, and woven fiber composites (further classified into the orthogonal, cylindrical structure, 5D woven structure, braided structure, orthogonal interlock structure, angle interlock structure woven fiber reinforced composites). Titanium has some salient properties like low density, high strength, very strong, extremely durable, but is expensive. The addition of reinforcements to composites incurs additional expenditure, and hence reinforced composites may lead to much more expenditure. But in today's time, scientists of material sciences are looking for creating composites with required properties, like those that are not easily deformable and are not impacted by variation in temperature, and at the same time stress resistant, at a much less cost. This has given rise to the development and usage of functionally graded materials (FGMs), which have the desired properties of expensive composites but can be produced at a much cheaper cost.

FGMs are composite materials that are designed for specific functions and applications to show a graded variation in the materials properties. With the help of FGM, we can reduce the thermal stress and increases the strength of the material which results in maximizing its performance. These materials have different properties in different directions. The properties of FGM are ideal for the creation of future high-speed space crafts which can withstand extreme temperature and pressure conditions of space tours. FGM is used to avoid necking, corrosion cracking, fracture, fatigue. FGM is the most important nonhomogeneous material formed for high-temperature application. These days, many industries are interested in finding materials that are adaptive to very high thermal gradients by using ceramic layers along with metallic layers. The transition between the layers in a FGM is totally different than conventional material composites, and the FGM transitions are usually found with the help of power series (approximate solutions). The behavior of FGM can be commonly analyzed through finite element method (FEM), boundary element method, analytical/semianalytical methods with boundary conditions, meshless methods, higher order diskretization methods [24]. The ceramic in an FGM shows thermal barrier effects and protects the metal from corrosion. In Figure 12.1, it is seen that in FGM ceramic, metal, fiber, and micropore are uniformly designed to bear the stress, temperature, and many more other external factors acting on the material [25, 26]. On the other hand, in non-FGM, the place of constituent elements is not placed uniformly, in fact, they are arbitrarily placed, leading to nonachievement of required functional property.

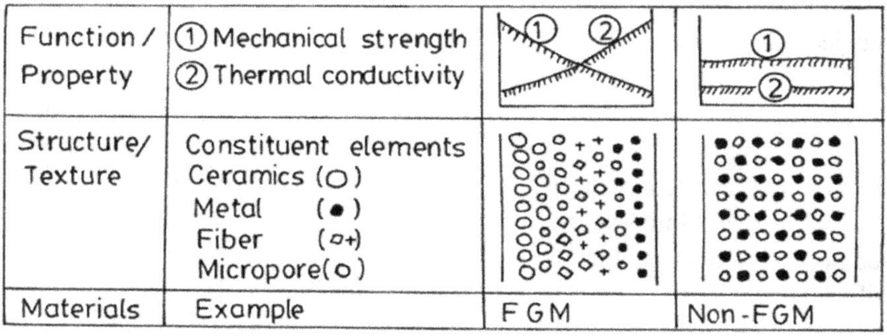

Function / Property	① Mechanical strength ② Thermal conductivity		
Structure/ Texture	Constituent elements Ceramics (O) Metal (•) Fiber (o+) Micropore(o)		
Materials	Example	FGM	Non-FGM

FIGURE 12.1 Schematic representation of the characteristics of functionally gradient material.

When stress is applied to a material, the material changes its shape, and after removing the load, if the material returns to its original shape, then such deformations are called elastic deformation, for example, sponge, elastic rubber band, etc. In plastic deformations, materials will be permanently deformed on the application of external load, that is, the object can never return to its original shape after unloading. Plastic deformation is not a reversible process like an elastic deformation, for example, coil springs used in shockers of cars get deformed after some years. In plastic deformations, Hooke's law is not valid on the deformations/strains of the material. Deformations also depend upon the temperature and intensity of the load. We get to know when the material is deformed through the yield criterion.

Most of the engineering components that undergo fast wear and tear resemble the shape of a rotating disk or a rotating cylinder. The two shapes have been actively studied by various researchers over the past eight decades, to develop materials that are stronger, stiffer, and are less prone to creep. A good amount of phenomenological study has been conducted in the field of creep analysis of rotating disks and rotating cylinders. The review paper as follows is an attempt to encapsulate them in short for a broader understanding of phenomenology of creep analysis for people working in this field.

12.2   FGM AND NANOTECHNOLOGY

Engineers developing materials of their interest and utility have developed a keen interest in FGMs and composites because of the reasons that the FGM and composites yield desired results as per the requirement and that too at a very less expensive cost compared to conventional usage of materials. This has led to the increased popularity of FGM as the choicest material for the development of new components. With the advent of nanotechnology, use of nanomaterials fast started picking up for the making of engineering components from nanotech-based FGMs. In civil engineering, nanotechnology has made a remarkable advent in the field of structural engineering in ensuring the strength and integrity of high-rise structures. Nanotechnology has also led to creation of nanomaterials that have increased strength and durability. Some of the examples of nanocomposites and FGM are nanosilica, nanotitania, nanoalumina, carbon nanotubes, graphene oxides, and graphene nanoplatelets [32].

The desired outcomes from the use of nanotechnology and nanomaterials depend on whether it is possible to tailor the structures of materials at nanoscale to achieve the actually desired properties. Nanotechnology can render unique and desired physical as well as chemical properties in a material or composite, namely, strength, durability, light, heavy, reactive, resistant, conduction, induction, etc. [33].

In electronic engineering and computer sciences, nanomaterials are being used in making computers and electronic systems faster and more ergonomic. Nanomaterials are increasingly being researched to develop random access memory and central processing units faster and which are capable of carrying more data [33].

Similarly, nanotechnology has usage in medical and environmental sciences also. In the field of medicine, the use of nanomaterials is fast becoming in vogue keeping in consideration the desired results obtained by the use of nanomaterials which were not possible by the use of conventional materials.

The manufacturing techniques of FGM and composites are classified into two categories called—thin FGM technique and bulk FGM technique [34]. In the thin FGM technique, the methodology used is the spray method or vapor condensation method. However, the thin FGM method has drawbacks like it requires high energy intensity for making the product and as the method involves spray, spraying of nanoparticles can be extremely harmful to the human body, and also they are not environment friendly [34]. The common methods are powder metallurgy, centrifugal method, and additive method.

One of the most common methods of making FGMs or composites using the bulk FGM technique is the centrifugal solid particle method. However, for making nanomaterials, centrifugal solid particle method will yield proper results [35]. The author suggests that centrifugal mixed particle method be more suited for making FGM with nanoparticles. These methods are, however, not comprehensive enough to account for all types of complex material distributions.

To further study and develop new FGM materials modeling using computer-based software is done using AutoCAD and CAM. Modeling is followed by fabrication as per the modeled design. Modeling in itself is very challenging due to inherent bottlenecks faced in computer-aided designing software in conceptualizing 3D multilayered materials and other complex material distributions [35].

In all the above, utility of nanomaterials-based composites and FGM, strength, and durability of the material can be measured also by the study of creep in the structures. The decision of using such materials for future structures and future equipment in the field of engineering is very much dependent on the creep analysis of the materials.

Leo Razdolsky in his book [36] has highlighted the two important approaches called deterministic and probabilistic approaches for creep modeling of composites and nanomaterials. This book reviews creep at high temperatures on crystalline composites and nanomaterials and diskusses two different creep models to describe a range of behavior of materials. The author demonstrates the usage of Newtonian nonlinear viscoelastic creep theory to be incorporated into the general theory of mechanics to design and analyze composite materials.

A lot of research has been done in the field of nanocomposites; however, the domain of nanotechnology provides a huge scope of further research in this field of nanocomposites. It is understood that demand for neocomposites with the most accurate features, as required, is always going to be there in the market. Thus, the study in this direction is always going to be of help to the mankind in future.

12.3 YIELDING CRITERIA

Yielding criterion is the hypothesis concerning the limit of elasticity under any possible combinations of stresses. Tresca's yield criterion and von Mises are two of the very important yielding criterion. Through these, we get to know when the material operating under the same load conditions start getting deformed in a plastic manner. These yield criteria are the functions of principal stresses σ_1, σ_2, σ_3, in such a way that $\sigma_1 > \sigma_2 > \sigma_3$.

12.3.1   TRESCA'S YIELDING CRITERION (OR MAXIMUM SHEAR–STRESS CRITERION)

$$\tau_{max} = \frac{\sigma_1 - \sigma_3}{2}$$

The material is deformed when τ_{max} (maximum shear stress) reaches a critical value (k).

Conditions for uniaxial tension are $\sigma_1 = \sigma_e, \sigma_2 = \sigma_3 = 0, \tau_e = \frac{\sigma_e}{2}$. Solving these expressions, we get

$$\tau_{max} = \frac{\sigma_1 - \sigma_3}{2} = \tau_e = \frac{\sigma_e}{2}$$

$$\sigma_e = \sigma_1 - \sigma_3$$

Conditions of pure shear state,

$$\sigma_1 = -\sigma_3 = k, \ \sigma_2 = 0$$

Using these conditions in the above expression, we get

$$\sigma_1 - \sigma_3 = 2k = \sigma_e \text{ implies } k = \frac{\sigma_e}{2}$$

Note: Tresca's yielding criterion intermediate principal stress is neglected, therefore, the following von Mises yielding criterion is preferred by researchers in most of their research paper.

12.3.2 VON MISES (OR DISTORTION-ENERGY) YIELDING CRITERION

In 1913, this criterion was proposed by von Mises for isotropic material that yielding exists when stress deviator exceeded some critical value

i.e., $\frac{1}{6}\left[\left(\sigma_1 - \sigma_2\right)^2 + \left(\sigma_2 - \sigma_3\right)^2 + \left(\sigma_3 - \sigma_1\right)^2\right] = k^2$

Now, the condition of yielding in uniaxial tension is $\sigma_1 = \sigma_e$, $\sigma_2 = \sigma_3 = 0$, $\sigma_e^2 + \sigma_e^2 = 6k^2$. After solving both the equations, we get von Mises yielding criterion as

$$\sigma_e = \frac{1}{\sqrt{2}}\left[\left(\sigma_2 - \sigma_3\right)^2 + \left(\sigma_3 - \sigma_1\right)^2 + \left(\sigma_1 - \sigma_2\right)^2\right]^{\frac{1}{2}}$$

12.3.3   HOFFMAN'S YIELDING CRITERION

For isotropic material, Hoffman's yielding criterion utilizes uniaxial compression (f_c) and tensile yield stresses (f_t) and is formulated as follows:

$$\left(\sigma_{11}^{2}+\sigma_{22}^{2}+\sigma_{33}^{2}\right)-\left(\sigma_{11}\,\sigma_{22}+\sigma_{22}\,\sigma_{33}+\sigma_{33}\,\sigma_{11}\right)+\left(f_c-f_t\right)\left(\sigma_{11}+\sigma_{22}+\sigma_{33}\right)-f_c f_t=0$$

12.3.4   HILL'S YIELDING CRITERION

In 1948, Hill proposed this criterion for anisotropic material by modifying the von Mises yielding criterion (isotropic material).

$$2f = F\left(\sigma_{22}-\sigma_{33}\right)^{2}+G\left(\sigma_{33}-\sigma_{11}\right)^{2}+H\left(\sigma_{11}-\sigma_{22}\right)^{2}=1$$

Here, the potential function $2f$ can be found from the effective stress expression given as

$$\sigma_e = \frac{\sqrt{2f}}{\sqrt{G+H}}$$

In Hill's yielding criterion, it is assumed that the material is orthotropic. This means that there are three planes of symmetry that are mutually intersecting each other at 90 degrees at every point inside the material, and these planes are known as the principal axes of anisotropy. When principal axes of anisotropy are taken in x, y, and z directions, Hill's yielding criterion converts into the following form:

$$F\left(\sigma_2-\sigma_3\right)^{2}+G\left(\sigma_3-\sigma_1\right)^{2}+H\left(\sigma_1-\sigma_2\right)^{2}=1$$

where F, G, and H are anisotropic constants. If $F = G = H = 1$, then this reduces to von Mises yielding criterion.

12.3.5   SINGH AND RAY'S YIELDING CRITERION

In 2003, Singh and Ray proposed this criterion for anisotropic material using uniaxial compression (f_c) and tensile stress (f_t). This criterion is developed using Hill's criterion and Hoffman's criterion

$$2f = F\left(\sigma_{22} - \sigma_{33}\right)^2 + G\left(\sigma_{33} - \sigma_{11}\right)^2 + H\left(\sigma_{11} - \sigma_{22}\right)^2 + (k_1\sigma_{11} + k_2\sigma_{22} + k_3\sigma_{33}) = 1$$

where $k_i = \left[\dfrac{f_c - f_t}{f_c\, f_t}\right]_{i\text{th principal direction}}$ $i = 1, 2, 3$

Since the difference between tensile stress and compressive stress has been calculated by thermal residual stresses, so $k_1 = k_2 = k_3 = \left[\dfrac{f_c - f_t}{f_c\, f_t}\right]$

12.4   CONSTITUTIVE EQUATIONS

Constitutive equations of a material are a relationship between the applied stresses and strain. These are combined with other equations called associated flow rules to solve physical problems or real problems related to creep behavior and deformation. These equations are a set of differential equations that are almost impossible to be solved without assumptions. As a result, various approximation schemes are used to solve the constitutive equations and obtain the values of stresses and strains in creep and deformations problems of various solid materials and composites.

12.4.1   GENERALIZED ASSUMPTIONS IN YIELDING CRITERIA

In the classical theory of plasticity, the material region is assumed to be divided into two parts, elastic and plastic region, which is separated by a yield surface depending on the symmetry and other physical considerations such as yield condition for elastic–plastic deformation, and yield condition and creep laws for creep deformation [38]. Some of the generalized assumptions used in defining yielding criteria of materials are as follows:

1. Material is incompressible, that is, $\dot{\varepsilon}_r + \dot{\varepsilon}_\theta + \dot{\varepsilon}_z = 0$.
2. The principal axes of stress coincide with the axes of anisotropy i.e., $\tau_{xy} = \tau_{yz} = \tau_{zx} = 0$
3. The principal axes of strain rates coincide with the same axes i.e., $\dot{\gamma}_{xy} = \dot{\gamma}_{yz} = \dot{\gamma}_{zx} = 0$

4. The creep rate depends upon some function of the stress and vice-versa

    i.e., $\dot{\varepsilon} = f(\sigma) \, or \, \sigma = f(\dot{\varepsilon})$

5. The axial strain is zero or the true axial strain is constant w.r.t. radius

    i.e., $\varepsilon_z = 0 \, or \, \varepsilon_z = k$.

6. The material is isotropic and homogenous and after deformation, there is no change in the shape of the cylinder.

7. The volume of the material is constant, that is, $\varepsilon_r + \varepsilon_\theta + \varepsilon_z = 0$.

12.4.2 STRAIN–DISPLACEMENT RELATION FOR INFINITESIMAL DISPLACEMENT

In the classical theory of elasticity, the displacements are assumed to be so small that the squares and products of displacement gradients are neglected and the measure of strain thus becomes linear [38].

$$\text{Radial strain rate} = \dot{\varepsilon}_r = \frac{du_r}{dr},$$

$$\text{Tangential strain rate} = \dot{\varepsilon}_\theta = \frac{u_r}{r}$$

where r = radius of the cylinder, $u_r = \dfrac{du}{dt}$ = radial displacement rate , u = radial displacement.

Compatibility equation is obtained by eliminating u_r from radial strain rate and tangential strain rate equations

$$r \frac{d\dot{\varepsilon}_\theta}{dr} = \dot{\varepsilon}_r - \dot{\varepsilon}_\theta$$

In thick-walled cylinders, the stress variation across the thickness of walls is also significant. To solve such problems axis symmetry about the z-axis is considered and the stress equilibrium-related differential equations are solved in polar coordinates [39]. Generally, the stress equations of equilibrium without any external forces can be given as

$$\frac{\partial \sigma_r}{\partial r} + \frac{1}{r}\frac{\partial \tau_{r\theta}}{\partial \theta} + \frac{\partial \tau_{rz}}{\partial z} + \frac{\sigma_r - \sigma_\theta}{r} = 0$$

$$\frac{\partial \tau_{\theta r}}{\partial r} + \frac{1}{r}\frac{\partial \sigma_{\theta}}{\partial \theta} + \frac{\partial \tau_{\theta z}}{\partial z} + 2\frac{\tau_{\theta r}}{r} = 0$$

$$\frac{\partial \tau_{zr}}{\partial r} + \frac{1}{r}\frac{\partial \tau_{z\theta}}{\partial \theta} + \frac{\partial \sigma_z}{\partial z} + \frac{\tau_{zr}}{r} = 0$$

For axis symmetry about the z-axis $\frac{\partial}{\partial \theta} = 0$, thus

$$\frac{\partial \sigma_r}{\partial r} + \frac{\partial \tau_{rz}}{\partial z} + \frac{\sigma_r - \sigma_{\theta}}{r} = 0$$

$$\frac{\partial \tau_{\theta r}}{\partial r} + \frac{\partial \tau_{\theta z}}{\partial z} + 2\frac{\tau_{\theta r}}{r} = 0$$

$$\frac{\partial \tau_{zr}}{\partial r} + \frac{\partial \sigma_z}{\partial z} + \frac{\tau_{zr}}{r} = 0$$

In a plane stress situation, axial stress of the cylinder is equal to zero and due to uniform radial deformation and symmetry $\tau_{rz} = \tau_{\theta z} = \tau_{r\theta} = 0$. The equation of equilibrium becomes $r\frac{\partial \sigma_r}{\partial r} = \sigma_{\theta} - \sigma_r$.

The Levy–Mises equations for an orthotropic material in the footsteps of Hill can be written as

$$(G+H)\sigma_x - H\sigma_y - G\sigma_z = \frac{2\sigma}{d\varepsilon}d\varepsilon_x,$$

$$(H+F)\sigma_y - F\sigma_z - H\sigma_x = \frac{2\sigma}{d\varepsilon}d\varepsilon_y,$$

$$(F+G)\sigma_z - G\sigma_x - F\sigma_y = \frac{2\sigma}{d\varepsilon}d\varepsilon_z,$$

$$N\tau_{xy} = \frac{\sigma}{d\varepsilon}d\gamma_{xy},$$

$$L\tau_{yz} = \frac{\sigma}{d\varepsilon}d\gamma_{yz},$$

$$M\tau_{zx} = \frac{\sigma}{d\varepsilon}d\gamma_{zx}.$$

where F, G, H, L, M, N are constants of the material, σ_x, ..., τ_{xy} are stress components, $d\varepsilon_x$, ..., $d\gamma_{xy}$ are the increments of plastic strain components, and $d\varepsilon_1$ is the plastic strain increment invariant.

For complete isotropic (taking $L = M = N = 3F = 3G = 3H$), these equations reduce to the Levy–Mises equations for isotropic material.

The fundamental constitutive equations for orthotropic theory of creep are given by replacing the strain increments with the strain rates in the above Levy–Mises equations, we get

$$\dot{\varepsilon}_x = \frac{\dot{\varepsilon}}{2\sigma}\left[(G+H)\sigma_x - H\sigma_y - G\sigma_z\right],$$

$$\dot{\varepsilon}_y = \frac{\dot{\varepsilon}}{2\sigma}\left[(H+F)\sigma_y - F\sigma_z - H\sigma_x\right],$$

$$\dot{\varepsilon}_z = \frac{\dot{\varepsilon}}{2\sigma}\left[(F+G)\sigma_z - G\sigma_x - F\sigma_y\right],$$

$$\dot{\gamma}_{xy} = \frac{\dot{\varepsilon}}{\sigma}N\tau_{xy},$$

$$\dot{\gamma}_{yz} = \frac{\dot{\varepsilon}}{\sigma}L\tau_{yz},$$

$$\dot{\gamma}_{zx} = \frac{\dot{\varepsilon}}{\sigma}M\tau_{zx}$$

Fundamental constitutive equations for orthotropic theory of creep given by Bhatnagar and Gupta [4] are as follows:

$$\dot{\varepsilon}_r = \frac{\dot{\varepsilon}}{2\sigma}\left[(G+H)\sigma_r - H\sigma_\theta - G\sigma_z\right],$$

$$\dot{\varepsilon}_\theta = \frac{\dot{\varepsilon}}{2\sigma}\left[(H+F)\sigma_\theta - F\sigma_z - H\sigma_r\right],$$

$$\dot{\varepsilon}_z = \frac{\dot{\varepsilon}}{2\sigma}\left[(F+G)\sigma_z - G\sigma_r - F\sigma_\theta\right]$$

The stress invariant σ (also known as von Mises stress invariant when $F = G = H = 1$) is

$$\sigma = \frac{1}{\sqrt{2}} \left[F\left(\sigma_\theta - \sigma_z\right)^2 + G\left(\sigma_z - \sigma_r\right)^2 + H\left(\sigma_r - \sigma_\theta\right)^2 \right]^{\frac{1}{2}}$$

The true strain rate corresponding to the above von Mises stress invariant is given by

$$\dot{\varepsilon} = \frac{\sqrt{2}}{3} \left[\left(\dot{\varepsilon}_\theta - \dot{\varepsilon}_z \right)^2 + \left(\dot{\varepsilon}_z - \dot{\varepsilon}_r \right)^2 + \left(\dot{\varepsilon}_r - \dot{\varepsilon}_\theta \right)^2 \right]^{\frac{1}{2}}$$

12.5 CREEP

Creep is a time-dependent deformation in response to a constant load. Creep is defined as the slow rise of plastic deformation under the action of stress below the yield strength of the material. Its dictionary meaning is "move slowly and carefully to avoid being heard or noticed."

For example, a tensile material under a load will undergo strain. The material under the same stress/load condition behaves in the fashion as shown in Figure 12.2, wherein over a period of time the strain with the same load condition, increases very slowly, leading to plastic deformation in the material such that on the removal of the load, the material undergoes nonelastic changes, which may or may not impact the physical and other properties of the material.

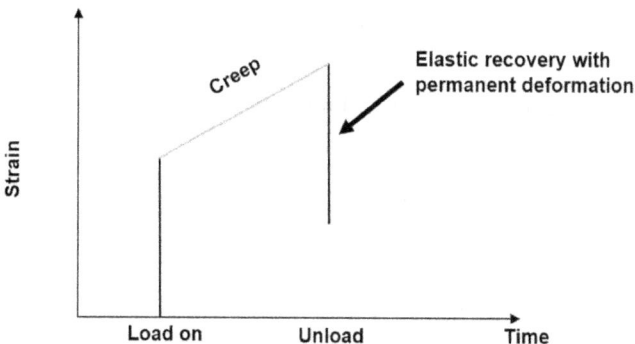

Creep often takes place in three defined stages in which creep rate changes with increasing time:

1. Primary creep or transient creep: in this creep, rate decreases with time. It exists due to dislocation movement. Here the effect of working hard is more than the recovery process. Strain rate is a function of time.
2. Secondary creep or steady-state creep: here, strain rate is constant. Rates of work hardening and recovery are equal. It is generally plastic in character depends on the surrounding temperature.
3. Tertiary creep or fracture stage creep: in this, creep rate increases until failure or fracture due to which necking occurs. Time shortens at elevated temperature.
4. These stages of creep behavior are shown in the curve in Figure 12.4.

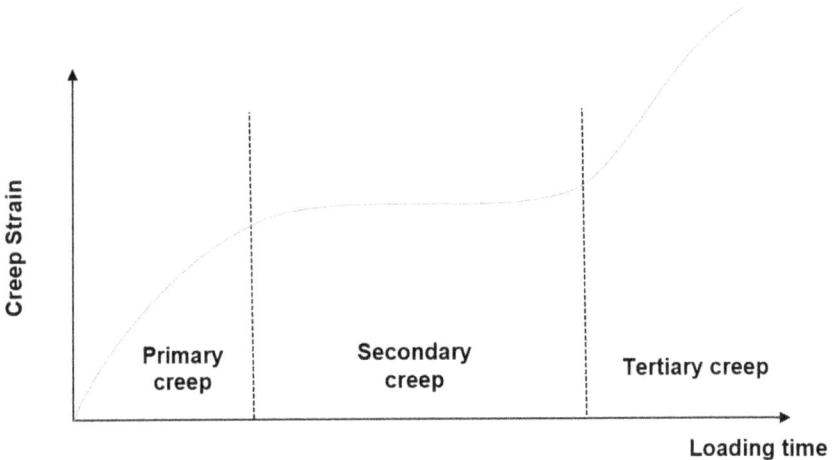

FIGURE 12.2   Creep strain versus loading time.

Creep laws assumed by different authors:

(1) Norton's law: $\dot{\varepsilon} = A\sigma^n$

(2) Time-hardening law: $\dot{\varepsilon} = B\sigma^n t^{m-1}$

(3) Strain-hardening law: $\dot{\varepsilon} = C\sigma^{\frac{n}{m}} \varepsilon^{-1} \, 1-\left(\frac{1}{m}\right)$

(4) Marin's law: $\dot{\varepsilon} = De^{-q}\sigma^m + E\sigma^n$

where A, B, C, D, E, n, m, q are the experimental constants in uniaxial cases by Finnie and Heller.

(5) For steady-state creep, Norton's law: $\dot{\varepsilon} = A\dfrac{n}{\sigma}$

(6) In aluminum-based composite, the effective creep rate is related to effective stress through threshold stress (σ_0) is given as

$$\dot{\varepsilon} = A\left(\frac{\sigma - \sigma_0}{E}\right)^n \exp\left(\frac{-Q}{RT}\right)$$

where A, n, Q, E, R, T are parameter, true stress exponent, true activation energy, Young's Modulus (depends on temperature), gas constant, and operating temperature, respectively [18].

It is alternatively written as

$$\dot{\varepsilon} = \left[M(\sigma - \sigma_0)\right]^n$$

$$\text{where } M = \frac{1}{E}\left(A\exp\left(\frac{-Q}{RT}\right)\right)^{\frac{1}{n}}$$

(7) Sherby's law is defined as

$$\dot{\varepsilon} = \frac{AD_L\lambda^3}{\left|\vec{b_r}\right|^5}\left(\frac{\sigma - \sigma_0}{E}\right)^n$$

where $A, D_L, \lambda, \left|\vec{b_r}\right|, E,$ and σ_0 are constant, the lattice diffusivity coefficient, subgrain size, magnitude of the burger's vector, Young's modulus, threshold stress, respectively.

12.6 REVIEW OF WORK DONE

12.6.1 CREEP IN CYLINDER

Rimrott et al. [1] concluded that in any hollow cylindrical vessel under constant inner pressure, the vessel will continuously expand at an accelerated rate until the wall thickness has decreased to such an extent that the strength of the material is no longer sufficient to hold the load of the constant internal pressure felt at the inner radius, and the vessel fractures due to application of two simultaneous effects, that is, constant internal pressure at the inner radius of the cylindrical vessel and reducing the strength of the material due to decreasing thickness due to expansion of the vessel. In this paper, creep-failure time (the time where the true strain approaches infinity) is examined for thin (i.e., $\sigma_r = 0$) and thick (i.e., the value of b/a is very large so that $a/b = 0$) walled cylindrical vessels of circular cross-section with closed ends under constant internal pressure. Assumptions made in this paper are the impact of primary creep is neglected. The behavior of the material is expressed by the creep relation in the secondary and tertiary creep stages. Material is isotropic, homogeneous, and the shape of the vessel does not change its shape during creep. The density of the material is constant (i.e., $2\pi ah = 2\pi a'h'$, where a and a' are original and actual radii and h and h' are original and actual wall thickness, respectively).The ratios of principal shear–strain rates to the shear–stresses are constant. that is,

$$\frac{\dot{\varepsilon}_\theta - \dot{\varepsilon}_r}{\sigma_\theta - \sigma_r} = \frac{\dot{\varepsilon}_r - \dot{\varepsilon}_z}{\sigma_r - \sigma_z} = \frac{\dot{\varepsilon}_z - \dot{\varepsilon}_\theta}{\sigma_z - \sigma_\theta} = \frac{3}{2K}$$

where K = creep modulus, σ_r, σ_θ, σ_z are radial, tangential, axial stresses, ε_r, ε_θ, ε_z are radial, tangential, axial strain rates, respectively.

In this paper, Norton's law is used. According to the writers, the mathematical solution of the problem is complicated so they represented it with the help of graphs.

Rimrott and Luke [2] analyzed the behavior of creep for a rotating cylinder using Norton's law and found the values of strain rates, strains, and stresses for creep exponent $(n) = 1$. The duo showed that we can find the values of strains, strain rates, and stresses for different values of creep exponent. The authors said that the cylinders can be made from any material and the findings are independent of inner and outer radius. The author derived the exact solution for isotropic material by assuming that strains are large. For this, the authors used the finite- strain theory to predict the time after which the cylinder will be deformed.

Schweiker and Sidebottom [3] developed an incompressible creep theory to predict deformations in a thick-walled cylinder under internal pressure and axial load at any given time and presented it in the form of dimensionless design curves that do not depend upon the properties of the material. The strain–stress–time expression for the material is approximated by a hyperbolic sine function. The general assumptions used here are that, the cylinder is closed-ended, material is incompressible, deformations are infinitesimal, and the total strain theory is valid. The experiment is observed on a bar made of high-density polyethylene having diameter of 1.5 inches. The authors deduce that deformation in the cylinder cannot be changed by redistribution of stress with time.

Bhatnagar and Gupta [4] formulated constitutive equations for multiaxial stress based on orthotropic theory of creep. In this paper, two applications are diskussed, that is, tension of a bar (principal axes of stress do not coincide with the axes of anisotropy) and the other one is variation in stresses taking load as constant with time. These constitutive equations are based on anisotropic plastic model of Hill.

Pai [6] obtained the solutions for steady-state creep of a thick-walled orthotropic cylinder under internal pressure by using a piecewise linear model (stress–strain rate equations) and the flow rule associated with von Mises criterion. He found the expressions of stresses and creep rate under the assumptions of plane strain, closed-ended, and open-ended. In all three axial conditions, creep anisotropy of material affected the strain rates. The paper shows the difference in creep behavior between tensile and compressive creep tests in addition to accounting for the directional preference of the material subjected to the tensile creep tests. He observed that plane strain condition is equivalent to closed-end conditions. The solution helps to predict creep rate (strain rate) measurements more accurately. During the steady-state creep conditions, isotropic materials in the elastic range or plastic range may become highly anisotropic.

King and Mackie [7] presented the study of creep deformation of a thick-walled cylinder with internal pressure and load on isotropic material. In this paper, author analyzes the tensile data and observed that tensile stress will increase continuously due to decrease in the cross-sectional area during extension, but this can be calculated by finding the value of effective stress where the creep rate is minimum. The authors concluded that where the permanent strains quickly reach the level of the elastic strains, the simplest steady-state theory accounting for cylinder deformation gives a reasonable correlation with the experimental diametral measurements. The other analyses in which some attempt was made to introduce the effect of time on the creep behavior of the materials did not result in improved predictions. This would seem to be due to the inability of the simple time-hardening law chosen to reflect adequately the stress dependence of the material over the range of tensile stresses involved. This stress range is much wider than that covering the significant stresses encountered in an individual tube test.

Bhatnagar and Gupta [8] found the different values of stresses and creep rates for a thick-walled cylinder of an orthotropic material subject to the internal pressure under the following assumptions:

1. Plane strain, that is, $\varepsilon_z = 0$.
2. Generalized plane strain: here cylinder is taken sufficiently long and plane section normal to the axes of cylinder before creep will remain the same after creep, that is, ε_z is not a function of r.
3. Plane stress, that is, $\sigma_z = 0$.

They used the constitutive equations of anisotropy creep theory and Norton's law and compared their results with those obtained by Pai [6]. In plane strain assumption, results came out that axial stress is dependent on material anisotropy. Results give two independent anisotropic constants. In plane stress case, the effect of anisotropy is small on radial stress. Lastly, they found that creep anisotropy has a significant effect on cylinder behavior and cylinder made from anisotropic material will be less deformed.

Bhatnagar and Arya [9] concluded that after applying the external pressure on the thick-walled cylinder, it results in decreasing the strain rate and increases the life of the cylinder, and is beneficial in designing a cylinder so that creep deformation will be less. In this paper, they analyzed that the anisotropic behavior of the material affects the creep behavior of a thick walled cylinder, that is, the strains, stresses, and strain rates are affected by the anisotropic material when internal pressure is applied. At the inner radius, axial stress is compressible and it is tensile at the outer radius. The paper concludes that the strain rate increases with increasing deformation (strain). The authors also diskussed the behavior of isotropic material and compared their results with Rimrott for an isotropic cylinder.

Arya and Bhatnagar [10] investigated the creep analysis of thick-walled cylinder made of anisotropic material under internal and external pressure, both considering the elastic strains. Authors obtained fundamental equations of an orthotropic cylinder for the behavior of creep by assuming that strain rate is a function of stress times a function of time (time-hardening law), the total axial strain is a function of time not radius, displacement is radial (a function of radius), the axial stress is uniform throughout. Solutions of these equations are diskussed for planar anisotropy and these results are compared with the isotropic case. Conditions for planar anisotropy are taken from Gupta [11] and Lekhnitskii [12] as

$$F = G, \quad S_{11} = \frac{1}{E} = S_{22}, \quad S_{12} = \frac{-v}{E},$$

$$S_{13} = \frac{-v^1}{E^1} = S_{23}, \quad S_{13} = \frac{1}{E^1}$$

where E and E^1 are Young's moduli, v and v^1 are Poisson's ratio, F and G are anisotropic creep constants, and S_{11}, S_{12},..., S_{33} are anisotropic elastic constants.

Authors estimated the values of radial stress and tangential stress by taking different values of r (ranges from 1.0 inches to 2.0 inches) and time t ($t = 100$ h, 1000 h, 10,000 h) and presented radial strain rate and tangential strain rate versus time and radius graphically. The values of strain rates is more in cases of isotropic material than in cases of anisotropic materials.

Bhatnagar et al. [13] investigated the steady-state creep analysis of a hollow thick-walled rotating cylinder using Norton's power law by finding the values of strain rates and stresses in an anisotropic material. Rotating cylinders are used in power generations and various industries. Assumptions made by the authors are that the material of the cylinder is homogeneous and orthotropic, no change in the volume after creep deformations (i.e., $\acute{\varepsilon}_r + \acute{\varepsilon}_\theta + \acute{\varepsilon}_z = 0$), condition of plane–strain occurs ($\acute{\varepsilon}_z = 0$), there is a relationship between effective stress, and effective strain rates in such a way that for small values of effective strain rate for the preceding anisotropic material than those for the isotropic material will be obtained. In numerical computations, the material of the cylinder is taken as steel or steel alloys implemented on three cases of anisotropy. They concluded that anisotropy of material and creep exponent (n) both affected the strain rates and stresses. The authors calculated the values of effective stress for anisotropic material for which the ratios of axial to tangential strain rate and of radial to tangential strain rate is equal to 1.2 and is more than the corresponding values for isotropic material (i.e., 1.0). Authors suggested the anisotropic material is beneficial for the formulation of cylinders, giving reasons that the life of the cylinder will be longer (because strain rate is less) and it allows the cylinder to support large forces without a risk of failure under creep. Graphical representation of radial stress, tangential stress, axial stress, effective stress, and tangential strain rate for three cases of anisotropy and taking creep exponent $n = 1.0$, 3.0, 6.9, etc.

Bhatnagar et al. [14] developed the application of finite strain theory and carried out that the structure of the cylinder will not be safe if creep analysis is done on the basis of small strain theory. In this paper, author worked on creep analysis of orthotropic thick-walled rotating cylinder, and the cylinder is made up of anisotropic material. In this paper, authors assumed that the material of cylinder is orthotropic, incompressible (i.e., $\dot{\varepsilon}_r + \dot{\varepsilon}_\theta + \dot{\varepsilon}_z = 0$), and the axial strain is taken as constant k (i.e., $\dot{\varepsilon}_z = k$), Norton's law is used to define the relation between true strain rate and true stress and constitutive equations given by Bhatnagar and Gupta [4] for orthotropic materials are used. Here equilibrium equation can be obtained by using tangential strain and radial strains and by evaluating this equation, the expression for the radial stress is found. After that authors used boundary conditions by taking axial force improvised on the cylinder as zero and tested for creep exponent $n = 1$. To find the values of stresses, strain rates and yielding criterion, the value of anisotropic constants F, G, and H are necessary. Authors used five cases (obtained by equating an equation using Simpson's rule) for finding the values of anisotropic constants.

- Case 1: Ratio of axial strain rate and tangential strain rate = 1.2 = ratio of tangential strain rate and radial strain rate.
- Case 2: Ratio of axial strain rate and radial strain rate = 1.2 = ratio of radial strain rate and tangential strain rate.
- Case 3: Radial strain rate = tangential strain rate = axial strain rate.
- Case 4: Ratio of radial strain rate and axial strain rate = 1.2 = ratio of tangential strain rate and radial strain rate.
- Case 5: Ratio of tangential strain rate and axial strain rate = 1.2 = ratio of radial strain rate and tangential strain rate.

Authors have construed that cylinder has the longest collapse (failure) time with anisotropy type as mentioned in case 1 and shortest collapse time with anisotropy of material of type as diskussed in case 5. The results were represented graphically by taking various radii ratio. They concluded that cases 1 and 2 will be better than the isotropic material of the cylinder and, cases 4 and 5 are not as good as the isotropic material of rotating cylinder. This means isotropic cylinders results lie in between cases 1 and 5 results. For nonlinear materials the value of true strain rates will be more for anisotropic material as in cases 4 and 5.

Chen et al. [15] compared the estimated solution of higher order equation with the results of finite element analysis using the software ABAQUS because a higher-order term helps to get an exact solution for the time-dependent behavior of a functionally graded cylinder. Author used the fifth-order equation which is suitable to calculate the creep stress distribution (as terms higher than fifth order can be neglected in the calculation)and diskussed the creep behavior of functionally graded cylinders under internal pressure as well as external pressure. The effect on this material is symmetric about an axis and depends on the radial coordinate, approximated solution is derived using Taylor's expansion series. Asymptotic solutions for the radial stress, circumferential stress, and axial stress under internal pressure and external pressure are plotted in this paper. More attention is on the calculation when internal pressure is applied to the functionally graded cylinder.

Abrinia et.al. [16] applied an analytical solution for calculating radial stress and tangential stress of a thick-walled functionally graded cylindrical vessel under internal pressure and applied temperature. Assumption has been made as to the modulus of elasticity and thermal coefficient of expansion are related to radius with a power law

i.e.,
$$E(r) = E_0 r^\beta$$
$$\alpha(r) = \alpha_0 r^\beta$$

where E_0, α_0 are the modulus of elasticity and thermal coefficient of expansion, respectively, r is the ratio of the value lying between inner radius and outer radius, and outer radius ($r = R/b$). This power-law relation is used to check the changing properties of material along the radius and developed by Tutuncu et al. [17]. It is assumed that Poisson's ratio value is constant in the elastic range. The property of nonhomogeneity in thick FGM cylinder is applied by selecting a parameter that affects the value of stresses in the cylindrical vessel. Authors used Maple 9.5 software for plotting variations of normalized radial stress, tangential stress in a cylindrical vessel under the loading of pressure, under the loading of temperature. Further, they plotted the graphs of distribution of normalized radial stress, normalized circumferential stress, relative radial stress, relative hoop stress in a vessel under the combined loading of temperature and internal pressure for different values of the parameter ranging from −2 to 2. To obtain the optimum value of parameter, radial stress and circumferential stress of FGM are compared by the stresses of homogeneous material using same boundary conditions. Authors concluded that if the value of parameter is equal to zero, then radial stress and circumferential stress expressions of FGM and homogeneous material are same. Lastly, authors suggested that this paper can be further extended by the researcher in the yielding region to find the elastic–plastic solution.

Singh and Gupta (2009) [18] investigated the steady-state creep analysis in thick-walled cylinder made of Al-SiC$_p$ composite under internal pressure. The stresses and strain rates are examined for different values of size, content of the reinforcement (SiC$_p$) and applied temperature and concluded that there is very less impact of size, content of reinforcement, and applied temperature on stresses in the cylinder. Moreover, the value of radial strain rate and tangential strain rate in the cylinder can be reduced by decreasing the size of SiC$_p$ and applied temperature, and increasing its content. For steady-state creep, the effective strain rate is a function of effective stress by Sherby's law in aluminum-based composites. Sherby's law is given by

$$\dot{\varepsilon}_e = A^1 \left(\frac{\sigma_e - \sigma_0}{E} \right)^n \exp\left(\frac{-Q}{RT} \right)$$

where A^1, n, Q, E, R, T are parameter, true stress exponent, true activation energy, Young's Modulus (depends on temperature), gas constant, and operating temperature, respectively.

Sherby's law is altered by the author as

$$\dot{\varepsilon}_e = \left[M\left(\sigma_e - \sigma_0 \right) \right]^n \text{ taking creep exponent } n = 5.$$

where

$$M = \frac{1}{E} \left(A^1 \exp\left(\frac{-Q}{RT} \right) \right)^{\frac{1}{n}}$$

The value of threshold stress will be found using linear extrapolation technique. Assumptions made in this paper are material incompressible, isotropic, and SiC$_p$ and is distributed uniformly in aluminum matrix. Initially, less pressure is applied and after some time, it will be constant. At any point, stresses remain constant in the cylinder with time (i.e., steady-state condition of stress). Elastic deformation is less and neglected. Effective stress is taken by von Mises yielding criterion given in Dieter [19]. Variation in strain rates, stresses, effective stress in the composite cylinder for various sizes of the SiC, content of reinforcement, and operating temperature. Authors concluded that stresses are increasing when applied from inner to outer radius of a thick-walled composite cylinder. In an isotropic thick-walled composite cylinder, radial and tangential strain rates keep on decreasing when moved from inner to outer radius.

Habibi et al. [21] defined a numerical method for investigating the creep and elastic behavior of FGM in rotating cylinders. Authors assumed that the cylinder is made up of finite width layers with constant thermodynamic properties in each layer. Displacement equations are used to find the compatibility equation for thermoelasticity theory. Norton's law (also steady-state creep of cylinder) is used in this paper as

$$\dot{\varepsilon} = K\sigma^{\varsigma}\tau^{q}$$

where ζ and q are experimental constants.

Equilibrium equations for a rotating functionally graded thick-walled cylinder based on thermos-elasticity theory are derived and nonlinear heat transfer equation in radial direction are used to get temperature distribution in a rotating cylinder. The solution obtained by using three different cases on boundary conditions: hollow cylinder with free edges (Inner and outer surfaces are free from any constraint (i.e., $\sigma_{rr} = 0$, $t =$ inner radus and outer radius), hollow cylinder with fixed-free edges (i.e., $u = 0$ when $r =$ inner radius and $\sigma_{rr} = 0$, $t =$ inner radus and outer radius), filled cylinder with free edges (i.e., $u = 0$ when $r = 0$; $\sigma_{rr} = 0$, $t =$ outer radius). Here, radial stress and u is the radial displacement, r is the radius of cylinder. Expression of the radial stress is calculated by using stress–strain relation for plain strain conditions. Authors validated their results with the help of examples and compared these results with the results generated by Singh and Ray graphically. Comparison between radial displacement distribution and FEM, dimensionless radial, and tangential strain distribution in the radial direction for different gradients and for different values of n are presented graphically. Authors concluded that increase in the value of n, cylinder properties approaches the properties of metal base (has a smaller modulus of elasticity) and radial strain changes from positive values to negative values and tangential strain increases toward the positive direction. Increase in thermal gradient in cylinder would result in increase in tangential strain rate in the positive direction and radial strain increases in the positive direction in the middle portion of the cylinder when observed from inner radius to outer radius. In FGM cylinders, when creep started the stress and strain rate for disk and cylinder reaches to a steady-state value.

Tejeet et al. [22] carried out the creep behavior analysis in a thick-walled isotropic cylindrical vessel made of aluminum matrix reinforced with silicon carbide by using Sherby's law in which threshold stress is included. The law is described as

$$\dot{\varepsilon}_e = \left[M\left(\sigma_e - \sigma_0\right)\right]^n$$

where M is the creep parameter, ε_e, σ_e, σ_0 are the effective creep rate, effective stress, threshold stress, respectively, n is the creep exponent and the values selected for creep exponent are 3, 5, and 8. von Mises yielding criterion is used with constitutive equations and equilibrium equations to obtain the stress distribution, creep stress, and strain rates in thick-walled composite cylindrical vessels under internal pressure. Authors observed that radial stress, tangential stress, and axial stress increases, and the strain rates decreases along the radius. Radial stress, tangential stress, axial stress, effective stress, effective strain rates, radial and tangential strain rates in the composite cylinder are represented graphically. Authors concluded that for isotropic material, radial stress is maximum at inner radius and zero at outer radius, and the steady-state effective strain rate, radial strain rate, and tangential strain rate decreases everywhere in the composite cylinder with decrease in the value of creep exponent n from 8 to 3. Radial stress is compressible along the radius of the cylinder. Tangential stress is minimum at the inner radius and maximum at the outer radius. Axial stress increases as moved along inner radius to outer radius.

Kohli et al. [23] described that small strain theory leads to unsafe cylinder designs by using the finite strain theory. Small strain theory helps in designing a cylinder so that the cylinder can work under hard mechanical and thermal loadings. This paper analyzed the behavior of creep of a thick-walled aluminum-based composite reinforced with silicon carbide cylinder under internal pressure and large strains are assumed here (because it helps us to use the finite strain theory). Sherby's law is used and Norton's law is not used because of large creep exponent and higher activation energy. Material is homogeneous and isotropic, volume of the material remains constant, effective stress is related to creep strain rate by $\sigma_e = f(\varepsilon_e)$, cylinder is closed from both the ends (plain strain condition exists). For an isotropic material, the effective stress is given by von Mises criterion. Authors calculated creep rate at inner radius, effective stress, and then plotted radial stress, tangential stress, effective stress, axial stress, and strain rates at different radii of a cylinder. They concluded that at the center of the cylinder, radial stress shows a slight change and tangential stress, axial stress, effective stress, and strain rates show a large change throughout the cylinder when the strain increases at inner radius.

12.6.2   CREEP IN ROTATING DISC

Wahl et al. [27] presented steady-state creep behavior and stress distribution in a rotating disk made of anisotropic material. The validity of theoretical results was observed experimentally. To study the behavior of creep von Mises and Tresca yield criterion is used with a power law. Authors compared the results formed from Tresca yield criterion and von Mises criterion and revealed that in case of Tresca yield criterion stresses in the rotating disk are in better agreement than von Mises criterion.

Wahl [28] calculated stress distributions with the help of Tresca and associated flow rules on inner rotating disks having variable thickness. The stress distributions are calculated on disks undergoing steady-state creep at very high temperatures. The results obtained were compared with available data and are found to be reasonable. The methodology of Wahl when applied for transient state, where stress distribution changes are from starting period up to steady-state conditions, the results so obtained are also found satisfactory.

Wahl [29] simulated real conditions of blades by subjecting the disks to peripheral load. He finds that the steady-state creep rate is directly proportional to the power function of stress and time. Wahl formulated various design charts of stress distribution for different values of stress and creep exponent at various diameters of disks.

Saroja [30] analyzed the creep deformation in a thin rotating disk with a central circular hole with radii 1.25 inches and 6 inches of hole and disk, respectively, and constant thickness. Assumptions made are biaxial state of stress exists, axial stress is zero. Author compared her results of stress distribution with the results found by Wahl and presented it graphically. Author represented the creep at inner and outer diameter of a rotating disk at a temperature of 1000 °F and an initial speed of 15,000 rpm. It is concluded that we can eliminate the assumption that radial strain rate is equal to zero made by Wahl.

Ma [31] presented the creep analysis in rotating solid disks with variable thickness and temperature both and finally compared the results with stress distributions in uniform thickness and variable temperature disks. Tresca's yield criteria, associated flow rule, and exponential function creep law at steady-state conditions are used for calculation. In this paper, author concluded that we can solve a complicated problem with simpler way by the closed form. Ma represented temperature distributions in rotating solid disks, creep stresses at different values of ratio of radius at any point on the disk with outer radius of the disk graphically. With the help of Tresca's criteria, associated flow rule, exponential function creep law, equilibrium equations, the incompressibility equations and compatibility equations, the stress equations are easily derived.

Venkatraman and Patel [37] analyzed creep in annular plates with uniform pressure by using maximum shearing stress criterion. Authors assumed that creep rate is a power function of moment multiplied by a function of time. They diskussed four mathematical problems and concluded that in three problems, the nature of the moment distribution is affected by the relative size of the hole. Further, the results show that the choice of stress profile depends upon the size of the hole and creep exponent. However, in fourth mathematical problem, the nature of stress profile does not depend upon either size of the hole or the creep exponent.

Wahl [40] described the effects of the transient period in evaluating rotating disk tests under creep conditions using time-hardening and strain-hardening relations. Changes in tangential stress distribution during transient periods (Tresca criterion and time-hardening assumed) are represented graphically. He compared the test results and theoretical values calculated using the Tresca and Mises criteria in rotating disks and concluded that the use of the Mises criterion may yield low calculated creep deformation. Finally, author proposed that the use of the Mises criterion and associated flow rule may lead to low values of creep deformation when applied to states of biaxial tension stress.

Bhatnagar et al. [41] presented the analysis of the torsion problem for a rod made of orthotropic material. Authors considered a uniform, long, solid cylindrical rod of the circular cross-section of the radius a, with its axis parallel to the z-axis of anisotropy, subjected to two equal and opposite torques, applied at the ends. The origin is taken on the axis of the rod. The analysis is based on Norton's law generalized for a multiaxial state of stress for an orthotropic medium. The fundamental constitutive equations for creep of an orthotropic material given by Bhatnagar and Gupta [5] were used. The components of displacement have been assumed to be similar to those given by the theory of elasticity. The assumption was that the resultant stress has to be a function of θ. An expression giving the twist as a function of time has also been derived.

Chang [42] analyzed the anisotropic behavior of the material and stresses in rotating disks or orthotropic cylinders. The radial and tangential stresses in a rotating circular disk are functions of radial coordinates only and shear stress is equal to zero. However, the deformed shape does not remain circular. The stresses and displacements found in a rotating disk are axially symmetrical. Since the success of this method is hinged on the fact that the boundary condition integral is integrable, the same technique can also be used to solve some other type of boundary value problem as long as the boundary conditions can be integrated. The semi-inverse method is used to obtain a closed-form solution for the stresses in rotating circular and elliptical disks (cylinders) made of orthotropic material. It is remarkable to note that the radial and circumferential stresses of the circular disk are only functions of radial coordinate and shear stress is identically equal to zero. The displacements are axially symmetric and the deformed disk is noncircular. However, the stress components in a rotating elliptical disk are axially symmetric. The method employed can be used to solve certain classes of boundary value problems as long as the boundary conditions integral is integrable.

Arya and Bhatnagar [43] presented the creep deformation of orthotropic rotating disks by assuming that the material is incompressible, axial stress is negligible, creep rate is a function of stress multiplied by a function of time which is the time-hardening law. Expressions of strain rates and stresses were obtained with the help of time-hardening law (where effective stress for orthotropic material defined by Hill is used), constitutive equations given by Bhatnagar and Gupta for orthotropic theory of creep, and the equilibrium equation for a disk with constant thickness. The basic equations are obtained using a method of successive approximations. Authors compared steady and nonsteady state solutions. To study the effect of anisotropy, the following ratios of the anisotropic creep constants were chosen.

- Case 1. Greatest Anisotropy: $G/F = 0.5$, $H/F = 1.5$
- Case 2. Smaller Anisotropy: $G/F = 0.75$, $H/F = 1.25$
- Case 3. Anisotropy vanishes (Isotropic case): $G/F = 1.0$, $H/F = 1.0$

Tangential stress versus radius (ranging from 1 inch till 6 inches) for all the three cases diskussed above is represented graphically. The figures show that tangential stress at any radius decrease at all times for an anisotropic material.

Guowei et al. [44] developed a unified yield criterion (UYC) for plastic limit analysis of a rotating annular disk with variable thickness. Upper and lower boundary conditions of the solutions are derived by selecting a weight coefficient in the UYC.

The UYC is a piecewise linear function of principal stresses as follows:

$$\sigma_1 - \frac{1}{1+b}(b\sigma_2 + \sigma_3) = \sigma_0 \quad \text{when} \quad \sigma_2 \leq \frac{1}{2}(\sigma_1 + \sigma_3),$$

$$\frac{1}{1+b}(\sigma_1 + b\sigma_2) - \sigma_3 = \sigma_0 \quad \text{when} \quad \sigma_2 \geq \frac{1}{2}(\sigma_1 + \sigma_3)$$

Because of piecewise linearity and versatility, the UYC is extremely useful in engineering application. The formula derived here based on unified field criterion gives the exact solutions of rotating disks for different yield criteria. Stress distribution and limit angular velocity with respect to Tresca criterion, von Mises criterion and Yu criterion are illustrated, compared and represented graphically.

Singh and Ray [45] investigated the steady-state creep response in a particle-reinforced isotropic FGM disk with linear variation of particle distribution along the radial distance and compared with that of a disk containing the same amount of particle distributed uniformly. Authors contributed tremendously by analytically constructing a distribution function first time for an engineering component made of functionally gradient material. They investigated the results on a disk made of composite materials containing silicon carbide particles in a matrix of 6061 aluminum alloy. Steady-state creep behavior of the composites was explained by using Norton's law. The material parameters of creep vary along the radial distance in the disk due to varying composition, and this variation has been estimated by regression fit of the available experimental data. Singh and Ray analyzed that the tangential stress increases due to increased density caused by higher particle content in the region near the inner radius of the FGM disk. By lowering the value of creep parameters due to increased particle content, the steady-state creep rate decreases. Thus, for the assumed linear particle distributions in isotropic rotating disks, the steady-state radial, and circumferential creep rates are smaller than those in an isotropic disk with uniform particle distribution.

Singh and Ray [46] in the research paper state that many a times anisotropy is introduced due to whiskers and short fiber formation during manufacture of isotropic composite materials. The authors thus modeled anisotropy and creep in orthotropic aluminium silicon carbide composite rotating disk to show that anisotropy due to formation of whiskers and short fiber formation actually helps in significantly lowering the tangential creep rate apart from reduction in compressive radial strain. The results thus obtained in the paper then were significant from the point of view of manufacturing of composite materials in the context of real-life engineering.

Singh and Ray [47] published two research papers in the year. The one which is related to creep analysis in an isotropic FGM rotating disk of Aluminum Composite. The authors have assumed that there is a linear particle distribution in the composite for it to be isotropic. The authors deduced that steady-state tangential and radial creep rates are smaller by almost an order of magnitude compared to that of an isotropic disk with uniform particle distribution. The authors also concluded that for isotropic rotating FGM disk with linearly decreasing particle content with increasing radius, steady-state creep response is significantly greater than compared to the isotropic rotating FGM disk with uniform particle content distribution.

Singh and Ray [48] in their second paper of the year proposed new yield criteria for residual stress and steady-state creep in anisotropic composite rotating disk. The authors felt that many a times, residual stress remains the object in the process of manufacturing which may be because of the process involved like extrusion or because of sudden cooling of the object from high to low temperature. This residual stress yield different yield stresses in tension and compression. The authors modified the Hill Criterion following Hoffman isotropic yielding criterion to take into account the difference in yielding behavior due to residual stresses. The authors compared the results of steady-state creep obtained thus for anisotropic rotating composite having residual stress following the new criteria of yielding which without residual stress (i.e., limiting case) reduces to Hill's anisotropic or Hoffman's isotropic yielding criteria. The authors found that the results are comparable when the residual stress is assumed to be negligible. The authors were able to thus establish that usage of aluminum composites under rotation leads to the formation of residual stresses. The formation of residual stresses in rotating machine objects further aggravates steady-state tensile creep response and thus may significantly limit the life of the object under consideration and hence study of residual stress is very significant from the point of view of steady-state creep.

Chen et al. [49] analyzed the creep response for an FGM subjected to both internal and external pressure. The authors found that when the properties of FGM are considered to be axis-symmetric and dependent on radial coordinates an asymptotic solution can be found using Taylor's expansion series for time-dependent creep response. The approximate solutions thus found from Taylor series expansion were compared with the results of FEM. It was found that although the use of higher order terms in Taylor series may help in obtaining more accurate results, a fifth-order form of solution is sufficiently accurate to calculate the creep stress distribution with satisfactory approximation.

Deepak et al. [50] felt that the experimental studies conducted thus far reveal that steady-state creep rate in metal matrix composites based on aluminum and its alloys reinforced with ceramic particles is reduced by several orders of magnitude when compared with pure aluminum and alloys. They studied that in studies done by Singh and Ray steady-state creep response of aluminum and aluminum alloy matrix composites was done using Norton's power equation and were highly contested due to high and often variable values of apparent stress exponent and activation energy. Following which Gupta et al. conducted studies pertaining to steady-state creep responses of aluminum silicon carbide composites with Sherby's constitutive creep law with stress exponent of 8. The study was again found to be untenable because it led to consistently higher values of activation energy for creep compared with those anticipated for lattice self-diffusion in aluminum materials. Thus Deepak, Gupta, and Dham decided to analyze the impact of stress exponent at 3, 5, and 8 of the creep law on creep performance of the aluminum–silicon matrix composites made rotating disks. The authors concluded that steady-state creep in Al-SiC$_p$ is explained in a better way by assuming the stress exponent of creep law at 5, rather than at 3 or 8. They also found that distribution of stress varies with varying values of stress exponent and thus also the strain rates are significantly affected with varying stress exponent. The authors found that radial stress corresponding to stress exponent 3 is highest and for stress exponent 8 is lowest. Tangential stress is highest at inner radius for stress exponent 3 and gradually becomes lowest at outer radius. Similarly, for stress exponent 8 tangential stress is lowest at inner radius which gradually becomes highest at the outer radius. Radial and tangential strains are highest for stress component 8 and lowest for stress component 3. Values for stress exponent 5 lie in between the two stress exponent values for both stress and strain.

Singh and Gupta [51] in their research have given a mathematical model to describe steady-state creep response in an isotropic FGM composite cylinder. The paper studies the effect of linear gradient based reinforcement on the variation of creep stresses and creep rates in the FGM composite cylinder. The paper concluded that radial stress in the composite cylinder decreases throughout with increase in gradient in the distribution of reinforcement in the aluminum composite. The authors found that with increase in reinforcement gradient the tangential, axial, and effective stress are significantly affected. The radial stress is highest for the region where the particle gradient of reinforcement is highest. The tangential stress is highest for highest particle gradient at inner radius but becomes the lowest at outer radius. Similar behavior is seen for axial stress and effective stress. The strain distribution curves show that strain is highest with lowest reinforcement density and vice versa. It was also observed that with increased density of particle gradient the strain rate becomes more uniform thus increasing the life of the cylinder.

Chamoli et al. [52] studied the steady-state creep behavior of an anisotropic rotating disk made of aluminum silicon carbide particulate composite (Al-SiC$_p$). The creep behavior is described using Sherby's law. The study found that anisotropy of a material significantly impacts the creep behavior of the rotating disks.

Vandana and Singh [53] analyzed the steady-state creep in a rotating disk made of Al-SiC$_p$ composite with variable thickness using Sherby's law. The paper analyzed creep response for varying thickness as constant, linear variation, and hyperbolic variation. The change in radial stress was not found to be very significant in the three cases. The authors concluded that material anisotropy can help in restraining creep response in a rotating disk. They also found that as the ratio of anisotropic constants basically depend on tensile and compressive yield stresses of composite hence care should be taken in introducing anisotropy. The study also established that anisotropic constant where G and H are greater than F is more favorable for long-lasting of the rotating disks.

Vandana and Singh [54] analyzed the effect of residual stress on composite rotating disks with variable thickness. In the creep model given by the authors, it was concluded that stress and strain distribution get affected by thermal residual stress in both isotropic and anisotropic rotating disks. The authors observed that the effect of thermal residual stress on creep response is less for anisotropic material than for isotropic materials. The authors proposed that designers of components thus must keep in mind the residual stress in their designs.

Sharma and Sahni [55] analyzed creep response of thin rotating disks having variable thickness and variable density with edge loading. The authors established that edge loading is safer in cases where density and thickness decrease with increasing radius of a rotating disk. The authors also found that deformation is much more significant in rotating disks that have edge loading than those without edge loading.

Thakur et al. [56] studied the impact of linear thermal gradient on steady-state creep behavior in rotating disks made of isotropic aluminum-silicon carbide composite. The study is significant considering the fact that many engineering components resembling the rotating disks are exposed to varying temperature and thus the creep response of such component can be best studied if the rotating disks are studied on a temperature gradient. The authors found that the variation of radial strain rate is similar in all the disks whereas the magnitude of radial strain rate first increases rapidly and then started decreasing. It reaches a minimum before increasing again toward the external radius in the presence of a temperature gradient. The radial strain rate is slowest somewhere in the middle of the thickness of the cylinder. The authors also established that the tangential strain rate decreases significantly over the entire radius of the disk operating under thermal gradient than compared with the disk operating without any thermal gradient. The results are quite significant which show that the strain rates decrease significantly over the entire radius in the isotropic disk operating under a temperature gradient and is to be considered for safe designing of a rotating disk under elevated temperature.

Thakur et al. [57] presented the creep behavior of a rotating disk in the presence of load and thickness by using Seth's transition theory. The proposed model is used in mechanical and electronic devices. The model is also used in steam and gas turbines, turbo generators, flywheel of internal combustion engines, turbojet engines, and brake disks.

Thakur and Sethi [58] analyzed the creep deformation and stress distribution in a transversely isotropic material disk subjected to the rigid shaft by using Seth's transition theory. Authors observed that radial stress has the maximum value at the inner surface of the disk as compared to the hoop stress. At the inner surface of the disk, the strain rates have also maximum values.

KEYWORDS

- creep
- functionally gradient material
- nanotechnology
- strains and stresses

REFERENCES

1. Rimrott, F. P. J., Mills, E. J. and Marin, J., "Prediction of Creep Failure Time for Pressure Vessels", Journal of Applied Mechanics, Vol. 27, pp. 303–308, (1960).
2. Rimrott, F. P. J., and Luke, J. R. "Large Strain Creep of Rotating Cylinders", ZAMM, Vol. 41, pp. 485–500, (1961).
3. Schweiker, J. W. and Sidebottom, O. M. "Creep of Thick Walled Cylinders subjected to Internal Pressure and Axial Load", Experimental Mechanics, Vol. 5, pp. 186–192, (1965).
4. Bhatnagar, N. S. and Gupta, R. P., "On the Constitutive Equations of the Orthotropic Theory of Creep", Journal of the Physical Society of Japan, Vol. 21, No. 5, pp. 1003–1007, (1966).
5. Bhatnagar, N. S. and Gupta, R. P., "On the Constitutive Equations of the Orthotropic Theory of Creep", Wood Science and Technology, Vol. 1, pp. 142–148, (1967).
6. Pai, D. H. "Steady State Creep Analysis of Thick Walled Orthotropic Cylinders", International Journal of Mechanical Sciences, Vol. 9, pp. 335–348, (1967).
7. King, R. H. and Mackie, W. W., "Creep of Thick-Walled Cylinders", Journal of Basic Engineering, Vol. 89, pp. 877–884, (1967).
8. Bhatnagar, N. S. and Gupta, S. K., "Analysis of Thick Walled Orthotropic Cylinder in the theory of creep", Journal of the Physical Society of Japan, Vol. 27, No. 6, pp. 1655–1661, (1969).
9. Bhatnagar, N. S. and Arya, V. K., "Large Strain Creep Analysis of Thick Walled Cylinders", International Journal of Non-Linear Mechanics, Vol. 9, pp. 127–140, (1974).
10. Arya, V. K. and Bhatnagar, N. S., "Creep of Thick Walled Orthotropic Cylinders Subject to Combined Internal and External Pressures", Journal Mechanical Engineering Science, Vol. 18, No. 1, pp. 1–5, (1976).
11. Gupta, R. P., "Some Problems on the Theory of Creep", PhD Thesis, University of Roorkee, (1967).
12. Lekhnitskii, S.G. "Theory of Elasticity of an Anisotropic Body", Holden & Day, (1963).
13. Bhatnagar, N. S., Arya, V. K. and Debnath. K. K. "Creep Analysis of Orthotropic Rotating Cylinder", Journal of Pressure Vessel Technology, Vol. 102, pp. 371–377, (1980).
14. Bhatnagar, N. S., Kulkarni, P. S., and Arya, V. K., "Creep Analysis of Orthotropic Rotating Cylinders Considering Finite Strains", International Journal of Non-Linear Mechanics, Vol. 21, No. 1, pp. 61–71, (1986).
15. Chen, J. J., Tu, S. T., Xuan, F. Z., and Wang, Z. D. "Creep Analysis for a Functionally Graded Cylinder Subjected to Internal and External Pressure", Journal of Strain Analysis, Vol. 42, pp. 69–77, (2007).
16. Abrinia, K., Naee, H., Sadeghi, F. and Djavanroodi, F. "New Analysis for the FGM Thick Cylinders Under Combined Pressure and Temperature Loading", American Journal of Applied Sciences, Vol. 5, No. 7, pp. 852–859, (2008).
17. Tutuncu, N. and Ozturk, M., "Exact Solutions for Stresses in FG Pressure Vessels", Journal of Composites, Part B: Engineering, Vol. 32, pp. 683–686, (2001).
18. Singh, T.. and Gupta, V. K., "Effect of Material Parameters on Steady State Creep in a Thick Composite Cylinder Subjected to Internal Pressure", The Journal of Engineering Research, Vol. 6, No. 2, pp. 20–32, (2009).
19. Dieter, G. E. (1988), Mechanical Metallurgy. McGraw-Hill: London.

20. Finnie, I. and Heller, W. R. (1959), Creep of Engineering Materials. McGraw-Hill Book Co. Inc., pp. 117, 228.

21. Habibi, N., Samawati, S. and Ahmadi, O., "Creep Analysis of the FGM Cylinder Under Steady State Symmetric Loading", Journal of Stress Analysis, Vol. 1, No. 1, pp. 9–21, (2016).

22. Singh, T. and Singh, I., "Modeling to Steady State Creep in Thick-Walled Cylinders under Internal Pressure", International Journal of Mechanical and Mechatronics Engineering, Vol. 10, No. 5, (2016).

23. Kohli, G. S., Singh, T., Singh, H., "Creep Analysis in Thick Composite Cylinder Considering Large Strain", Journal of the Brazilian Society of Mechanical Sciences and Engineering, Vol. 42, pp. 1–8, (2020).

24. Ahmed, J. A. "Analytical Solutions and Multiscale Creep Analysis of Functionally Graded Cylindrical Pressure Vessels", Research Dissertation, Faculty of the Louisiana State University and Agricultural and Mechanical College, (2017).

25. Vinson, J. R. and Sierakowski, R. L. "The Behaviour of Structures Composed of Composite Materials", 2nd edition, Kluwer Academic Publishers, New York, Boston, Dordrecht, London, Moscow, (2004). Ebook ISBN: 0-306-48414-5

26. Singh, S. B. and Ray, S. "Steady-State Creep Behaviour in an Isotropic Functionally Graded Material Rotating Disc of Al-SiC Composite", Vol. 32A, pp. 1679–1685, (2001).

27. Wahl, A. M., Sankey, G. O., Manjoine, M. J., and Shoemaker, E. "Creep Tests of Rotating Disks at Elevated Temperatures and Comparison with Theory", Journal of Applied Mechanics, Vol. 21, pp. 225–235, (1954).

28. Wahl, A. M., "Analysis of Creep in Rotating Disks based on the Tresca Criterion and Associated Flow Rule", Journal of Applied Mechanics, Vol. 23, pp. 231–238, (1956).

29. Wahl, A. M., "Stress Distributions in Rotating Disks Subjected to Creep at Elevated Temperature", Journal of Applied Mechanics, Vol. 24, pp. 299–306, (1957).

30. Saroja, B. V., "The Analysis of Creep in a Thin Rotating Disk with a Central Circular Hole", Aircraft Engineering and Aerospace Technology, Vol. 32, No. 2, pp. 34–36, (1960).

31. Ma, B. M. "Creep Analysis of Rotating Solid Disks with Variable Thickness and Temperature", Journal of the Franklin Institute, Vol. 271, pp. 40–55, (1961).

32. https://www.intechopen.com/books/new-uses-of-micro-and-nano-materials/ nano-materials-in-structural-engineering by Małgorzata Krystek and Marcin Górski (accessed November 5, 2018).

33. https://www.nano.gov/you/nanotechnology-benefits

34. Mahamood, R. M., Akinlabi, E. T., Shukla, M. and Pityana, S. "Functionally Graded Materials: An Overview", Proceedings of the World Congress on Engineering, (2012).

35. Watanabe, Y., Inaguma, Y., Sato, H. and Miura-Fujiwara, E. "A Novel Fabrication Method for Functionally Graded Materials under Centrifugal Force: The Centrifugal Mixed-Powder Method", Published online 2009. doi: 10.3390/ma2042510.

36. Razdolsky, L. (2019), Phenomenological Creep Models of Composites and Nanomaterials, Deterministic and Probablistic Approach, CRC Press, 1st Edition.

37. Venkatraman, B. and Patel, S. A., "Creep Analysis of Annular Plates", Journal of the Franklin Institute, Vol. 275, pp. 13–23, (1963)

38. Temesgen, A., Singh, S. B. and Pankaj T., "Classical and Non Classical Treatment of Problems in Elastic–Plastic and Creep Deformation for Rotating Discs". Materials

Physics and Chemistry, Apple Academic Press, (2020). Ebook ISBN: 9780367816094 and DOI: 10.1201/9780367816094

39.   https://nptel.ac.in/content/storage2/courses/112105125/pdf/module-9%20lesson-2.pdf.

40.   Wahl, A. M., "Effects of the Transient Period in Evaluating Rotating Disk Tests Under Creep Conditions", Journal of Basic Engineering, Vol. 85, pp. 66–70, (1963).

41.   Bhatnagar, N. S., Gupta, S. K. and Gupta, R. P., " The Torsion of an Orthotropic Rod in the Theory of Creep", Wood Science and Technology, Vol. 3, pp. 167–174, (1969).

42.   Chang, C. I., "The Anisotropic Rotating Disks", International Journal of Mechanical Sciences, Vol. 17, pp. 397–402, (1975).

43.   Arya, V. K. and Bhatnagar, N. S., "Creep Analysis of Rotating Orthotropic Disks", Nuclear Engineering and Design, Vol. 55, pp. 323–330, (1979).

44.   G. Ma, H. Hao and Y. Miyamoto, "Limit Angular Velocity of Rotating Disc with Unified Yeild Criterion", International Journal of Mechanical Sciences, Vol. 43, pp. 1137–1153, (2001).

45.   Singh, S. B. and Ray, S. "Steady-State Creep Behavior in an Isotropic Functionally Graded Material Rotating Disk of Al-SiC Composite", Metallurgical and Materials Transactions A, Vol. 32A, pp. 1679–1685, (2001).

46.   Singh, S. B. and Ray, S. "Modeling the Anisotropy and Creep in Orthotropic Aluminum-Silicon Carbide Composite Rotating Disc", Mechanics of Materials, Vol. 34, pp. 363–372, (2002).

47.   Singh, S. B. and Ray, S. "Creep Analysis in an Isotropic FGM Rotating Disc of Al-SiC Composite", Journal of Materials Processing Technology, Vol. 143–144, pp. 616–622, (2003).

48.   Singh, S. B. and Ray, S. "Newly Proposed Yield Criterion for Residual Stress and Steady State Creep in an Anisotropic Composite Rotating Disc", Journal of Materials Processing Technology, Vol. 143-144, pp. 623–628, (2003).

49.   Chen, J. J., Tu, S. T., Xuan, F. Z. and Wang, Z. D. "Creep Analysis for a Functionally Graded Cylinder Subjected to Internal and External Pressure", Journal of Strain Analysis, Vol. 42, pp. 69–75, (2007).

50.   Deepak, D., Gupta, V. K. and Dham, A. K., "Impact of Stress Exponent on Steady State Creep in a Rotating Composite Disc", Journal of Strain Analysis, Vol. 44, pp. 127–135, (2009).

51.   Singh, T. and Gupta, V. K., "Creep Analysis of an Internally Pressurized Thick Cylinder Made of a Functionally Graded Composite", Journal of Strain Analysis, Vol. 44, pp. 583–594, (2009).

52.   Chamoli, N., Rattan, M. and Singh, S. B., "Effect of Anisotropy on the Creep of a Rotating Disc of Al-SiC$_p$ Composite", International Journal of Contemporary Mathematical Sciences, Vol. 5, No. 11, pp. 509–516, (2010).

53.   Vandana and Singh, S. B., "Modeling Anisotropy and Steady State Creep in a Rotating Disc of Al-SiC$_p$ Having Varying Thickness", International Journal of Scientific & Engineering Research, Vol. 2, No. 10, pp. 1–12, (2011).

54.   Gupta, V. and Singh, S. B., "Creep Modeling in a Composite Rotating Disc with Thickness Variation in Presence of Residual Stress", International Journal of Mathematics and Mathematical Sciences, Volume 2012, Article ID 924921, 14 pages, doi:10.1155/2012/924921.

55.   Sharma, S. and Sahni, M., "Creep Analysis of Thin Rotating Disc Having Variable Thickness and Variable Density with Edge Loading", *Annals of Faculty Engineering*

Hunedoara—International Journal of Engineering, Tome XI- Fascicule-3, pp. 279–296 (2013).

56. Thakur, P., Gupta, N. and Singh, S. B., "Thermal Effect on the Creep in a Rotating Disc by Using Sherby's Law", Kragujevac Journal of Science, Vol. 39, pp. 17–27, (2017).

57. Thakur, P., Sethi, M., Shahi, S., Singh, S., Emmanuel, B., F. S., and Sajic, J. L. "Modelling of Creep Behaviour of a Rotating Disc in the Presence of Load and Variable Thickness by using Seth Transition Theory", Structural Integrity and Life, Vol. 18, No. 2, pp. 153–160, (2018).

58. Thakur, P., Sethi, M. K. "Creep Deformation and Stress Analysis in a Transversely Material Disk Subjected to Rigid Shaft", Mathematics and Mechanics of Solids, Vol. 25, No. 1, pp. 17–25, (2020).

INDEX

For Product Safety Concerns and Information please contact our EU
representative GPSR@taylorandfrancis.com
Taylor & Francis Verlag GmbH, Kaufingerstraße 24, 80331 München, Germany

www.ingramcontent.com/pod-product-compliance
Lightning Source LLC
Chambersburg PA
CBHW060814220326
41598CB00022B/2611